In many animals, from bees to bulls, peacocks to people, the sex of the individual can be determined at a glance. How and why have these differences in appearance and behaviour developed? What is the nature and extent of sex differences in humans? Why do individuals in some species change sex?

In a series of lively and highly readable articles, the word's leading experts in the field of sex differences have reviewed the latest molecular, genetic, hormonal, anatomical and behavioural data from a wide range of species, bridging the gap between geneticists, biologists and socio-biologists. Such an overview has never been attempted before and it should have wide appeal, especially to undergraduates and graduates in the biological and medical sciences.

D1166328

THE DIFFERENCES BETWEEN THE SEXES

THE DIFFERENCES
BETWEEN THE SEXES

Edited by

R. V. SHORT
Monash University, Australia

and

E. BALABAN
Harvard University, USA

Published by the Press Syndicate of the University of Cambridge
The Pitt Building, Trumpington Street, Cambridge CB2 1RP
40 West 20th Street, New York, NY 10011-4211, USA
10 Stamford Road, Oakleigh, Melbourne 3166, Australia

First published 1994

Printed in Great Britain at the University Press, Cambridge

A catalogue record for this book is available from the British Library

Library of Congress cataloguing in publication data
The Differences between the Sexes/edited by R. V. Short and E. Balaban.
p. cm.
Papers presented at the 11th International Conference on
Comparative Physiology, Sept. 12–14, 1992, Crans-sur-Sierre, Switzerland.
Includes bibliographical references and index.
ISBN 0 521 44411 X (hardback). – ISBN 0 521 44878 6 (pbk.)
1. Sexual dimorphism (Animals) – Congresses.   2. Sex differences – Congresses.
I. Short, Roger Valentine, 1930–   .   II. Balaban, E. (Evan)
III. International Conference on Comparative Physiology
(11th: 1992: Crans, Switzerland)
QP81.5.D54   1994
591.1′6 – dc20   93-27413 CIP

ISBN 0 521 44411 X   hardback
ISBN 0 521 44878 6   paperback

KW

Contents

Contributors

Evan Balaban
Department of Organismic and Evolutionary Biology, Harvard University, 26 Oxford Street, Cambridge MA 02138, USA

Tim H. Clutton-Brock
Department of Zoology, University of Cambridge, Downing Street, Cambridge CB2 3EJ, UK

Matthew Cobb
Mécanismes de Communication, CNRS URA 1491, Université Paris-Sud, 91000 Orsay Cédex, France

David Crews
Institute of Reproductive Biology and Departments of Zoology and Psychology, The University of Texas, Austin, TX 78712, USA

Gisèle Desvages
Département Dynamique du Génome et Evolution, Institut Jacques Monod, CNRS et Université Paris VII, 2 Place Jussieu – Tour 43, 75251 Paris Cédex 05, France

Andrea Dettling
Anthropologisches Institut und Museum, Universität Zürich-Irchel, Winterthurerstrasse 190, 8057 Zürich, Switzerland

Mireille Dorizzi
Département Dynamique du Génome et Evolution, Institut Jacques Monod, CNRS et Université Paris VII, 2 Place Jussieu – Tour 43, 75251 Paris Cédex 05, France

Jean-François Ferveur
Mécanismes de Communication, CNRS URA 1491, Université Paris-Sud, 91000 Orsay Cédex, France

Karl Fredga
Department of Genetics, University of Uppsala, PO Box 7003, S-750 07 Uppsala, Sweden

Manfred Gahr
Max-Planck Institut für Verhaltensphysiologie, Post Starnberg, 8130 Seewiesen, Germany

Marc Girondot
Département Dynamique du Génome et Evolution, Institut Jacques Monod, CNRS et Université Paris VII, 2 Place Jussieu – Tour 43, 75251 Paris Cédex 05, France

Jennifer A. Marshall Graves
Department of Genetics and Human Variation, Le Trobe University, Bundoora, Victoria 3083, Australia

Jan-Åke Gustafsson
Department of Medical Nutrition, Karolinska Institute, Huddinge University Hospital, F60 Novum, 14186 Huddinge, Sweden

Paul H. Harvey
Department of Zoology, University of Oxford, South Parks Road, Oxford OX1 3PS, UK

Jean-Marc Jallon
Mécanismes de Communication, CNRS URA 1491 Université Paris-Sud, 91000 Orsay Cédex, France

Gerald A. Lincoln
MRC Reproductive Biology Unit, Centre for Reproductive Biology, 37 Chalmers Street, Edinburgh EH3 9EW, UK

Mary F. Lyon
MRC Radiobiology Unit, Chilton, Didcot, Oxfordshire OX11 0RD, UK

Robert D. Martin
Anthropologisches Institut und Museum, Universität Zürich-Irchel, Winterthurerstrasse 190, 8057 Zürich, Switzerland

Yuzura Oguma
Mécanismes de Communication, CNRS URA 1491, Université Paris-Sud, 91000 Orsay Cédex, France

Claude Pieau
Département Dynamique du Genome et Evolution, Institut Jacques Monod, CNRS et Université Paris VII, 2 Place Jussieu – Tour 43, 75251 Paris Cédex 05, France

Joyce H. Poole
Kenya Wildlife Service, PO Box 40241, Nairobi, Kenya

Marilyn B. Renfree
Department of Zoology, University of Melbourne, Parkville, Victoria 3052, Australia

John D. Reynolds
School of Biological Sciences, University of East Anglia, Norwich NR4 7TJ, UK

Noëlle Richard-Mercier
Département Dynamique du Génome et Evolution, Institut Jacques Monod, CNRS et Université Paris VII, 2 Place Jussieu – Tour 43, 75251 Paris Cédex 05, France

Jon Seger
Department of Biology, University of Utah, Salt Lake City, UT 84112, USA

Douglas Y. Shapiro
Department of Biology, Eastern Michigan University, Ypsilanti, MI 48197, USA

Roger V. Short
Department of Physiology, Monash University, Clayton, Victoria 3168, Australia

J. William Stubblefield
CERA, Charles Square, 20 University Road, Cambridge MA 02138, USA

Lesley A. Willner
17 Church Crescent, London N20 0JR, UK

Jean D. Wilson
Department of Internal Medicine, The University of Texas, Southwestern Medical Center, 5323 Harry Hines Boulevard, Dallas, TX 75235–8857, USA

John C. Wingfield
Department of Zoology, University of Washington, Seattle, WA 98195, USA

Patrick Zaborski
Département Dynamique du Génome et Evolution, Institut Jacques Monod, CNRS et Université Paris VII, 2 Place Jussieu – Tour 43, 75251 Paris Cédex 05, France

Foreword

The fact is that animals, if they are subjected to a modification in minute organs, are liable to immense modifications in their general configuration. The phenomenon may be observed in the case of gelded animals; only a minute organ of the animal is mutilated, and the creature passes from the male to the female form. We may infer, then, that if in the primary conformation of the embryo an infinitesimally minute but essential organ sustain a change of magnitude, the animal will in one case turn to male and in the other to female; and also that, if the said organ be obliterated altogether, the animal will be of neither one sex nor the other . . .

The female is softer in disposition, is more mischievous, less simple, more impulsive, and more attentive to the nurture of the young; the male, on the other hand, is more spirited, more savage, more simple and less cunning. The traces of these characteristics are more or less visible everywhere, but they are especially visible where character is the more developed, and most of all in man.

The fact is, the nature of man is the most rounded off and complete, and consequently in man the qualities above referred to are found most clearly. Hence woman is more compassionate than man, more easily moved to tears, at the same time is more jealous, more querulous, more apt to scold and to strike. She is, furthermore, more prone to despondency and less hopeful than man, more void of shame, more false of speech, more deceptive, and of more retentive memory. She is also more wakeful, more shrinking, more difficult to rouse to action, and requires a smaller quantity of nutriment.

Aristotle, *Historia Animalium*

Preface

The Eleventh International Conference on Comparative Physiology was held on September 12–14, 1992, in the beautiful mountain setting of the Hotel du Golf et des Sports, Crans-sur-Sierre, Switzerland, under the sponsorship of the Interunion Commission on Comparative Physiology. All the participants are deeply indebted to Professor C. Liana Bolis, the Secretary-General, and Professor Ewald Weibel, the Co-Chairman, for the excellent local arrangements.

The organizers were able to bring together a small international group of scientists from all four corners of the globe, many of whom had never met before and were often unaware of one another's work. The theme that united us was the endlessly fascinating topic of *The Differences Between the Sexes*; we began by signing our names on the flyleaf of a copy of Charles Darwin's *The Descent of Man, and Selection in Relation to Sex* to pay homage to his penetrating original observations on this subject.

We discussed the advantages of sexual reproduction as compared to parthenogenesis, and went on to review the nature and extent of the host of differences in body size, shape, structure, function, colour, smell, sound and behaviour that distinguish male from female in insects, fish, amphibians, reptiles, birds and mammals. We then turned our attention to the adaptive significance of these sex differences in terms of lifetime reproductive success, looking at species as diverse as fruit flies, deer, birds and elephants. Finally we focused on the environmental, hormonal, chromosomal and genetic mechanisms responsible for producing these striking differences. We concluded with a few speculations about the differences which separate, but which may also help to unite, man and woman.

The conference was made possible by very generous financial

contributions from the Arien Foundation, the Interunion Commission on Comparative Physiology, Organon (Australia) Pty Ltd, Schering Research Foundation, the Swiss Academy of Sciences, and the Swiss National Science Foundation.

We would like to give our special thanks to Dr Jane Barrett of Cambridge University Press for her meticulous attention to a host of editorial details, and for her felicitous turn of phrase that clarified many obscurities in the text.

We hope that the published proceedings of this conference will appeal to anybody with a natural curiosity about life, be they molecular biologist or mammologist, behaviourist or biochemist, clinician or cytogeneticist, student or specialist, or just the ordinary man or woman.

R. V. S. and E. B.

Overview

1

Why sex?

ROGER V. SHORT

Sex is a subject that has proved to be endlessly fascinating to generations of biologists, but recent advances in our understanding of animal behaviour, embryology, molecular biology and genetics have shed new light on some of the age-old questions. How and when did sexual reproduction first evolve? What are the advantages and the disadvantages of reproducing sexually, the costs and the benefits, and what are the consequences?

Sex is usually seen as one of the cornerstones of natural selection. The process of meiosis, in which the chromosomes in the diploid germ cells exchange genes by crossing over and then shuffle the maternal and paternal chromosomes to deal a random haploid set to each of the gametes, produces an enormous amount of genetic variability. This is increased still further when maternal and paternal gametes from different individuals meet at fertilization and fuse their two sets of chromosomes at syngamy to produce a new, genetically unique diploid organism. This variability is the substrate on which natural selection can act.

But maybe there is also a down-side to sexual reproduction and its attendant natural selection, since the whole process will always act to favour the selfish interests of the individual. Richard Dawkins has even gone so far as to propose the concept of a 'selfish gene' which dominates the life of the organism that plays host to it. But, since the gene is no more than an inert fragment of DNA until it is able to translate its hidden chemical message into some cellular function, it is difficult to see how the interests of the untranscribed gene in the nucleus could ever differ from those of the organism it helped to produce. If one acted against the other, the gene would always be the loser; if the organism suffered, the gene would not be transmitted to succeeding generations.

Even if we may have difficulty in believing in selfish genes, selfish individuals are certainly a reality; natural selection ensures that

3

self-interest will always reign supreme in the struggle for survival. But where will this progressive evolutionary drive lead us? Natural selection will normally ensure that a balance is maintained between the organism and its environment, and it is the environment, such as the availability of food or shelter, that is the ultimate constraint. But there is a fatal flaw in the design that could lead to runaway selection and eventual destruction of the species. This will occur if natural selection eventually results in the evolution of an organism that can assume control of its environment and can escape from these natural constraints.

Our new-found ability to transmit knowledge across time has catapulted us into this league. Suddenly, we are able to profit from the sum total of all knowledge that mankind has ever possessed; uniquely amongst all species, we have been able to escape from the slow, blind, groping, random trial-and-error of natural selection, exchanging it for the explosive potential of a Lamarckian style of evolution. The acquisition of knowledge has given us the ability to inherit acquired characteristics through non-genetic means. This accumulated wisdom has enabled us to dominate our environment; we now have the power to change it in the twinkling of an eye and Nature cannot prevent us, it can only punish us after the event. The voice of reason may make us try to exercise restraint individually by limiting our food consumption or our fertility, but reason is unlikely to prevail; we may be able to *think* rationally as individuals, but corporately we seem unable to *act* in a rational manner. Even in a democracy, our leaders are selected largely because of their own self-interest; they want the job. Put a group of selfish individuals in charge and they will usually tend to act out of self-interest, not altruism towards those whom they allegedly seek to serve, or concern for the world in which their offspring must live.

We should all have shuddered a little at the recent announcement by the world's radioastronomers that they can find no evidence of other intelligent lifeforms within 40 million light years of us. Perhaps the lifespan of a species is inversely proportional to its degree of intellectual development? The probability that a species that has evolved to be as intelligent and all-conquering as ours could survive for long is remote indeed. We may live in a silent universe for a very good reason. Paradoxically, evolution may have ensured that we have one of the shortest survival times of any species, since it has made us, effectively, our own executioner.

But these grim forebodings about our future as a species did not concern Charles Darwin. When he first published *The Origin of Species by Means of Natural Selection, or the Preservation of Favoured Races in the Struggle for Life* in 1859, he certainly saw evolution as a dynamic process, but his immediate concern was the mechanism by which the favoured individual propagated his or her desirable traits to others. Darwin realized that evolution would not occur unless the chosen few were fertile and could pass on their more desirable characteristics to their offspring. So, in *The Descent of Man, and Selection in Relation to Sex* (1871), he went on to develop his ideas about sexual selection. It is worth quoting him on the distinction between natural selection and sexual selection, since these ideas still form the basis for our understanding of the differences between the sexes, the theme of this book:

Sexual Selection depends on the success of certain individuals over others of the same sex, in relation to the propagation of the species; whilst Natural Selection depends on the success of both sexes, at all ages, in relation to the general conditions of life. The sexual struggle is of two kinds; in the one it is between the individuals of the same sex, generally the males, in order to drive away or kill their rivals, the females remaining passive; whilst in the other, the struggle is likewise between the individuals of the same sex, in order to excite or charm those of the opposite sex, generally the females, which no longer remain passive, but select the more agreeable partners.

Clearly, Darwin appreciated that in polygynous species, where the males mate with more than one female, the competition between males for access to the females is greater than the competition between females for access to the males, leading to an exaggerated development of those secondary sexual characteristics used in inter-male aggressive encounters (see Figure 1.1). As Darwin wrote: 'That some relation exists between polygamy and the development of secondary sexual characters appears nearly certain'.

He even appreciated the potentially self-destructive nature of sexual selection, which he thought might sometimes act counter to natural selection, instead of in concert with it. Probably thinking of the massive antlers of the great Irish elk (which in fact are not out of proportion to its overall body size; see Chapter 6 by Gerald Lincoln, and Figure 1.2) or even the tail of the peacock, he said: 'The development however of certain structures – of the horns, for instance, in certain stags – has been carried to a wonderful extreme; and in some cases to an extreme which, as far as the general conditions of life are concerned, must be slightly injurious to the male.'

But Darwin went to his grave still baffled by the nature of the mechanism that enabled desirable characteristics to be transmitted from one generation to the next. He was unaware of the publication in 1866

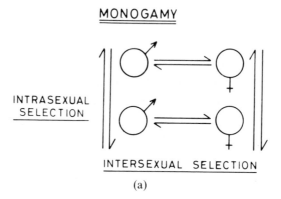

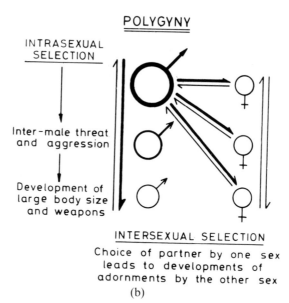

Figure 1.1. Differential effects of intersexual and intrasexual selection in (a) monogamous and (b) polygynous mating systems. (From Short, R. V. (1979).)

of his contemporary, Gregor Mendel, working away in the monastery gardens of Brno, Czechoslovakia, on the particulate nature of inheritance in plants, although Mendel had a copy of Darwin's *Origin of Species*. Mendel's work was almost completely ignored until it was rediscovered by others at the turn of the century. If only Darwin had known about genes, chromosomes and meiosis! Instead, he was forced to come up

Figure 1.2. The great Irish elk. (From Short, R. V. (1976) The origin of species. In *Reproduction in Mammals*, Book 6, *The Evolution of Reproduction*, p. 112. Eds. C. R. Austin and R. V. Short. Cambridge University Press.)

with a concept he named pangenesis to explain how characteristics might be inherited. He developed this idea at length in his 1868 publication *The Variation of Animals and Plants under Domestication*, but few people were attracted to it, for obvious reasons. This caused him bitter disappointment. He summarized his pangenesis hypothesis in *The Descent of Man, and Selection in Relation to Sex*: 'According to this hypothesis, every unit or cell of the body throws off gemmules or undeveloped atoms, which are transmitted to the offspring of both sexes, and are multiplied by self-division. They may remain undeveloped during the early years of life or during successive generations; and their development into units or cells, like those from which they were derived, depends on their affinity for, and union with other units or cells previously developed in the due order of growth'.

Darwin's 'gemmules' were thus the product of the cell, unlike our concept of a gene as a chemically coded message that goes to make the cell in the first place. And in the absence of any understanding of hormones, Darwin was forced to explain the development of sexual dimorphisms in secondary sexual characteristics as being due to the dormancy of the appropriate gemmules in one sex or their development in the other.

Darwin's ideas on intrasexual and intersexual selection were developed further by the English geneticist and statistician R. A. Fisher in his monumental 1930 publication *The Genetical Theory of Natural Selection*. Fisher was intrigued by the genetic control of sexual dimorphisms. Although an obvious solution might seem to be to locate the genes controlling all mammalian male secondary sexual characteristics on the only male-specific chromosome, the Y, Fisher was quick to see that Nature had usually opted for a far more subtle strategy. The Y chromosome was used simply to code for the presence of a testis; once made, the testis could then produce hormones that could act both locally and systemically to induce the development of a spectacular array of male secondary sexual characteristics. Therefore, the genes for these male secondary sexual characteristics did not need to be located on the Y chromosome, but could be on any of the autosomes. They would be expressed only under the activating influence of male sex hormones (see Chapter 8 by Jean Wilson). Thus antler growth is usually confined to male deer, but the genes regulating the antler do not need to be sex-linked and confined to the Y chromosome. We know that they are present, but dormant, in the female; if you implant a female with the male sex hormone testosterone, she too will develop antlers. Thus, the genes for most sexually dimorphic characteristics are hormonally sex-limited, not genetically sex-linked. This has made it much easier to develop a wide array of different sexually dimorphic characteristics in different species, since it is easier to develop hormone-dependent expression of an autosomal gene than to translocate it onto the Y chromosome. No wonder we see such spectacular and diverse forms of male secondary sexual characteristics such as body size, brain development, hair growth, teeth, horns or antlers in different species, whereas the Y chromosome appears to be highly conserved across all mammals and almost devoid of genetic information, apart from genes which determine spermatogenesis and the presence of the testis.

Since exceptions to rules demand an explanation, it is fascinating to discover that in marsupial mammals some sexually dimorphic structures such as the scrotum, pouch and mammary gland are under direct genetic, rather than indirect hormonal, control. The genes in question seem to be on the X rather that the Y chromosome, so that their differential expression is dosage dependent – one X chromosome codes for a scrotum, whereas two X chromosomes result in the development of a pouch and mammary gland instead (see Chapter 9 by Marilyn Renfree). Why did marsupials choose to do things differently when they sailed away from the rest of the world on Gondwanaland, some 150 million years ago?

Just as Fisher championed and expanded Darwin's view of sexual selection, so Julian Huxley was able to use his wide practical experience as an ornithologist and field naturalist to clothe the theories of Darwin and Fisher with practical examples. He realized that all aspects of the male and female reproductive tract were likely to be involved in inter-sexual selection, from the gametes themselves to the gonads, the male and female copulatory organs, ducts and accessory glands, any structures facilitating the discovery or recognition of one sex by the other, and any behavioural display that stimulated reproductive activity. Huxley gently chided Darwin for his modesty in not referring to the copulatory organs themselves, and re-emphasized the point made by Darwin and Fisher that polygynous mating systems would maximize sexual dimorphisms, whereas monogamous systems would minimize them.

Biologists were slow to develop ideas about the biological costs and benefits of sexual reproduction, although this has now become a happy hunting ground for sociobiologists (see Chapter 15 by Tim Clutton-Brock). It was the American, R. A. Trivers, who first pointed out that an animal's reproductive strategy in terms of the type of mating system it adopts is dictated by the relative energy investment of the two parents in rearing their young. Thus in birds, where the absence of lactation enables both sexes to play an equal role in feeding the young, monogamous mating systems are generally desirable, therefore sexual dimorphisms in body size are unlikely to develop. However, even in monogamous species, the sexes may still be distinguishable by plumage characteristics; recent evidence from DNA fingerprinting of progeny suggests that they may not be as monogamous as we had imagined and this is further borne out by the relatively large size of the testes in so-called monogamous birds. The smallest relative testis size is actually found in birds that lek, where the females seldom mate with more than one male.

It is also of interest that in birds, where the male is the homogametic sex (ZZ), the neutral or default state appears to be the gaudy plumage of the male, on which femaleness is superimposed by ovarian oestrogens. Many people still find it difficult to believe that in spectacularly dimorphic species like chickens, peacocks, pheasants and ducks, an ovariectomized female will develop the male's gaudy plumage, which is normally suppressed in the female by her oestrogens; her dowdy appearance is presumably of adaptive significance, since it will provide better camouflage when she is sitting on the nest. Peahens have been shown to select as a mate the peacock with the largest number of eye-spots in his train, which is a measure not of his androgen-induced virility, but of his lack of oestrogen-induced femininity. Since peahens with diseased ovaries can develop the male's plumage, the train is not a very reliable indicator of the male's reproductive potential. Oestrogen also seems to be the factor responsible for transforming the indifferent avian gonad into an ovary, since female chicken embryos, given an aromatase inhibitor, develop testes rather than ovaries. But the testosterone secreted by the cockerel's testes is nevertheless still responsible for his aggressive and sexual behaviour and for the development of such male secondary sexual characteristics as the comb and spurs.

Why should birds have evolved a completely different sex-determining mechanism from mammals? Not only are females the heterogametic sex (ZW) and oestrogen the dominant hormone, but the Z and W sex chromosomes are not even homologous to the X and Y chromosomes of mammals (see Chapter 18 by Jennifer Graves and Chapter 20 by Claude Pieau and colleagues). Perhaps we should be looking for an avian W-linked ovary-determining gene, SRW, to compare with the mammalian Y-linked testis-determining gene, SRY.

Sexual selection in humans and the great apes

When Darwin first put forward the concept of sexual selection, he considered the way in which it acted to develop general bodily characteristics, but was too modest to extend his argument to the genitalia themselves. Therefore, I undertook a study of gonadal and genital development in humans and the great apes. This showed clearly that somatic size and genital development are not necessarily related to one another. Male gorillas weighing around 250 kg, although twice as big as females, nevertheless have minute 10 g testes, whereas 50 kg male chimpanzees that are only slightly larger than females nevertheless have

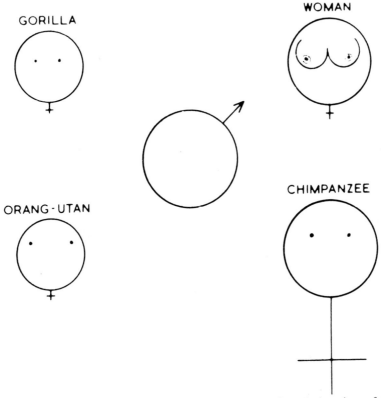

Figure 1.3. The male's view of the female, illustrating the relative sizes of the nulliparous breast and the external genitalia. (From Short, R. V. (1979).)

enormous 60 g testes. In order to explain this apparent paradox, we have to understand the different mating systems of the species in question, since these can have quite different effects on somatic and genital development (see Figures 1.3 and 1.4).

Somatic selection for body size is concerned with successful inter-male competition for a mate, as Darwin suggested originally. Genital selection, on the other hand, is far more complex; although influenced by the mating system, it is ultimately a reflection of the frequency of copulation and the number of sexual partners. Thus, in monogamous species that pair for life there is little or no somatic dimorphism in overall body size; since copulatory frequencies are also low, the testis is relatively small and the female does not have to advertise her sexual state to all-comers by developing pronounced sexual swellings. Marmoset monkeys and gibbons are good examples of monogamous primates that illustrate these points.

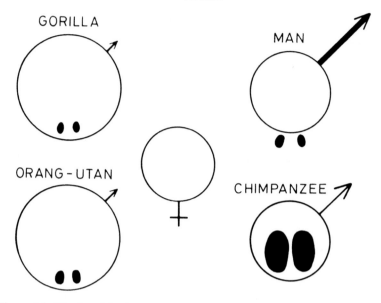

Figure 1.4. The female's view of the male, illustrating the relative sizes of the penis and the testes. (From Short, R. V. (1979).)

In polygynous species, where one male has exclusive access to several females, enhanced inter-male competition will result in the males becoming much bigger than the females. But the copulatory frequency may still be low and, in the absence of any need for increased sperm production, the testes may still remain relatively small. (Testicular size is determined almost exclusively by the volume of the seminiferous tubules; the testosterone-producing Leydig cells only account for a small proportion of overall testicular size. Thus, testicular size is an excellent predictor of spermatogenic capacity but a poor predictor of androgenic activity.) Females of polygynous species do not necessarily need to develop pronounced sexual swellings to advertise oestrus, since the dominant male is constantly in attendance.

The gorilla provides a good illustration of these polygynous attributes. A silverback male may have a harem of three or four females, but each adult female will come into oestrus only once or twice every four years or so, since for the rest of the time she is either pregnant or in lactational anoestrus. Thus, the male may only get an opportunity to copulate about once a year, since mating in this species is confined to the brief time when the female is in oestrus. This accounts for the relatively minute testes of this great ape, the high proportion of morphologically abnormal spermatozoa in the ejaculate, and the absence of any pronounced sexual swelling in the female at the time of oestrus.

In so-called promiscuous or multi-male mating systems, things are very different. Inter-male competition for access to females may still result in males being larger than females, although if all males have unrestricted access to the oestrous female, there may be little size dimorphism. The fact that several males mate with one female when she is in oestrus introduces sperm competition. The male who deposits the greatest number of good spermatozoa in the female's reproductive tract is most likely to sire the offspring, so there will be strong selection pressure for increased sperm quality, increased spermatogenic capacity and, hence, increased testicular size. In addition, males may have a relatively high copulatory frequency if they have reproductive access to all the adult females in the troop whenever they come into oestrus, and this will also favour increased testicular size. As for the female, if she is going to mate with several males, it pays her to advertise her oestrous state, leading to the development of pronounced sexual swellings at the time of ovulation that can be seen at a distance by the males. Chimpanzees illustrate these features very clearly.

And how about humans? The fact that men are 15–20% bigger than women suggests that we are not inherently monogamous. The relatively small size of a man's testes, which each weigh around 20 g, the high proportion of morphologically abnormal spermatozoa, and the lack of any cyclical sexual swelling in the female at ovulation suggests that neither are we adapted to a multi-male promiscuous mating system. A woman's ability to initiate copulation at any stage of her reproductive cycle, rather than just around the time of ovulation, has greatly increased the frequency of copulation. But since the vast majority of human acts of intercourse are unrelated to reproduction and are probably for social bonding, the size of a man's testes has not increased accordingly. The best guess would be that we are basically a polygynous primate in which the polygyny usually takes the form of serial monogamy. Since there is no male menopause, men have a longer fertile life than women, so a man is likely to produce more offspring in his lifetime than a woman. This is accentuated by the fact that if older men remarry, almost invariably they marry a younger woman. Humans have developed a variety of secondary sexual characteristics like a large penis and well-developed breasts that further enhance intersexual bonding, and women are unique amongst mammals in being covert ovulators, concealing the event not only from others but often from themselves as well.

One of the reasons why Darwin did not concern himself with the gonadal and genital aspects of sexual selection may be because he was writing at a time when the details of mammalian fertilization were still largely unknown. For example, in *The Variation of Animals and Plants under Domestication* when discussing mammals he stated that: 'More than one spermatozoon is requisite to fertilize an ovum' and 'The rate also of the segmentation of the ovum is determined by the number of spermatozoa'.

In *The Descent of Man, and Selection in Relation to Sex,* when discussing human beginnings, he merely states that: 'Man is developed from an ovule, about the 125th of an inch in diameter, which differs in no respect from the ovules of other animals'. Spermatozoa are not mentioned anywhere in the book and fertilization is discussed only briefly in the context of plants and lower animals. Although there was the presumption that spermatozoa fertilized the eggs of mammals, the first worthwhile description of mammalian fertilization was not made until 1875, four years after the publication of *The Descent of Man, and Selection in Relation to Sex.*

Germ cells

By the century's end, when mammalian fertilization had become an established fact, the German embryologist August Weismann proposed the concept of 'the continuity of germ plasm'. He believed that an immortal germ plasm, concentrated in the germ cells and transmitted from generation to generation by the gametes, budded off a mortal soma at each generation. In other words, a chicken was merely an egg's way of making another egg. Today, thanks to the techniques of experimental mammalian embryology, we find no evidence for the existence of germ plasm in mammals. The cells of the early cleaving embryo are totipotent; one embryonic stem cell taken from a mouse embryo $4\frac{1}{2}$ days after fertilization and injected into a recipient embryo, can give rise to both germ cells and somatic cells in the resultant chimaeric individual.

But even if germ plasm does not exist, the germ cells do show certain remarkable peculiarities that distinguish them from somatic cells. They are thought to originate in the embryonic ectoderm and then to migrate out of the embryo into the extraembryonic mesoderm around the tail bud. Subsequently, they literally wriggle their way back into the embryo again, invading first the mesoderm of the primitive streak and then the endoderm before finally coming to rest in the developing embryonic gonad. This brief sojourn outside the embryo may be to escape the DNA methylation that overtakes all the somatic cell lineages of the embryo during gastrulation. The germ cells are the only cells that have to preserve their totipotency throughout life; DNA methylation may be the reason for the loss of totipotency of the somatic lineages resulting in progressive differentiation and commitment.

The germ cells also appear to be under a different genetic control to the somatic tissues of the gonad. The first evidence for this came from experimental mouse fusion chimaeras in which XX and XY embryos were combined into one individual that usually developed as a phenotypic male. However, progeny testing showed invariably that the spermatozoa were all derived from the XY parent; XX germ cells seemed constitutionally incapable of undergoing spermatogenesis, even when placed in a normal testicular environment (but see Chapter 19 by Karl Fredga for some interesting exceptions). Similarly, in spontaneously occurring XX male mice, goats and humans, where one of the X chromosomes may carry a translocated segment of the Y chromosome's sex-determining region, the somatic tissue of the testis develops normally but the germ cells are unable to survive. Even when the recently discovered testis-determining gene SRY is injected into a genetic female mouse, testicular formation and male phenotypic development occurs normally, but the animal is sterile because none of the germ cells survive. SRY certainly appears to be the gene responsible for somatic differentiation of the testis, acting in a cell-autonomous manner to induce the supporting cells of the undifferentiated gonad to beome Sertoli cells (see Chapter 17 by Mary Lyon and Chapter 18 by Jennifer Graves). But it is emphatically not the gene responsible for inducing the germ cells to undergo spermatogenesis.

If XX germ cells are genetically committed to develop into oocytes, it may be because they contain two functional X chromosomes which escaped methylation and X-inactivation whilst hiding outside the embryo. XO germ cells, in contrast, appear to be capable of undergoing either spermatogenesis or oogenesis, depending on whether they are in a testicular or ovarian environment. However, XO oogonia are abnormal because they undergo accelerated atresia immediately after birth, and XO spermatogonia produce grossly abnormal spermatozoa that are incapable of fertilizing an egg. Again, this points to the fact that the female germline probably needs a double dose of X-linked genes for normal oogenesis and the male germline needs a gene or genes on the Y chromosome (other than SRY) for normal spermatogenesis.

There is evidence to show that, occasionally, XY germ cells can undergo oogenesis when in an ovarian environment. Once again, this highlights the fact that the presence of SRY does not commit a germ cell to male development. Perhaps there are X chromosome homologues of the pairing segment of the Y chromosome that facilitate oogenesis or maybe the XY oocyte is comparable to an XO oocyte.

In all mammals studied to date, XXY Klinefelter's syndrome results in testicular development and a normal male phenotype, but no germ cells survive in the testes. Presumably, this is because the two functional X chromosomes commit them to oogenesis, but there are no follicular support cells, since they have already differentiated into Sertoli cells under the influence of SRY.

The action of SRY, inducing Sertoli cell development in the somatic tissue of the gonad, is one of the first signs of somatic sexual differentiation, but soon it is followed by a striking sexual dimorphism in the behaviour of the germ cells. Following an initial wave of mitotic divisions in the primordial germ cells in both sexes when they enter the undifferentiated gonad, in the seminiferous tubules the male germ cells undergo mitotic and meiotic arrest that is withdrawn only under hormonal influences at the time of puberty, when spermatogenesis commences. It is believed that the Sertoli cells secrete a mitotic and/or meiotic inhibitor, since any germ cells lying outside the testis, for example in the developing adrenal gland, immediately enter meiosis as if they were oogonia. In the female, there is no mitotic arrest and, after a wave of mitotic divisions in early embryonic life, all the germ cells spontaneously enter meiosis at the same time, regardless of whether they are in the adrenal gland or the ovary. This is convincing evidence that, in the early stages of gonadal differentiation, it is the somatic tissue which controls the development of the germinal tissue. Later on, it is the genotype of the germ cell, rather than the phenotypic environment in which it finds itself, that decides its fate.

Why should the germ cells show such a dramatic sexual dimorphism in their mitotic and meiotic behaviour so very early in embryogenesis? In order to understand the reasons for this, we need to look at the mitochondria.

The transmission of mitochondrial DNA

Mitochondria are found in the cytoplasm of all somatic cells of eukaryotes, but they contain a form of DNA that is distinctly different from that in the cell's nucleus. It is now generally agreed that the mitochondria arose from a bacterium that invaded the ancestral prokaryotic cell around 1.5 billion years ago and eventually established a mutually beneficial symbiotic relationship with it. The mitochondria have since become the 'power stations' of all eukaryotic cells, and are responsible for much of their energy production from substrates in the cytoplasm. There is also growing evidence to suggest that an accumulation of deleterious mutations in the mitochondrial DNA (mtDNA) during an organism's lifetime may be a major contributory factor to the ageing process, leading ultimately to the death of the host. The more metabolically active the cell, perhaps the higher the mutation rate of its mtDNA; this is because of the high rate of production of oxygen free radicals in the mitochondria. These radicals chemically damage the molecules in their immediate vicinity, especially mtDNA and proteins. The nuclear DNA, on the other hand, may be protected from these mutagenic metabolites by the cell membrane that surrounds the nucleus and is the distinguishing feature of eukaryotes. Mitochondrial DNA, which lacks this membranous defence, certainly has a much higher mutation rate than nuclear DNA.

The exciting thing about this mitochondrial parasite within the cytoplasm is that it is not involved in sexual reproduction. Maybe it has never learned how or maybe it has been excluded by selection, but the end result is that the mtDNA of the embryo is almost exclusively maternally derived. This is of profound importance for our understanding of reproduction and evolution.

The mtDNA of all vertebrates seems to have the same set of 37 genes specifying 13 proteins, all of which are components of the mitochondrial respiratory chain and oxidative phosphorylation system, as well as two ribosomal RNA and 22 transfer RNA molecules; these genes are tightly packed in about 16.5 kb of a double-stranded circular mtDNA molecule. But unlike the nuclear DNA, which can cleanse itself of undesirable mutations at meiosis and fertilization by Mendelian segregation and recombination, mtDNA has no known repair mechanism. We do not even know whether the mitochondrion represents a single, elongated thread or a series of discreet cytoplasmic organelles; nor do we know how it replicates itself. Such knowledge is essential if we are to understand how mtDNA mutations accumulate during an individual's lifetime.

The mammalian egg is packed with mtDNA molecules, containing about 10^5 of them, compared to only a few thousand mtDNA molecules in each somatic cell. But the spermatozoon contains only around 100 mtDNA molecules, entirely confined to the mitochondrial sheath that surrounds its midpiece and provides the energy for the beating of the tail. There is no mtDNA in the head of the spermatozoon. Although the midpiece of the spermatozoon does penetrate the egg cytoplasm at

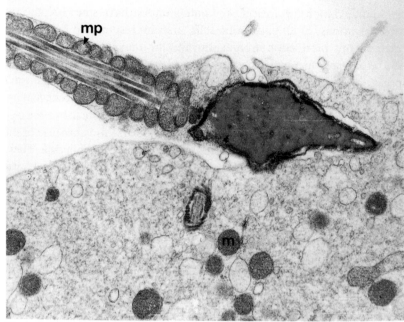

Figure 1.5. Electron micrograph of a spermatozoon entering the human egg at fertilization and showing abundant mitochondria (m) in the egg cytoplasm, and the mitochondrial midpiece of the fertilizing spermatozoon (mp) which will not contribute to the mitochondrial makeup of the resultant embryo. (From Santhananthan, A. H., Ng, S. C., Edirisinghe, R., Ratnam, S. S. and Wong, P. C. (1986) Human sperm–egg interaction *in vitro. Gamete Research,* **15** 317–26.)

fertilization (Figure 1.5), the genetic evidence suggests that it contributes little or no mtDNA to the embryo.

From the time of puberty onwards, male germ cells proliferate mitotically and then enter meiosis to produce sperm. Since these meiotic divisions provide an opportunity for removal of deleterious mutations from the germ cell's nuclear DNA, increasing paternal age has relatively little effect on the incidence of deleterious genetic mutations in the sperm. In contrast to this, mitochondrial mutations in male germ cells would be expected to accumulate over time. Sperm bearing these mutations in the mitochondrial sheath of the midpiece might show reduced motility and hence reduced fertilizing capacity, but even if they did fertilize an egg no harm would be done in view of the lack of paternal inheritance of mtDNA. Fertilization is therefore a very effective intergenerational barrier to the transmission of mtDNA mutations.

Contrast this situation to that of the female primordial germ cells.

A succession of mitotic divisions ensures that there are several million oogonia present in the fetal ovaries. Then all mitosis ceases, never to be resumed; this finite stock of oogonia cannot be replenished in later life. All the oogonia then enter meiosis; this occurs well before birth in most species. But instead of completing meiosis to produce mature, haploid oocytes, meiosis becomes arrested halfway through the first reduction division at the diplotene or dictyate stage, when homologous chromosomes have paired and undergone crossing-over, but have not yet separated. The oocyte remains in this dormant resting state from fetal life until the time of ovulation, which may be many years later. Immediately before ovulation the first and second meiotic divisions are completed in a matter of hours and a mature, haploid oocyte is formed. The oocyte is therefore the oldest cell in the female's body. After increasing periods of time in this arrested dictyate stage, the homologous pairs of chromosomes apparently become more tightly adherent to one another, so that when meiosis is finally completed, the homologous chromosome pairs may fail to separate, resulting in the production of an oocyte with too many or too few chromosomes, leading to the formation of a genetically defective embryo; this is thought to be the origin of the trisomy characteristic of Down's Syndrome, a condition whose incidence increases markedly with increasing maternal age.

This prolonged period of relative metabolic inactivity of the dictyate oocyte may be vital for the protection of its mtDNA from harmful mutations. Since mtDNA lacks the possibility of repair during meiosis, it is essential that the ovulated egg contributes a healthy complement of mtDNA to the future embryo. If the mtDNA mutation rate is dependent on the metabolic activity of the cell in which it resides, the female germ cell needs to be kept in a metabolically quiescent state until the time of ovulation.

One of the other characteristics of the female germline is the very high rate of death through atresia of dictyate oocytes; of the 7 million or so that are formed in the embryonic human gonad, under half a million survive to puberty and only a small fraction of these will ever be ovulated. Maybe oocyte atresia is a way of selecting out defective oocytes; it could act as a cleansing filter for mtDNA.

The final and most obvious criterion which distinguishes male from female gametes is their size. The selection for a small number of very large, immotile gametes is the characteristic by which we arbitrarily define the female sex. The male has opted for the production of a larger number of very small, highly motile gametes that have to be transported to the female for fertilization. Almost all sexually reproducing forms of life have indulged in this process of anisogamy. Whether large eggs are necessary to ensure the transmission of sufficient mutation-free maternal mtDNA

Figure 1.6. The egg of Aepyornis, the extinct elephant bird from Madagascar. The adult bird is thought to have weighed about $\frac{3}{4}$ tonne and it was probably exterminated by early man, like the closely related giant moas that once inhabited New Zealand. The nearest living relatives of these giant Rattites are the ostrich from Africa, the rhea from South America, the emu from Australia, the cassowary from Australia and New Guinea, and the kiwi from New Zealand.

to the embryo is an intriguing thought. The mammalian egg does not need to contain large amounts of yolk to nourish the embryo, and it is curious that all eutherian eggs, from mouse to elephant, are of comparable size. Avian eggs increase with the size of the species because of the need to provide sufficient yolk for the development of the chick. Perhaps the largest cell that has ever existed was the large yolky egg of Aepyornis, the now-extinct elephant bird from Madagascar, which was many times bigger than even the eggs of the largest dinosaurs (Figure 1.6), and yet Aepyornis spermatozoa would have been invisible to the naked eye.

Could it be that this, the most striking and fundamental of all sexual dimorphisms, the behaviour of the germ cells and the size of the gametes, reflects the constraints of that selfish parasite within us, our mitochondria? Our germ cells give us the potential for immortality, but perhaps it is our mother's mtDNA that tempers it with mortality. Whilst selfish genes seem to be theoretically implausible, selfish mitochondria are just another example of host–parasite interaction, which has had profound consequences for the evolution of sexual reproduction.

It is all rather reminiscent of the fairytale of the Sleeping Beauty and the Prince. The female germ cell enters into a long metabolic sleep from the moment of its formation in the embryo, thereby preserving the beauty of its mtDNA. It is awakened only by the kiss of the Prince, in preparation for the ensuing ovulation and fertilization. But there the fairytale ends, for it is difficult to believe that we humans are destined to live happily ever after.

Further reading

Austin, C. R. (1961) *The Mammalian Egg.* Blackwell, Oxford.

Birkhead, T. R. and Møller, A. P. (1992) *Sperm Competition in Birds.* Academic Press, London.

Darwin, C. (1871) *The Descent of Man, and Selection in Relation to Sex.* John Murray, London.

Dawkins, R. (1976) *The Selfish Gene.* Oxford University Press.

Hastings, I. M. (1992) Population genetic aspects of deleterious cytoplasmic genomes and their effect on the evolution of sexual reproduction. *Genetic Research Cambridge*, **59**, 215–25.

Linnane, A. W., Marzuki, S., Ozawa, T. and Tanaka, M. (1989) Mitochondrial DNA mutations as an important contributor to ageing and degenerative diseases. *Lancet*, **1**, 642–45.

Linnane, A. W., Zhang, C., Baumer, A. and Nagley, P. (1992) Mitochondrial DNA mutation and the ageing process: bioenergy and pharmacological intervention. *Mutation Research*, **275**, 195–208.

Margulis, L. and Sagan, D. (1986) *Origins of Sex.* Yale University Press, New Haven.

McLaren, A. (1991) Sex determination in mammals. *Oxford Reviews in Reproductive Biology* (ed. S. R. Milligan), **13**, 1–33.

Møller, A. P. (1988) Ejaculate quality, testes size and sperm competition in primates. *Journal of Human Evolution*, **17**, 479–88.

Petrie, M., Halliday, T. and Sanders, C. (1991) Peahens prefer peacocks with elaborate trains. *Animal Behavior*, **41**, 323–31.

Short, R. V. (1979) Sexual selection and its component parts, somatic and genital selection, as illustrated by Man and the Great Apes. *Advances in the Study of Behavior*, **9**, 131–58.

Wilson, A. C., Caum, R. L., Carr, S. M., George, M., Gyllensten, U. B., Helen-Bychowski, K. M., Higuchi, R. G., Palumbi, S. R., Prager, E. M., Sage, R. D. and Stoneking, M. (1985) Mitochondrial DNA and two perspectives on evolutionary genetics. *Biological Journal of the Linnaean Society*, **26**, 375–400.

2

Constraints to parthenogenesis

DAVID CREWS

The two products most prized by the poultry industry are meat and eggs. Today there are strains of domestic chickens that either produce large roosters destined for the Sunday pot or fecund hens that will lay eggs every day for their entire lifetimes. However, it is not possible to have both traits in the same strain. Despite repeated efforts, selection for bulk in roosters carries with it small hens that are poor layers, whereas selection for egg production yields puny males. Thus, one-half of the biomass produced (male or female chicks as the case may be) is without commercial value; indeed, often the unwanted sex is ground up into food for the desired sex. However, if each individual produced was the desired sex, there would be greater production, greater efficiency and, ultimately, greater profit for the producer and lower cost to the consumer. The impact of unisexuality can be seen in Figure 2.1. In this example we assume that only 10 breeding females exist and that each female produces 30 eggs each year. Let us assume further that the young produced become sexually mature in their third year. Finally, we will assume that a 1:1 sex ratio occurs in unmanipulated animals, there is no mortality, and all offspring have equal fecundity. (Clearly, these assumptions are unrealistic, but any decreases in production will be equivalent in the various scenarios.) If just females are produced, their number will increase exponentially, with 10 200 females being produced over a four-year period compared to 2700 females produced with no manipulation; at the end of seven years this difference becomes 633 100 versus 56 150! If just males were produced, the numbers would be 56 150 males in the manipulated strain versus 28 075 males in the strain with no manipulation.

Sex then can be a roadblock to production because the majority of animals consists of separate sexes (male and female) which must mate to produce young. This is called gonochorism. There are some species,

23

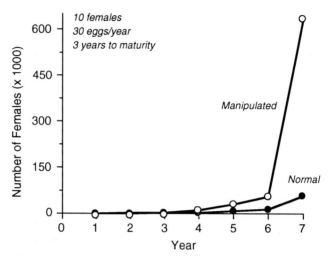

Figure 2.1. Hypothetical increase in the number of female individuals if just females are produced versus the number of female individuals produced in a gonochoristic species with a 1:1 sex ratio. (See text for further details.)

however, that are hermaphroditic, combining both male and female traits within the same individual at the same time or at different times in the life cycle. But hermaphrodites must also mate to produce young; self fertilization is rare. Hermaphroditism is common in the invertebrates, but there are relatively few hermaphroditic vertebrates and virtually all of them are fish. The only animal forms known to circumvent the problem of sex are parthenogenetic or all-female species.

Still another problem is that sex in many invertebrates and vertebrates is determined by various genotypic mechanisms. In fruitflies it is the number of X chromosomes that makes males. In mammals the male is heterogametic (meaning it has two different sex chromosomes), being XY compared to the female's XX genetic constitution. In birds it is the female who is the heterogametic sex (ZW compared to the ZZ bearing male). It has long been appreciated that if it were possible for a genetic female mammal to be sex-reversed so that it would breed as a male, just daughters would be produced. A genetic male bird breeding as a female would result in just sons. This has been achieved, but this avenue ends at one generation. (Exceptions to these rules are discussed in Chapter 19 by Karl Fredga.)

Current techniques in reproductive biology, combined with advances in the molecular genetics of development, may soon allow us to produce for commercial purposes a unisexual species that can perpetuate itself easily. But can self reproduction or 'parthenogenesis' be achieved economically?

Unisexual animals have been created by man, but with different results in each circumstance. A well-established technology in fish aquaculture is functional sex-reversal using treatments of sex steroid hormones. The primary objective is to produce lots of fish. In the United States, catfish have become common table fare. In catfish the male is the heterogametic sex. Treating the fry with androgen will result in genetic females (XX) that breed as males, thereby producing just female offspring (which immediately doubles the number of reproducing fish). Then there is the Beltsville turkey. Marlow Olsen of the USA Department of Agriculture noticed a low incidence of parthenogenetic development in certain strains of domestic turkeys. By selection, he was able to build this into a strain in which 30–40% of the unfertilized eggs would develop spontaneously. But, as in all birds, the female is the heterogametic sex and male offspring were produced, effectively preventing completion of the parthenogenesis. Moreover, only about 20% of the mature parthenogenetic offspring had viable sperm. The small numbers of animals produced, 26 fertile parthenogens in 20 years, make this marvellous discovery too costly from the commercial standpoint. The ultimate goal then is to create a species capable of clonal reproduction, or parthenogenesis, in which all individuals are female and generate genetically identical offspring: female to perpetuate the lineage and genetically identical because uniformity is a critical ingredient of any mass production effort.

There are but a handful of species that consist naturally of female individuals only. One group of these animals exhibits male-dependent parthenogenesis; that is, the species consists of just female individuals that must mate with the males of another species in order to reproduce. There are more invertebrate species than vertebrates of this type. Hybridogenesis occurs in the mole salamanders (genus *Ambystoma*), the European water frog (*Rana esculenta*), and the top minnow (*Poeciliopsis*). In the formation of the unisexual hybrid, the genome of one of the parental species (species A) is discarded by differential segregation, while only the genome of the other parental species (species B) is transmitted to the egg. Subsequent eggs produced by these hybrids are then fertilized by the sperm of species A males (Figure 2.2). Something similar called gynogenesis occurs in the Amazon molly (*Poecilia formosa*). In this all-female species the individual solicits the attention of males of a closely related sexual species and mates with them. However, the sperm only activate the egg to start developing. There is no nuclear fusion and, hence,

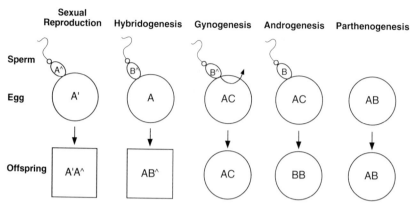

Figure 2.2. Unisexual modes of reproduction compared to sexual reproduction. In sexual reproduction, the mother produces a haploid egg (A′) that, due to recombination, contains a unique combination of genes from her mother and father; her male mate produces a haploid sperm (A^) in the same way; the resulting offspring are highly variable. In hybridogenesis, one ancestral genome (A) is transmitted to the egg without recombination while the other ancestral genome (B) is discarded; sperm (B^) from ancestral species B then fertilizes the egg, producing an offspring with the original A genome and a new B^ genome. In gynogenesis, sperm from a related bisexual ancestor is required to stimulate development of the egg, but the male pronucleus is eliminated after fertilization. In androgenesis, the eggs produced by a hybrid eliminate the maternal DNA, incorporating the paternal DNA upon fertilization. In parthenogenesis, the hybrid genome (one haploid genome from bisexual ancestor A, the other haploid genome from bisexual ancestor B) is transmitted to the egg complete and without recombination; the egg develops without sperm into an offspring genetically identical to the mother. It can be argued that the unisexual modes of reproduction represent successive steps in the evolution of parthenogenesis from sexual reproduction. The square indicates that the offspring have unique genetic identities (through recombination) whereas the circle indicates that the offspring are genetically identical to a parent. Letters indicate origins of individuals and not a pedigree.

no mingling of sperm and egg DNA. A third type of male-dependent parthenogenesis, androgenesis, has been described in the stick insects (*Bacillus* spp.). Here the hybrid female's egg, when fertilized by sperm from a male of one of the original ancestral species, eliminates the maternal DNA, substituting the paternal DNA; thus, the eggs develop under the instruction of the paternal DNA. Both male and female androgenetic offspring are fertile.

The other type of parthenogenesis is obligate parthenogenesis. The species that exhibit this consist only of female individuals that do not need to mate (and hence do not require sperm to activate embryogenesis) in order to reproduce. In obligate parthenogenesis there is no reduction in ploidy (indeed, usually there is an increase) and the egg is produced without genetic recombination. Parthenoforms can arise by mutation, through hybridization between sexual species or by artificial selection. (I might add that there are still other modes of reproduction such as that found in the aphid. In this animal, an individual can either be gonochoristic or parthenogenetic within its own lifetime, but the mode of reproduction can vary across generations depending upon the quality of the environment.)

Obligate parthenogenesis is rare in the vertebrates and appears to be restricted to the lizards; the Brahminy blind snake (*Ramphotyphlops braminus*) may also be unisexual, although this has not been confirmed. There are about 40 species of lizards including the Australian geckos (*Heteronotia*), eurasian rock lizards (*Lacerta*) and southwestern whiptail lizards (*Cnemidophorus*) that reproduce in this manner. In fact, in many instances we actually know which species hybridized to create the parthenogen (Figure 2.3).

Given the natural variety of reproductive mechanisms and considering the advances made in the molecular genetics of development in just the last few years, it is not far fetched to expect that we might soon understand the molecular basis of parthenogenesis. With this information, I predict that scientists by the year 2000 will have developed the methods necessary to create unisexual parthenogenetic species. Ann McLaren already has succeeded in making mouse embryos from two female pronuclei, but they have not developed into viable young. But, is the fact that there are so few parthenogenetic species in nature indicative of some built-in costs to sexuality that can become constraints to parthenogenesis? The purpose of this chapter is to identify some of these constraints and to explore the costs and consequences of creating parthenogenetic vertebrates. Namely, is it possible and, if so, what exactly will we have created? Will there be built-in limitations to such organisms?

Why are there so few parthenogens?

Many have questioned why sex even exists, particularly as sex involves a host of dangers and disadvantages. Even considering its consequence, mating is a risky business and the participants are often 'seemingly unaware' of their environment. In non-human animals the dangers are usually predators; in humans, there are social mores and sexually transmitted diseases.

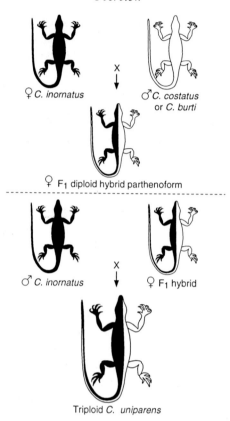

Figure 2.3. Evolution of the triploid *Cnemidophorus uniparens*, a unisexual vertebrate. The parthenogenetic desert grassland whiptail, *C. uniparens*, arose from the
hybrid union of a female little striped whiptail, *C. inornatus*, and either a male *C.
costatus* or *C. burti*, although which is a matter of dispute among systematists.
Whatever the paternal species, it is known that *C. uniparens* arose from a female
F$_1$ hybrid backcrossing with a male *C. inornatus*. Thus, the maternal species has
contributed two sets of chromosomes to the triploid genome characteristic of this
parthenogen. Orlando Cuellar was able to show how this feat of self reproduction
is accomplished: immediately prior to the first reduction division, the chromosomes duplicate themselves; meiosis then proceeds normally with the result that
each egg is chromosomally complete and female. Parthenogenetic organisms like
these may be particularly useful to the behavioral endocrinologist interested in
evolution. These species allow one to study brain–behavior evolution in a manner
impossible with conventional sexual species. This is because, in one sense, the
parthenogenetic animal and its related sexual species represent a 'snapshot' of
evolution, allowing comparison of the neuroendocrine mechanisms that control
reproductive behaviors in the descendant species with those of the ancestral
species. The larger size of *C. uniparens* reflects its polyploid nature. (Adapted from
David Crews, *Psychobiology of Reproductive Behavior: an Evolutionary Perspective*,
©1987, pp. 106, 108, 109, 112. Reprinted by permission of Prentice-Hall, Inc.,
Englewood Cliffs, N.J.)

It is these apparent problems as well as other theoretical issues that make parthenogenesis seem to be such an obvious alternative. Evolutionary biologists have made three distinct arguments as to why parthenogenetic species *should* have an evolutionary and ecological advantage over gonochoristic animals. The 'superwoman' argument: many parthenogens have highly heterozygous, and often polyploid, genomes because of their hybrid origins. The 'all for one and one for all' argument: parthenogens have greater colonizing ability because each individual can be a founder. The 'if it ain't broke, don't fix it' argument: parthenogenesis protects against a favorable gene combination being rearranged which might occur during normal meiosis. Despite the logic and persuasiveness of these arguments, ultimately they are unsatisfactory.

If parthenogenesis is such an effective means of reproduction, why are there so few examples in nature? With the possible exception of blue-green algae, all unisexual species existing today are thought to have arisen from sexual species. Perhaps we should ask not only 'why does sex exist?' but also 'why is parthenogenesis so rare?'

Although unisexuality is widely distributed among many different kinds of animals, parthenogenetic species represent only a small fraction of 1% of the total number of species. Because unisexuality is a derived state, any unisexual species is likely to retain features of the ancestral sexual state. Are there negative consequences of such evolutionary baggage and does this help to explain why there are not more parthenogenetic species in the world? Three classes of constraints inherent in the sexual process that appear to be relevant are: (1) the behavioral facilitation of reproduction; (2) the retention of the causal mechanisms of male-specific traits; and (3) the mechanics of sex determination. When these are considered, we gain some insight into why, relatively speaking, there are so few parthenogenetic species in the world.

The impetus to mate

Reproduction in a population tends to occur within a defined period or breeding season. Mating may occur once only, or many times, but the sexual activity is followed almost always by a period in which there is no breeding. Some animals have spectacular, albeit brief, breeding seasons, such as the red-sided garter snake (Figure 2.4). Other animals have very prolonged reproductive periods, as in the house mouse. Not only do there tend to be defined breeding seasons, but, within that time, both sexes undergo marked behavioral changes that are related to their cycle of

Figure 2.4. Emergence of male Canadian red-sided garter snakes at hibernaculum entrance during the spring. Top panel: males emerge first and *en masse*. Middle panel: females then emerge individually over a three-week period resulting in the formation of mating balls. The female is the individual with the large head in the center of the figure; the rest are courting males. In garter snakes, the females are about three times larger than males. Unlike most mammals, the testes inhibit body growth in male garter snakes. Bottom panel: a mating ball of red-sided garter snakes. There is only one female present; the rest of the snakes are males. (From Crews, D. (1992) Diversity of hormone–behavior relations in reproductive behavior. In *Behavioral Endocrinology*, pp. 141–86. Eds. J. B. Becker, S. M. Breedlove and D. Crews. The MIT Press, Cambridge, Massachusetts.)

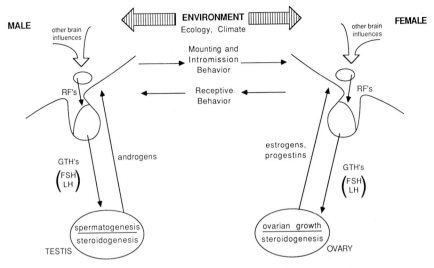

Figure 2.5. Dynamic relationship between the internal and external environments in the control of mating behavior in many vertebrate animals. The behavior of the male and the female help to synchronize the maturation and release of eggs and sperm so that fertilization occurs. Changes in climate, ecology or the behavior of other members of the species will initiate and modulate gonadal and hormonal changes required for reproduction. Thus, hormones regulate behavior in the individual animal and are themselves affected by other stimuli, including the behavior and, indirectly, the physiology of its mate. Abbreviations: RF's, releasing factors or hormones, GTH's, gonadotropins; FSH, follicle stimulating hormone; LH, luteinizing or ovulating hormone. (From Crews D. (1977) Integration of internal and external stimuli in the regulation of lizard reproduction. In *Behavior and Neurology of Lizards*, pp. 149–72. Eds. N. Greenberg and P. D. MacLean. DHEW Publication Number (ADN) 7-7-491.)

gonadal growth. The impetus to mate is regulated, therefore, by changes in the external and internal environments.

Reproduction is a symphony in physiology (Figure 2.5). The length of day, seasonal temperature changes, increases in humidity or flooding, or even the appearance of a particular food item, can be the actual trigger for initiating the sexual cycle. These environmental cues are transduced into neural stimuli that act in the hypothalamus to influence the neuro-hormones that regulate pituitary activity. The pituitary secretes protein hormones (called gonadotropins) that stimulate the gonad to grow. The growing gonad has two primary products: the gametes (eggs or sperm) and the steroid hormones (progestins, androgens and estrogens). The gonadal steroid hormones are carried in the circulation to act throughout the body. One site of action is the limbic system, where they serve to modulate further pituitary gonadotropin secretion. The limbic system is one of the 'core' brain structures shared by all vertebrates. It is an ancient network of brain nuclei that includes the preoptic area, anterior hypothalamus and ventromedial hypothalamus. Because of its involvement in sexual and aggressive behaviors, the limbic system is also called the emotional brain.

Gonadal steroids, then, activate sexual behavior in animals living in predictable environments that have long periods of benign environmental conditions conducive to reproduction. However, animals living in predictable environments with only brief periods for reproduction, or in environments that are harsh and completely unpredictable, have a particular stimulus (e.g. a specific display) or a suite of stimuli (e.g. rain and the color green) that triggers breeding activity. For example, in the Canadian red-sided garter snake, the warm spring temperatures stimulate emergence from underground winter hibernacula (Figure 2.4). A specific chemical or pheromone produced by the female then activates courtship behavior in males even though at this time the testes are fully regressed and not secreting androgens. In the deserts of Africa and western Australia, where there seems to be a perpetual drought, weaver birds and zebra finches will begin to pair within minutes of the first raindrops and then build a nest and lay eggs in a few days. Klaus Immelmann demonstrated that these birds are 'opportunistic breeders', maintaining almost fully active gonads at all times, thereby enabling them to respond immediately to the unpredictable rains.

Sexual synergism

One of the first major discoveries made by ethologists was the observation that the sexual and aggressive behaviors of all species tend to be stereotyped and characteristic of all individuals within that species. During the breeding season, sexual displays have profound effects on mating activity. This holds for the individual performing the display as well as for the individual who is the object of the display. The performer undergoes physiological changes – for example, in both male and female vertebrates the circulating concentrations of steroid hormones increase after a sexual interaction – perhaps as internal reinforcement of the immediately preceding event. The recipient of the display is also stimulated to undergo physiological changes. Thus, there is the relationship between the individual and the physical environment as well as the relationship between the individual and its internal milieu; the individual's behaviors and physiology influence and, in turn, are influenced by interactions with the opposite sex (Figure 2.5).

It is important to consider the sexual interactions of a mating pair as more than simply coordinating the transfer of gametes. That is, the behaviors displayed by each participant serve to integrate the reproductive cycle of the pair by acting as neuroendocrine primers. For example, in many species the female will not undergo normal ovarian growth if sexually active males are absent. Experiments with red deer, ring doves and green anole lizards have demonstrated that specific male courtship displays will facilitate the effects of a stimulatory environmental regimen, causing the courted female's ovaries to grow more rapidly than if she was exposed only to the environmental regimen. In cats, ferrets and mink, the ovarian follicles will develop under the appropriate environmental conditions, but if the female does not physically mate with a male she will not ovulate. In garter snakes and musk shrews, the ovaries will not even begin to grow until the female mates.

In sexual species then, female reproduction often depends absolutely on the presence and behavior of conspecific males. One result is that fertilization is assured. Another result is behavioral facilitation and coordination of reproductive cycles. This has important consequences. For populations with synchronized reproduction, it means that a greater number of young will be produced at a given time, making it less likely that one's own offspring will be killed by a predator. This is often cited as an example of the 'selfish herd' benefit. Hence, the behavioral facilitation of reproduction and its consequence for reproductive synchrony is a powerful force in sexual species. The necessity for male-like traits that stimulate female-typical reproductive physiology (as well as vice versa) can be viewed as a negative or, at best, a neutral constraint to parthenogenesis.

It would seem that parthenogenetic species would be emancipated from the need for male-typical sexual behavior. This is the case in the fruitfly, *Drosophila mercatorum*. Hampton Carson selected against sexual reproduction by raising those eggs of virgin females that went on to develop. This represented only a fraction of 1% of the eggs, but his persistence resulted eventually in the creation of a parthenogenetic, all-female strain. When Carson studied the behavior of these fruitflies, he found that they no longer displayed sexual receptivity. His interpretation was that the behaviors associated with sexuality (female receptivity) were lost, due to the absence of a selective force (the need for males for reproduction).

This all seems straightforward and reasonable. However, consider the case of the parthenogenetic whiptail lizard. Whiptail lizards are a modest-sized taxon of medium-to-large (5–25 g) animals. They are shaped like torpedoes and, as their shape suggests, can run at high speeds, but, fortunately for an aging biologist, only for short distances. Most whiptail lizard species are gonochoristic, yet, remarkably, one-third of the 45 species of whiptail lizards are unisexual, consisting only of female-like individuals that reproduce by obligate parthenogenesis.

'Sexual' behavior in unisexual species

In 1978, I observed serendipitously two unisexual whiptail lizards exhibiting behaviors remarkably similar to the courtship and copulatory behavior of the sexual ancestral species (Figure 2.6). (Serendipity often leads to new perspectives or paradigms. It reflects very much Louis Pasteur's admonition: 'Chance favors the prepared mind'. As such, serendipity must withstand constantly the myopia of the 'I will see it when I believe it' attitude.) The actual moment had its comic side. In my laboratory, photographic film is stored at $-20\,°C$ and there was some furious cursing as I tried to warm up the film and to load it into the camera to document what I then thought was a unique observation.

On seeing the animals engaged in this 'sexual' behavior it struck me that this display of both male-like and female-like behaviors *alternately by a single individual* demonstrated perfectly the fundamental bisexuality of the vertebrate brain; simultaneously hermaphroditic species make the same point in their behavior, but their gonads release both sperm and eggs (only rarely simultaneously) and male-typical and female-typical hormone secretions.

Since then, this chance observation has proved not to be a unique discovery, for others (e.g. Jay Cole and Orlando Cuellar) had seen this behavior but did not recognize its significance and so did not publish their observations. Furthermore, this behavior is expressed regularly and reliably during the reproductive season and is not restricted to *C. uniparens.* Such 'sexual' behavior has been observed now in at least five other species of parthenogenetic whiptails and in the parthenogenetic morning gecko.

Experiments indicate that engaging in 'sexual' behavior stimulates ovarian growth in the parthenogenetic whiptail, just as male courtship stimulates ovarian growth in its sexual ancestor, *C. inornatus.* In both species, the time to the first ovulation is decreased significantly if courtship

Figure 2.6. In gonochoristic whiptail lizards (a–c, *Cnemidophorus inornatus*) the male approaches and investigates the female with his bifid tongue, an action that presumably indicates involvement of chemical senses via the vomeronasal organ as occurs in snakes. If the female is sexually receptive, she stands still for the male, allowing him to mount her back. Usually just before the male mounts the female, he grips with his jaws either a portion of the skin on the female's neck or her foreleg. As the male rides the female, he scratches her sides and presses her body against the substrate. Then the male begins to maneuver his tail beneath the female's tail, attempting to appose their cloacal regions (the urogenital opening). During mating, one of two hemipenes (all lizards and snakes have two penises rather than one) is intromitted into the female's cloaca. With intromission, the male shifts his jawgrip from the female's neck to her pelvic region, thereby assuming a contorted copulatory posture I have termed the *doughnut*. This posture is maintained for five to ten minutes, after which the male dismounts rapidly and leaves the female. Unisexual whiptail lizards (e.g. d–f, *C. uniparens*) exhibit this same sequence of events. That is, one individual will approach and mount another individual and, after riding for a few minutes, the mounting (male-like) individual will swing its tail beneath that of the mounted (female-like) individual, apposing the cloacal regions. At the same time the mounting individual will shift its jawgrip from the neck to the pelvic region of the mounted individual, forming the doughnut posture. Since parthenogens are morphologically female, there are no hemipenes and intromission does not occur. (From Crews (1987), see Figure 2.3.)

occurs. Indeed, females of the sexual species will lay eggs only if a sexually active male is present; females housed with a male that fails to court, either because he has been castrated or has a low sex drive, do not ovulate and so do not produce eggs. Similarly, in the parthenogenetic whiptail, engaging in 'sexual' behavior increases the likelihood of ovulation. Over the course of a reproductive season this effect can be substantial. Isolated parthenogens will ovulate eventually, but it is rare that they will produce more than one clutch. Usually, two intact parthenogens housed together will each produce two clutches. If the parthenogen is caged with another parthenogen who has been treated hormonally so as to exhibit only male-like behavior, she will produce two and sometimes three clutches (this is the same number of clutches calculated to be produced in nature).

'Sexual' behavior in parthenogenetic whiptails is also related to the ovarian cycle (Figure 2.7). The behavioral roles during pseudocopulations are paralleled by differences in the circulating levels of sex steroid hormones. That is, individuals show primarily female-like behavior during the preovulatory stage when estrogen concentrations are relatively high and progesterone levels are relatively low; just the opposite is seen during the display of male-like mounting behavior which occurs most frequently during the postovulatory stages of the cycle when concentrations of estrogen are low and progesterone levels have increased.

But does 'sexual' behavior occur in nature? Much to my regret, I have never seen it, but there are at least three people who have observed this behavior in field studies. (How I envy these individuals.) I do have the second best evidence though. In 1990, Larry Young and I counted the number of whiptails (collected in the first two weeks of the reproductive season) that had bite marks on the back. (During copulation, the mounting animal clamps its jaws tightly on the side and back of the mounted animal. This jawgrip is so strong that a scar is formed that lasts for about two weeks). Significantly, the parthenogens were found to have copulation bite-marks. There was a difference in the frequency of bite marks between the species. Almost 80% of females and no males of the sexual species had bite marks, whereas they were present in only about half the parthenogens. Remember though that, because the partheno-genetic individuals alternate in their behavioral roles, at the beginning of the reproductive season only half the parthenogens would be expected to be behaving as females. This is also consistent with the observation that, in captivity, individuals will first establish a dominance relationship. In pairs, the dominant individual will complete its follicular growth, behaving as a female, while the other subordinate individual delays

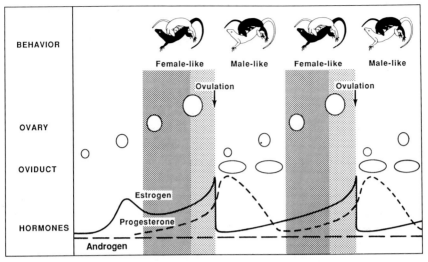

Figure 2.7. The psychobiology of 'sexual' behavior in the parthenogenetic whiptail lizard. Depicted is the relation among male-like and female-like 'sexual' behavior, ovarian state and circulating concentrations of sex steroid hormones during different stages of the reproductive cycle of parthenogenetic *Cnemidophorus uniparens*. The transition from receptive to mounting behavior occurs at the time of ovulation (arrow). The temporal patterns of ovarian hormone secretion are remarkably similar in the ancestral and descendant species. In both, the circulating concentrations of estradiol increase as the follicle grows, peaking around the time of ovulation. Progesterone levels begin to increase during the latter stages of follicular maturation, but are at their maximum at or immediately after ovulation; progesterone concentrations then decline as the corpora lutea are resorbed. In both female whiptails and in parthenogenetic whiptails, circulating concentrations of androgens are uniformly low and are not detectable by radioimmunoassay. (From Crews (1987), see Figure 2.3.)

ovarian growth and behaves initially as a male. With ovulation these behavioral and reproductive roles are reversed. Interestingly, the dominance relationship is stable throughout the season.

To return to the parthenogenetic fruitflies, do these animals also lack the ability to be stimulated by male courtship? Using a similar experimental design to that with the whiptail lizards, Carson and I evaluated the reproductive performance of sexual and parthenogenetic fruitflies by housing the parthenogen with or without a male; other individuals were isolated or placed in same-sex pairs as controls. In the sexual strain, there was a profound behavioral facilitation effect: virgin females housed with males produced 13 times more eggs than did isolated females and six times more eggs than did pairs of females. Since sterile (XO) males were as effective as fertile (XY) males, it must be the behavior and possibly the ejaculate that is the stimulatory agent. In the parthenogenetic fruitflies, however, there was no effect of social condition on egg production; the parthenogens laid the same number of eggs in the absence of males as in their presence.

The difference between the parthenogenetic whiptail lizards and the parthenogenetic fruitflies may lie in the origins of the species. The persistence of male-like 'sexual' behaviors and the behavioral facilitation of reproduction in parthenogenetic whiptails is perhaps due to the fact that the species arose from the hybrid union of two gonochoristic species. The loss of both these traits in the parthenogenetic strain of fruitfly may be due to the artificial selection of viable eggs of virgin females. Thus, in whiptails these traits are linked because of the historical behavioral interaction between a male and a female (hybridization), whereas in the fruitfly there was no such behavioral interaction. This suggests that the functional associations between gonadal sex and behavioral sex are not intrinsic but are dependent on how the species arose. Thus, how the unisexual strain is created may carry with it constraints associated with the stimulation of female-typical reproductive physiology.

The retention of male-specific pathways

If mating is going to occur, there have to be ways of distinguishing male from female. Some signals operate at a distance, such as odors, unique and/or exaggerated body parts, etc. (horns, crests, calls, scents). Sex-typical courtship behaviors, on the other hand, serve for close encounters and actually make fertilization possible. Just as evolution has selected for sex differences in morphology and physiology, so too has it shaped behaviors, and hence the brain. Indeed, sexual dimorphisms can be so extreme that in several instances female and male sexes were thought originally to be separate species

Today it is popular for evolutionary biologists (as has always been the case with comedians) to treat the sexes conceptually as being separate but dependent entities. By viewing sex differences in behavior and phenotype as a means of dividing up the resources in a species' niche, the costs and benefits of sex-specific traits and how they may have evolved can be appreciated.

The sexual brain

Among invertebrates, there are sex-specific structures in the brain. For example, the male tobacco hornworm moth recognizes the female by detecting, via his antennae, a pheromone she releases. The male antennae are associated with male-specific macroglomeruli in the central nervous system. Anne Schneiderman and John Hildebrand found that if the antennal imaginal discs (an imaginal disc is a 'bud' on the pupal body surface that will serve to create the specialized adult structures) of male pupae were grafted onto female pupae, the females developed male-like antennae upon metamorphosis. Further, the female's brain was altered, having male-like macroglomeruli, and they flew toward female pheromone sources! Similarly, if female antennal discs are transferred to male pupae, they develop female-typical antennae, and fly toward food and oviposition sources, something a male would otherwise never do.

As pointed out by Manfred Gahr (Chapter 12), vertebrates do not appear to have any sex-specific brain structures. Rather, they have more subtle differences, such as in the volume of specific nuclei, the complexity of certain connections, or the number of synapses made by or on a neuron. It has been known for more than 70 years that in sexual species each sex has the potential to become the opposite sex. Chromosomal anomalies or hormone treatment of zygotes can lead to functional sex reversal at times. Because of their sexual ancestry, unisexual vertebrates retain the potential for somatic and behavioral sexual dimorphisms.

This retention of male-specific pathways is seen particularly well in the embryos of hybridogenetic and gynogenetic vertebrates that have received treatment with exogenous androgens, resulting in the development of individuals with functional testes! Indeed, in my laboratory we have found that administration of the aromatase inhibitor CGS 16949A (courtesy Ciba-Geigy) to eggs of the parthenogenetic whiptails produces individuals that have distinct testes with no ovarian material, lacking oviducts, and possessing well-developed Wolffian ducts and hemipenes. This can only mean that male-specific traits are encoded by autosomal genes that are normally under the influence of male-typical steroid secretion and enzymatic patterns; in hybridogens, gynogens and true parthenogens, these genes normally are silent because of the female-like physiology.

Retention of male-specific pathways implies that, at a fundamental level, each individual vertebrate must be bisexual. The contemporary orthogonal paradigm of Richard Whalen is that maleness and femaleness represent separate developmental trajectories, not opposite ends of a single continuum (Figure 2.8). During a specific period of embryonic life, hormones that are secreted by the fetal gonad act to regulate the activity of those genes linked to reproduction, thereby sculpting those traits that later will differentiate male from female.

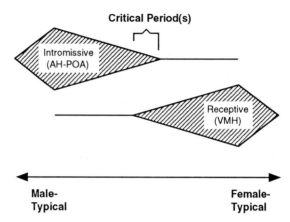

Figure 2.8. Schematic illustration of the current paradigm of the sexual differentia-
tion of brain areas mediating sexual behavior in vertebrates. Each individual has
residing within its brain the neural circuitry for both male *and* female sexual
behaviors. Abbreviations: AH–POA, anterior hypothalamus and preoptic area;
VMH, ventromedial hypothalamus. (From Crews, D. and Bull, J. J. (1987)
Evolutionary insights on reptile sexual differentiation. In *Genetic Markers of
Sexual Differentiation*, pp. 11–26. Eds. F. Haseltine, M. McClure and E. Goldberg.
Plenum Press, New York.)

If this sexual differentiation is going to occur, does maleness develop
independently of femaleness (and vice versa)? The concept of the male
as the organizing sex and the female as the default or neutral sex (that is,
if you are not a male, you must be a female), by its very nature, con-
tradicts this premise. Jim Bull predicted that a mutation that is beneficial
to one sex may be harmful to the opposite sex. William Rice has
demonstrated that this indeed occurred in fruitflies. The example cited at
the beginning of the chapter regarding size and fecundity in chickens is
another case in point.

At the level of the brain, specifically the neuroendocrine events that
underlie the expression of sexual displays, might one mechanism (e.g. that
controlling male-typical behaviors or physiology) interfere with the
adaptive evolution of another mechanism (that controlling female-typical
behaviors or physiology)? Whenever one has bothered to look for 'sexual'
behaviors in all-female species, pseudocopulatory displays are found that
are similar to the characteristic mating behaviors exhibited by their sexual
ancestors. Also, as seen above, these male-like behaviors can have a
definite function. But how could a mechanism that controls male-like
behaviors evolve in an all-female species? It turns out that the dependence
of male-typical sexual behaviors on testicular androgens is quite plastic.

Given the appropriate physiological constraints (we know of two: the length of time it takes to grow a gamete and the temperature necessary to sustain physiological processes like gametogenesis), sexual behaviors can evolve to become completely independent of steroid hormone activation. Recall the garter snake, the weaver bird and the zebra finch examples mentioned above. It is perhaps significant that this plasticity does not appear to extend to the primary and accessory sex structures.

However, the evolutionary loss of neural substrates, which are the physical connections in the brain that control behavior, is another matter. There is no doubt that peripheral sensory structures can be lost or gained, as is the case of the degenerate eyes of many cave organisms or the development of electroreception. But the loss of brain structures is rare: I can think of only a single instance among vertebrates. In fish, there are the Mauthner cells originating from the brain that control movement of the tail. The Mauthner cells are absent in the amniote vertebrates (mammals, birds, and reptiles); frogs have them as tadpoles but they have disappeared in the adult. For a long while, it was believed that, in humans, the vomeronasal organ (VNO), which is the primary filter for chemical pheromones in the brain, was present only in fetuses; adults were believed to lack the VNO. Now it is known that the structure is present in adult humans as well.

Given the apparent cost associated with an evolutionary loss of a central nervous system structure, it must be easier just to evolve another system of controls. That is, the structure remains but the agent activating that structure changes. In terms of the male-like 'sexual' behavior in parthenogenetic species, this would mean that a female physiology must become capable of stimulating specific brain areas to express male-like behaviors at the appropriate time. For this to happen, there must exist a predisposition for such novel functional relationships.

The evolution of new hormone–brain–behavior relationships

To date the only studies of the brain–behavior relations in a parthenogenetic organism have centered on the whiptail lizards. Early studies revealed that the display of male-like pseudocopulatory behaviors is not due to androgens secreted by the ovary. Radioimmunoassay of the circulating blood revealed that androgens are uniformly undetectable in the parthenogen throughout the reproductive cycle. In addition, the nature and pattern of sex steroid hormone secretions in females of the sexual species and the parthenogens are virtually identical. Other experiments using synthetic progestin agonists and progesterone receptor antagonists indicated that progesterone is not being metabolized into another hormone (e.g. androgen) in the brain. Together, this indicates that the evolution of parthenogenesis has *not* been accompanied by an alteration of the female-typical pattern of endocrine changes. What is the evidence for this claim?

Commonly, behavioral transitions mirror changes in hormone concentrations in the blood. There is an intriguingly close parallel between the transition from female-like to male-like 'sexual' behavior and the three-fold decrease in circulating levels of estrogen and the nine-fold rise in progesterone levels at ovulation (Figure 2.7). Is it possible that this shift in hormone concentrations plays a crucial role in controlling the expression of 'sexual' behavior? That is, could the postovulatory surge in progesterone have been exploited and become the new hormonal cue triggering male-like behaviors? To test this hypothesis, parthenogens had their ovaries removed and were then given an implant containing progesterone, estrogen, or nothing. Animals were paired with another ovariectomized parthenogen that had, or had not, received hormone treatment. The results were clear cut. Pseudocopulations occurred only in pairs in which both individuals were treated hormonally in a complementary fashion (e.g. in pairings of estrogen- and progesterone-treated individuals). Further, in all the pseudocopulations, the progesterone-treated parthenogen assumed the male-like role and animals treated with estrogen exhibited the female-like role. In the absence of the appropriate hormones, these behaviors are never exhibited.

Thus, unisexual and hermaphroditic vertebrates enable us to address this fundamental question about the evolution of the neural basis of sex-typical behaviors from a new perspective. Let us take the old idea that there are dual neural circuits in the brains of all vertebrates, one mediating mounting and intromission behavior and the other mediating receptive behavior (Figure 2.8). Researchers have long commented on males that exhibited female-typical sexual behaviors or, conversely, females that exhibited male-typical sexual behaviors. The bulk of modern research, however, has focused on the neuroendocrine mechanisms controlling homotypical behaviors, namely mounting behavior in gonadal males and receptive behaviors in gonadal females. In other words, each neural circuit has been studied extensively, but almost always in isolation from its complement. The parthenogenetic whiptails allow us to study these circuits operating together. Comparison studies allow us to say what is common to both and what is specific to each.

There appear to be two distinct neural circuits in the vertebrate brain that mediate sexual behavior (Figure 2.8). The anterior hypothalamus and preoptic area (AH–POA) are important integrative areas for the display of mounting and intromission behavior in all vertebrates studied to date. Similarly, the ventromedial hypothalamus (VMH) is an important integrative area for the display of receptivity. These areas are sensitive to sex hormones, as indicated by the now common finding that sex hormones are concentrated there. Furthermore, administration of sex hormones directly into these areas stimulate, whereas lesions in this areas abolish, male- and female-typical behaviors. Therefore, it is of particular interest that, in both the sexual and unisexual whiptails, intracranial implantation of androgens into the AH–POA elicits mounting behavior; androgen implants in the VMH not only fail to elicit mounting behavior, but they have no effect on receptive behavior. Conversely, implantation of estrogen in the VMH activates receptivity in both species, but estrogen implants in the AH–POA have no effect on receptive or mounting behavior.

A number of investigations have demonstrated sex differences in the vertebrate brain. But these are not differences of sex-specific nuclei like those described for the tobacco hornworm moth. Thus, I will argue that for vertebrates the sexual dimorphisms in behaviors cannot be attributed to the subtle dimorphisms in brain nuclei. My evidence will be the whiptail lizards. In the sexual whiptails only the males mount and only the females exhibit receptivity. The areas of the whiptail brain regulating these behaviors (AH–POA, VMH) are sexually dimorphic. The AH–POA, which we know to be involved in the control of mounting behavior, is larger in males than in females (Figure 2.9). The VMH, which controls female-typical receptivity, is larger in females. Juli Wade found that, during hibernation or following castration, the AH–POA and VMH of males became female-like in size (that is, the AH–POA shrank and the VMH enlarged). Recent studies using the Golgi silver impregnation technique to visualize individual neurons indicate that a similar relationship exists at the single-cell level. That is, the somata of neurons in the AH–POA of male *C. inornatus* are significantly larger than those in females, whereas in the VMH the somata of neurons is significantly smaller in males compared to females. These results indicate clearly that structural dimorphisms develop in the adult and, further, that testicular androgens control the seasonal growth of these areas.

However, in the parthenogenetic whiptail, which regularly and reliably exhibits both male-like and female-like 'sexual' behaviors, both the AH–POA and the VMH are similar in size to that seen in *females* of the

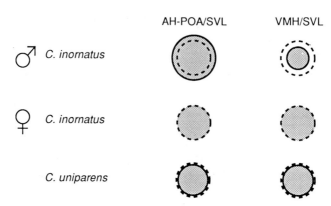

Figure 2.9. Schematic representations of the volumes of the sexually dimorphic areas in the brain relative to body size in sexual and parthenogenetic whiptails (*Cnemidophorus*). To aid in comparison, the volume of the AH–POA and the VMH of female *C. inornatus* is represented as a dashed outline in other drawings to indicate significant differences. The same relationships occur at the level of the soma of individual neurons in these brain areas. Abbreviations: AH–POA, anterior hypothalamus and preoptic area; VMH, ventromedial hypothalamus; SVL, snout-to-vent length. (From Crews (1992), see Figure 2.4.)

sexual ancestral species. This was found even in those individuals that were exhibiting male-like behavior. This same relationship applies to single neurons in these areas as determined by the Golgi method. Further, there was no difference in neuron soma size in those individuals exhibiting male-like behavior compared to those exhibiting female-like behavior. Even if the parthenogen is treated with androgen so that it exhibits both male-like behavior and coloration, the brain remains unchanged.

Such findings raise interesting questions about the meaning of sexual dimorphisms in the vertebrate brain. Clearly, the parthenogen retains the ability to express male-like behaviors. But it does so not because it has developed a masculinized AH–POA, but because it has coopted the naturally occurring progesterone surge to trigger the masculine behavioral potential that remains in a feminized brain. Taken together, these studies suggest that structural differences in the brain do not necessarily imply functional differences. Similarly, behavioral differences need not be paralleled by structural differences in the brain. Regarding the former postulate, Wade failed to find structural differences in brain area volumes between courting and non-courting males of the sexual whiptails during the breeding season or between courting and non-courting males castrated and given identical testosterone treatment.

Regarding the latter postulate, recall that the AH–POA and VMH do not differ significantly between parthenogenetic whiptails exhibiting male-like and female-like 'sexual' behaviors.

But how can progesterone, a 'female' hormone associated with ovulation, come to regulate a male-typical behavior? This is particularly puzzling given the well-known observation that androgen-dependent sexual behaviors in male mammals and birds are inhibited by treatments of progesterone. Indeed, progesterone has been used as a 'chemical castration' agent for convicted felony rapists. In the course of his studies of the male whiptail lizard, Jonathan Lindzey discovered that exogenous progesterone could stimulate sexual and copulatory behavior. That is, administration of exogenous progesterone *restores* the complete repertoire of male-typical sexual behavior in about one-third of castrated males. Indeed, these progesterone-sensitive males actively court and copulate with females with an intensity equal to that shown by castrates treated with androgen. I should mention also that while not all exogenous androgen-sensitive males are sensitive to progesterone, all progesterone-sensitive individuals are sensitive to both progesterone and exogenous androgens, suggesting that progesterone-sensitive males are a subset of exogenous androgen-sensitive males.

To test whether progesterone was acting directly to activate mounting behavior, Lindzey used various ligands, or synthetic agonists and antagonists that bind to progesterone receptors. The agonist R5020 induces progesterone receptor function, whereas RU486, a progesterone antagonist also known as the abortion pill, blocks progesterone receptor function. Administration of R5020 stimulated sexual behavior in castrated males, while RU486 abolished sexual behavior in castrated, progesterone-treated males. These experiments suggest that progesterone is probably exerting its stimulatory action as a progestin and not via conversion to other sex-steroid hormones and, further, acts via progesterone receptors and not via androgen receptors. This also suggests that the progesterone receptors may be functionally linked to neurons involved in male-typical sexual behavior.

How does an androgen-dependent male-typical behavior of the sexual ancestral species evolve to become a progesterone-dependent male-like behavior in the unisexual descendant species? Stephen Jay Gould and Elisabeth Vrba have pointed out that existing features may have been produced by two distinct historical processes, one of which is adaptation, or the gradual selection of traits resulting in improved functions. Some traits, however, evolved from features that served other roles, or had no function at all, and only later were coopted for their current role because they enhanced current fitness. This latter process may be termed exaptation. Thus, the difference between the two is that, in adaptation, traits are constructed by selection for their present functions, while exaptations are coopted for a unique use. In the present case, the variation among males in the sensitivity to progesterone may be the substrate on which selection operated, resulting in the novel hormone–brain–behavior relationship observed in the parthenogen. That is, the elevation of progesterone following ovulation presented an appropriate stimulus that, given the low circulating concentrations of androgens, was coopted to trigger mounting behavior in the parthenogen.

The story of the evolution of hormone–brain–behavior relationships does not end here. Larry Young found that progesterone can have both antiandrogenic and synandrogenic effects on sexual behavior in males of another reptile, the green anole lizard. How then in mammals and birds can progesterone serve as an antiandrogen and inhibit sexual behavior in males? A critical review of the published literature concerning this progesterone inhibition of male sexual behavior shows that all studies to date have used large pharmacological dosages. If administered at physiological levels, could progesterone stimulate male-typical sexual behaviors in genetic males?

Behavioral endocrinologists have not considered progesterone to be a hormone important to the regulation of male-typical sexual behavior as it is in female-typical sexual behavior. However, Pushpa and Satya Kalra found that male rats had a pronounced diurnal rhythm in progesterone secretion, where the peak in progesterone levels coincided with the period of greatest copulatory activity. Diane Witt, Larry Young and I collaborated on a study to test the hypothesis. Progesterone administered in physiological dosages (rather than the pharmacological dosages usually used) to castrated male rats causes them to mate with receptive females!

Thus, in vertebrates, the differences in brain circuitry that underlie these sex-typical behaviors are likely to be neutral constraints. However, it may be a negative constraint in humans, since the mind (versus the brain) operates within systems of social values.

The mechanics of sex determination

The retention of male-specific pathways, particularly those responsible for somatic differentiation, raises another interesting question. Namely, are the mechanics of sex determination a constraint in sexuality that potentially precludes the evolution of unisexuality? I have mentioned already the bipotential nature of the individual. Jean Wilson (Chapter 8) describes how the action of hormones during early development modifies the phenotype. This process is referred to as the organizational action of sex steroids. Studies with a variety of eutherian mammals demonstrate that perinatal castration results in adult individuals having female-typical morphology, physiology and behavior, whereas perinatal androgen treatment of females results in adult individuals with a masculinized phenotype. As pointed out by Marilyn Renfree (Chapter 9), the rule that sex differences result solely from gonadal secretions does not apply to marsupials. Indeed, the presence of XY fertile females and XX fertile males in wood lemmings (Chapter 19 by Karl Fredga) shows this not to be universal even in eutherian mammals!

Perhaps unisexual species can arise only from sexual species having male heterogamety. If a parthenogenetic vertebrate arose from sexual ancestors having female heterogamety this would lead to the production of sons, thereby failing to sustain the lineage. Recall the turkey and the fact that in birds the female is the heterogametic sex. Such a constraint does not appear to be the case in invertebrates.

Another constraint to parthenogenesis may be parental imprinting of the genome (Chapter 17 by Mary Lyon). Parental imprinting was discovered in mammals when developmental biologists observed that zygotes constituted from two haploid egg nuclei or from two haploid sperm nuclei would not develop to term, but that eggs constituted from one each of a haploid sperm nucleus and a haploid egg nucleus developed normally. The genome of an individual was thus being modified or 'imprinted' during gametogenesis differently in males and females, so that successful development requires complementary imprinted genomes. Thus, the male is necessary not only to provide half the chromosomes but also to provide chromosomes carrying the male imprint. Wherever this phenomenon exists, it poses yet another hurdle to parthenogenesis.

In species with environmental sex determination, each individual genotype is equally prone to develop into a functional male or a functional female. The extent to which environmental sex determination (versus genotypic sex-determining mechanisms) is compatible with parthenogenesis is not known. However, it would seem that, regardless of how the parthenogen was created (by selection or hybridization), the evolutionary advantages of environmental sex determination would not be visited upon the descendant parthenogen. Eric Charnov and Jim Bull have used empirical models to show that environmental sex determination is favored over genotypic sex determination when environmental resources are not distributed evenly but in patches and when the mother has no control over the environment in which the embryo will develop. These conditions are found in various reptiles, many of which have temperature-dependent sex determination (see Chapter 20 by Claude Pieau and colleagues).

Conclusion

Often, we hear of the two-fold cost of sex. First, there is the disadvantage of producing males. In many species, males represent half the population, but they cannot bear young. In all-female parthenogenetic populations, however, each individual is capable of reproducing. In small populations this can lead to a doubling in each generation compared to the slower growth rate of a sexual species. Second, when both sperm and egg are required for reproduction, there is the 'cost of meiosis' or the increased probability of losing advantageous gene combinations during the reduction divisions that accompany meiosis. If there are such advantages to unisexuality, why has parthenogenesis not replaced sexuality?

One possibility is that any modification of the sexual process usually results in other, lethal, developmental anomalies. However, it is possible to select for parthenogenesis, as demonstrated by the Beltsville turkey and the fruitflies. Another possibility is that the conditions necessary for parthenogenesis are rare. Many parthenogenetic vertebrates are known to have evolved from the hybrid matings of closely related sexual species. Recent studies by Peter and Rosemary Grant indicate that as many as 10% of bird species hybridize. Yet there are no known parthenogenetic birds (besides the turkey), suggesting that either this is too low a frequency to lead to parthenogenesis, or there are other constraints. Another possibility is that, in parthenogenesis, portions of the genotype are not 'tested and changed' by natural or sexual selection. Thus, because males are superfluous, male-specific genes in all-female species should accumulate mutations that may be deleterious if they are expressed again. The finding that treatment of embryonic hybridogenetic or gynogenetic fish with androgen will produce functional males who have viable sperm and mate, shows this not to be the case in at least these unisexual species.

Clearly, sexual reproduction must be advantageous, since it is the predominant mode in animals. But does the major advantage of sex, namely the reshuffling of genes in the face of environmental or genetic change, compensate for this two-fold cost of sex? Many outstanding evolutionary biologists and comedians do not think so, but no one has come up with a satisfactory explanation yet of 'why sex evolved'.

Consider that, since it exists, there may be great costs in abandoning sex. One cost often cited is that, in parthenogenetic reproduction, genetic variation is fixed and change occurs only by mutation. Thus, evolutionary success would require the creation of new clones with improved reproduction. Even though theory says that fixed genotypes may be an evolutionary dead-end, this does not appear to be the case when we study existing parthenogenetic species, although there are other costs. Because unisexuality is a derived state, any unisexual species probably retains features of the ancestral sexual state, creating negative evolutionary baggage. These features inherent to sexuality may consequently be constraints to unisexuality. If science is eventually to produce parthenogenetic stocks, we must explore these constraints in order to succeed.

Acknowledgements

I thank J. Bull, E. Prediger and K. Wennstrom for comments on the manuscript.

Further reading

Bell, G. (1982) *The Masterpiece of Nature. The Evolution and Genetics of Sexuality.* University of California Press, Berkeley, California.

Crews, D. (1987) Functional associations in behavioral endocrinology. In *Masculinity/Femininity: Basic Perspectives*, pp. 83–106. Eds. J. M. Reinisch, L. A. Rosenblum and S. A. Sanders. Oxford University Press, Oxford.

Crews, D. (1992) Behavioral endocrinology and reproduction: an evolutionary perspective. In *Oxford Reviews in Reproductive Biology*, pp. 303–70. Ed. S. Milligan. Oxford University Press, Oxford.

Dawley, R. M. and Bogart, J. P. (eds.) (1989) *Evolution and Ecology of Unisexual Vertebrates*, New York State Museum, Albany, New York.

Ghiselin, M. T. (1974) *The Economy of Nature and the Evolution of Sex.* University of California Press, Berkeley, California.

Goy, R. W. and McEwen, B. S. (1980) *Sexual Differentiation of the Brain.* The MIT Press, Cambridge, Massachusetts.

Smith, J. M. (1978) *The Evolution of Sex.* Cambridge University Press. Cambridge.

Somatic dimorphisms across the species

3

Sexual selection and the evolution of sex differences

JOHN D. REYNOLDS and PAUL H. HARVEY

Many differences between the sexes cannot be explained from the simple physiological necessities of successful reproduction. A male bird of paradise could father a successful clutch if his elaborate plume were plucked and a male red deer could sire many offspring if his antlers were sawn off. So why do male birds of paradise have elaborate plumes and why do male red deer grow antlers? After all, male sparrows and horses reproduce perfectly well without plumes and antlers. Inevitably, our answers involve a combination of causes, but it is useful from the start to distinguish between adaptation and contingency.

We might argue that plumes attract females, and antlers help in fights with males. Those species in which mating displays and competition are most important should have evolved the more elaborate traits. In that sense, the traits would be adaptations for gaining mating access to females. However, why have different kinds of ornaments and weapons evolved in different taxa? Why have male deer not evolved enormous claws, like those of fiddler crabs, and why do fiddler crabs not have antlers? Here we reach into the realm of contingency, although adaptation is not irrelevant. It so happens that particular evolving lineages already had weapons which were used in other contexts, or they had organs which selection could fine tune as weapons. Different lineages solved essentially the same problem in different ways, contingent upon the inheritance of appropriate precursors. Once we realize that enlarged claws and antlers are weapons used in male–male combat, or bright colours and loud songs attract females, we can also seek similarities of lifestyle in the different taxa that are associated with these traits. Can we identify ecological similarities between crabs with the largest claws and deer with the largest antlers? This trick of seeking instances of convergent evolution allows evolutionary biologists to interpret diversity in terms of a near-manageable number of selective processes.

This approach is the essence of Darwin's comparative method, instigated in *The Descent of Man, and Selection in Relation to Sex* (1871). Darwin used comparisons within and among species to suggest and to test reasons for life's diversity. We can apply this technique to understand why closely related species vary in the extent to which the sexes differ in morphology and behaviour.

Sexual selection versus natural selection

Before we attempt to classify the selective forces implicated in gender differences, we should make a caveat. Here, we are not concerned with the evolution of traits that organisms clearly need to survive and to reproduce as males or females. Testes, ovaries and associated plumbing are accepted as having evolved, at least initially, through natural selection, and are outside the scope of our discussion. We are interested in sex differences that cannot be explained so easily. However, like Darwin before us, we have to be careful when making a distinction between differences that have evolved through natural selection and those that have evolved through selection in relation to sex. The distinction, although sometimes slightly artificial, can be valuable. A male crab may need claws large enough to hold a female during copulation. If some crabs copulate in rougher waters than others, their claws may need to be larger to hold on to the female. Those claws would be organs necessary for reproduction and do not concern us here. However, some crabs, like the fiddler crabs we have mentioned, have claws which are far larger than necessary to clasp the female, and which may hamper their survival. These sex differences are of interest because they do not have a ready explanation. What causes variation among species in the degree of enlargement of male claws?

More subtly, only one successful sperm is necessary for an egg to be fertilized. Why then do the males of some species have much larger sperm-producing capacity than their close relatives? We might suspect that these are the males of species where the female needs to fertilize more than one egg or where the male gets the opportunity to fertilize the egg(s) of more than one female. We would be wrong, as we see below. If we had been right, the differences between related species would follow as a result of what Darwin called 'natural selection' rather than what he called 'selection in relation to sex' and what, more usually, we call 'sexual selection'.

Historically, sex differences arising from sexual selection have been considered to be of two types. First, differences can result from mating preferences: if males will mate only with females that have bright colours, then females may evolve bright colours. Second, differences can result from selection for success at combat among members of one sex to gain mating access to members of the other sex: if males are a limiting resource, those females that can win fights for them will be the ones that leave offspring. More recently, a number of sex differences have been ascribed to a more subtle form of combat involving sperm competition. Whenever females mate with several males, then those males might evolve characteristics to help them win at fertilizing the eggs.

Which sex will diverge farthest from the natural selection optimum? The answer depends upon which sex is a more limiting resource for reproduction by the other. Often, males provide very little for their young, and their rate of reproduction is limited by the number of eggs they can fertilize. The potential reproductive rate of females, on the other hand, tends to be limited by the number of eggs they can produce or by the number of young they can raise. If females are a limiting resource for males, Darwin argued that the consequent 'law of battle' among males was bound to favour the elaboration of behavioural displays and ornaments to attract females, as well as weapons and greater body size to fight with other males. The evolutionary trajectory of sexually selected traits will then depend on genetic variation for precursors in the lineage (contingency), and proceed to the point where the cost in terms of reduced viability is offset by the gains in reproductive success (adaptation). Where females contribute little or no parental care relative to the male, and male care cannot be shared easily among the offspring of many females, males become a limiting resource. Females are thus more competitive, and typically evolve more elaborate courtship displays and weaponry. Well-known examples include phalaropes and some species of pipefishes, dendrobatid frogs, midwife toads, giant water bugs and bush crickets. In all these cases, male parental care takes longer per brood than the rate at which females can produce eggs (see Clutton-Brock and Vincent for a review).

Sexual selection can be important in all sexually reproducing species, including plants (see Harvey and Bradbury), and it can have profound impacts extending beyond reproduction. For example, if males evolve a larger size than females, we should expect them to have a different diet and generally to behave in a way that accords with their larger size. In other words, they will come to occupy a different ecological niche from females. These side effects will be discussed at the end, after separate considerations of the three main ways in which sexual selection can produce differences between the sexes: mate choice, combat and sperm competition.

Mate choice

Males in many taxa spend a large amount of time displaying rather ornate characteristics, such as bright colours or long tails. Why have these evolved? Darwin suspected that such traits might appeal to an aesthetic sense of females. Unfortunately, Darwin lacked solid experimental proof of such preferences, although he did encourage a sceptical pigeon fancier, Mr W. B. Tegetmeier, to dye the feathers of male birds magenta. Apparently, the female pigeons were not impressed, and neither were Wallace and Huxley impressed with the general notion that female animals should prefer non-adaptive ornaments and displays.

Over 100 years later, David Semler and Malte Andersson tried again, with more encouraging results. Semler noticed that male three-spined sticklebacks in a Washington lake varied greatly in the extent of red coloration on their throats, whereas females were silver-brownish. After finding that females preferred to spawn with red males, he painted some dull-coloured males with red lipstick and others with transparent lip gloss as controls. In laboratory preference tests, females again preferred the red males (Figure 3.1A). This suggests that female choice may be at least partly responsible for the evolution of red colours in male sticklebacks.

Andersson used an even more spectacular animal, the African long-tailed widowbird, to test for female choice. The females are dull with short tails, while the males are black with red wing epaulets and tails measuring 50 cm. Andersson cut the tails of some males and used the feathers to extend the tails of other males. The result was that females nested in the territories of males with elongated tails at the expense of males with normal or shortened tails (Figure 3.1B).

Similar experiments have now been performed on a variety of species. In the normally monogamous barn swallow, Anders Møller found that not only did females prefer to pair with males that had elongated tails, but that the females that failed to get such males as partners attempted to engage in extra-pair copulations with them! It seems that the females wanted the genes of long-tailed males if they could not get the males themselves.

Although female choice may be at least partly responsible for the

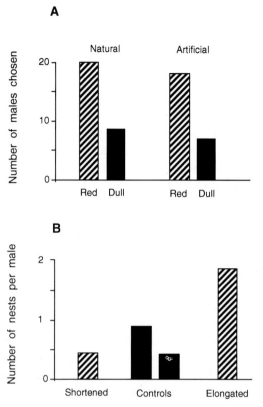

Figure 3.1. Female choice for extravagant male traits. (A) Female three-spined sticklebacks prefer redder males (Semler, 1971). (B) Female long-tailed widowbirds prefer males whose tails have been elongated, rather than ones with shortened tails, or controls (either cut and reglued, or uncut; Andersson (1982)).

evolution of elaborate male display traits, it is still not obvious *why* females prefer mating with such males. Several possibilities have been discussed in the literature on sexual selection, but at the moment we seem far from reaching a consensus.

R. A. Fisher suggested that if females develop a slight preference for males of a particular sort, say with tails longer than average, then those males will father a disproportionate number of offspring and, at the same time, the characteristic will become exaggerated. Fisher envisaged that females were selected to prefer males with longer tails because, originally, longer tails had a natural selection advantage, but that tail size would increase due to the mating advantage it conferred on sons until it reached beyond its optimum and had a natural selection disadvantage. Various models of Fisher's 'runaway' process demonstrate that it could be important until the benefits due to sexual selection balance the viability costs from natural selection, but only if there are no costs associated with female choice or if patterns of mutation are biased appropriately.

Others have argued that the displays indicate that their bearers possess 'good genes' of some sort. Amotz Zahavi suggested in 1975 that if a male can survive with a viability handicap, such as a long tail, females should choose to mate with him, since he has demonstrated his vigour in the face of adversity. Although this idea was treated with scorn for many years, it has gained increasing acceptance, and has produced some innovative spinoffs. For example, William Hamilton and Marlene Zuk argued that parasitized birds cannot develop extravagant secondary sexual characteristics. Consequently, if resistance to parasites is heritable, by choosing showy males as mates, females will produce offspring that are resistant to parasites.

There is some support for the 'good genes' hypothesis from a study of Trinidadian guppies. Females prefer to mate with larger males, and in laboratory studies these males sire faster-growing offspring (Figure 3.2). Larger size in daughters translates into higher fecundity. These results do not rule out potential Fisherian advantages from mating with larger males, since sons inherit their father's size, but they are consistent with a possible genetic benefit to females from choosing larger males.

More recently, Anders Møller and Randy Thornhill have argued that the degree of bilateral asymmetry in sexually selected traits might indicate quality. It is physiologically difficult for animals to produce symmetric traits, and any genetic abnormality might be reflected in differences in the lengths of the left and right elongated tail feathers of barn swallows, or in the size and shape of patches of colour on each side of a fish's body. Early tests have been encouraging for an assortment of traits, including the pincers on earwigs, wing lengths in scorpionflies, and whisker spots on lions, but we await rigorous genetic experiments.

A third possibility is that extravagant male characteristics could reliably indicate direct benefits to females or their offspring. In species in which males contribute parental care or territorial resources, perhaps male sexual traits reliably reflect such direct benefits in addition to any genetic quality they might indicate. Again, bilateral asymmetry of ornaments is one potential cue whereby animals could assess either genetic or environmentally caused variation in quality of prospective mates, and, hence, the resources they will provide for offspring. Ornaments might also indicate more subtle direct benefits to mates, even in species where members of the

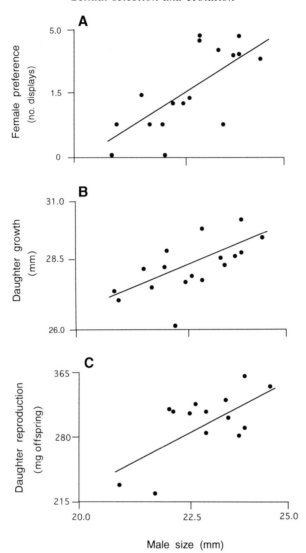

Figure 3.2. (A) Female guppies prefer larger males; (B) these males sire faster-growing daughters; (C) these daughters have higher fecundity. Each datum point represents the mean preference or offspring performance of two females tested per male. The scales are log-transformed (see Reynolds and Gross, 1992).

opposite sex appear to contribute little besides genes to the offspring. For example, if long tails demonstrate that males are healthy and dominant over rivals, females might choose such males because they run a lower risk of catching a sexually transmitted disease or of being harassed by other males when mating. If the costs of choice are low, these slight benefits could augment genetic benefits and explain why females are often highly choosy even in species where males have little to offer (see Reynolds and Gross).

There is a compelling body of experimental evidence showing the importance of mate choice in the evolution of sexually dimorphic traits. But, so far, few solid predictions have been upheld by comparisons among species of display characteristics, unlike the situation for comparisons of combat traits, which are described below. For example, several comparative studies have been performed on bird song and coloration, with attempts to relate extravagance to particular selective forces such as parasite load. The resulting correlations seem weak at best.

Why have comparative studies had so little success in explaining the evolution of displays related to mate choice? We suggest four reasons: wrong form of selection; wrong emphasis on benefits rather than costs of displays; wrong understanding of evolutionary precursors; and wrong measures of mate choice. Each possibility is illustrated with examples.

Wrong form of selection

Although bold colours may signal quality to prospective mates, perhaps they also signal quality to would-be predators, particularly the ability of their bearers to flee. Thus, perhaps sex differences in colours do not depend solely on mate choice (or on other forms of sexual selection), but also on differences in the importance of signalling to predators. When this 'unprofitable prey hypothesis' was first proposed by Robin Baker and Geoff Parker in 1979, it drew immediate and sustained fire from many quarters, partly because it ignored the sexual context of many displays. However, two recent studies suggest that it may deserve a second look. First, male skylarks have been observed singing bursts of their famed songs when hotly pursued by sparrowhawks. Second, an experiment by Frank Götmark showed that stuffed mounts of male pied flycatchers received fewer attacks from raptors than did female mounts, despite males having a conspicuous black and white pattern compared with the dull grey-brown colour of females. Could complex songs and bold coloration serve, at least in part, to advertise stamina and unprofitability to predators? These examples do not rule out the likelihood that animals also use such displays to demonstrate their quality to each other, especially where displays are limited to one sex. However, traditional comparative studies relating display characteristics to sexual selection alone might do well to consider other contexts.

Wrong emphasis on benefits rather than costs of displays

Perhaps some species are more sexually dimorphic than others because of differences in the costs of displays, due to natural selection, rather than in the benefits, through sexual selection. For example, some species of livebearing fish in the family Poeciliidae, such as the guppy, are strongly dimorphic in colour patterns. Females are a dull silver-tan colour, while male guppies sport a mosaic of colour patches including yellow, orange, blue, silver and black. In relatives, such as some species of mosquitofish, both sexes are cryptically coloured. The intensity of sexual selection could theoretically be the same in both species, hence the potential benefits of displaying colour patterns might not differ. The key might be found in the costs faced by males in both species during the evolution of sexual displays. If the habitats of mosquitofish contain more visually hunting predators than those of guppies, conspicuous males may pay a particularly high price, and this difference in the costs of displays could explain species differences in the degree of sexual dimorphism. Support for this comes from comparisons among populations of guppies in Trinidad, where John Endler and others have shown that the intensity of male coloration is related to differences in danger from local predatory fish.

The costs of many sexually selected traits such as colours are hard to measure. However, a recent comparative study of birds is encouraging. Based on aerodynamic models, Andrew Balmford, Adrian Thomas and Ian Jones knew that tails which were elongated into shallow forks were efficient for manoeuvering, whereas tails elongated more uniformly (like a magpie's) or only in the centre (a pintail) caused considerable drag. Thus, the authors predicted that birds that did not rely heavily on flight, such as ground feeders and non-migrating species, would be better able to afford the costs of long, graduated tails. This was supported by their comparative study (Figure 3.3A). Such species should also show greater sexual dimorphism in tail length, since the elongation must be due to sexual selection, rather than natural selection (which could explain forked tails). This prediction was also supported (Figure 3.3B). These results suggest that costs of sexually selected traits may be important in the diversity of sexual dimorphism found in birds. It remains to be seen whether ecological differences among species control such traits directly, or indirectly by influencing mating patterns, and hence the benefits of displays.

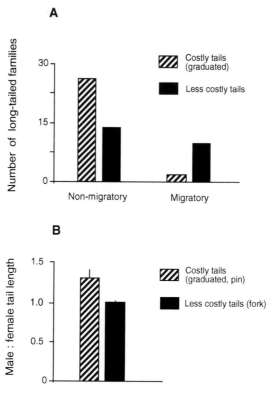

Figure 3.3. Costs of long tails and sexual dimorphism in birds. (A) The number of long-tailed families of birds according to whether they migrate, and the aerodynamic expense of different tail shapes. (B) Average sexual dimorphism (± 1 SE) in tail lengths of nine evolutionarily independent pairs of species, according to aerodynamic expense. Tail lengths were controlled for body size (Balmford *et al.*, 1993).

Wrong understanding of evolutionary precursors

In the guppy/mosquitofish example above, female guppies might prefer brightly coloured males because their ancestors were more sensitive to bright colours than were the ancestors of mosquitofish. Perhaps they specialized in bright foods. This could have provided the evolutionary template for sexual section to build upon. The potential importance of pre-existing biases in sensory systems has been promoted by Michael Ryan and John Endler. The hypothesis is important because it places sexual selection into a clear phylogenetic framework, and it leads to several testable predictions. For example, one could compare the colour sensitivities of related species with the colours used in sexual displays.

Studies of frogs (Ryan and colleagues) and swordtail fishes (Alexandra Basolo) have confirmed that females may have pre-existing preferences for traits which males of their own species do not possess. The implication is that if such traits do arise, they may evolve rapidly by female choice, perhaps because initially their bearers are easy for females to detect, and later, as the traits become costly, because they signal male quality.

Wrong measures of mate choice

The application of DNA fingerprinting and other molecular genetic techniques has revolutionized our understanding of paternity in wild animals. In turn, this has overthrown our earlier efforts at measuring the intensity of sexual selection. No longer can we assume that males have sired every offspring born to their putative mates. In the apparently monogamous indigo bunting, for example, David Westneat found that one-third of all offspring were actually fathered by someone other than the female's 'mate'! Interestingly, this species exhibits marked sex differences in colour patterns. This offers a clue to the puzzle of why such species should be so dimorphic, despite their apparent monogamy. There is much more scope for variation in male mating success than was assumed previously. Anders Møller and Tim Birkhead compiled studies of paternity in birds, and matched them with measures of sexual dimorphism. They showed that when one uses the correct measure of intensity of sexual selection (variation in number of mates fertilized), the differences among closely related species in extent of plumage dimorphism are indeed related to differences in sexual selection. Males of dimorphic species perform relatively more extra-pair fertilizations than males of monomorphic species, as predicted from sexual selection theory.

Combat

In many species, the struggle to obtain mates can be rather violent. Darwin suggested that fights between members of the same sex for dominance or resources which allow them to attract members of the opposite sex may be responsible for the evolution of many costly traits, such as antlers, tusks or large body size. The potential gains in reproductive success are likely to be particularly high whenever there is the opportunity to mate with numerous members of the opposite sex.

Since the late 1970s, several studies, particularly of various vertebrate groups, have demonstrated clear relationships between the extent of male weaponry and the degree of polygyny, as measured (e.g. Tim Clutton-Brock and colleagues) by female group size. For example, male primate canine size and male deer antlers are both reasonably well related to female group size. Of course, the fact that female deer (Cervidae) do not usually possess antlers accords with the idea that antlers have evolved as a consequence of sexual selection among males. However, many female antelopes (Bovidae) have horns which may be the same length as those of the males. In an illuminating comparative study, Craig Packer attempted to distinguish between the effects of sexual versus natural selection on dimorphism in the horns of African antelopes.

Male antelopes use their horns in vigorous head-to-head clashes when competing for access to mates. Both sexes use their horns in self-defence against predators, and females also use their horns to defend their calves. Packer compared three components of horn morphology: total length, basal area and shape. Among those species in which both sexes possess horns, males and females have horns of equal length, but male horns have double the basal area and tend to be more curved with tips pointing back to the base rather than forward as in females.

Packer interpreted his results, with additional supporting evidence, as follows. The increased basal area of male horns allows them to withstand twice the force without breaking during a butting or shoving match. Evidence that males do indeed need strong horns is provided by the fact that, despite their greater basal area, the frequency of observed horn breakage is much higher among males than among females, and is highest of all among those males with the thinnest horns. Packer goes on to argue that the more complex horn shape in the male serves to catch the blows of an opponent's horns and to gore the neck or belly with a hooking movement of the head. Since females use their horns to ward off attacks by predators, they do not need such thick horns, as is evidenced by their lower rate of horn breakage. Furthermore, small-bodied species tend to flee from predators rather than standing to fight, and it is the females of small-bodied species ($<25\,\mathrm{kg}$) that are almost invariably hornless.

A complementary study of the role of weaponry in intrasexual combat comes from Peter Jarman's review of the distribution of dermal shields (patches of thickened skin) on the bodies of mammals. Such shields develop in areas that are likely to receive blows from the weapons of conspecifics during combat. Jarman concludes that there is a repeated matching of distributions of probable blows and thickened skin in mammals. Furthermore, dermal shields are distributed more extensively among males than among females, pointing to the importance of sexual selection in the evolution of both male weaponry and dermal shields.

These studies show that to understand differences among species in weapons and body size we should pay attention to the details of *how* animals fight with one another. Shoving, biting, grappling, or stabbing can each select for different combative and defensive traits. Whichever traits are favoured, they tend to be more pronounced in males, compared to females, in mammal species where females live in groups, and where males have potential mating access to more than one female. During the period when behavioural ecologists and evolutionary biologists were demonstrating these correlations, they also began to understand better why females live in groups of different sizes in the first place. Usually, combinations of the need for defence against predators and the clumping of food resources in space and time select for differences in female group size.

Although comparative studies have provided important advances in our understanding of sexual dimorphism through combat, until recently they shared a common limitation for making evolutionary inferences – lack of independence among related species. Whenever species are compared with one another, it is necessary to consider similarities shared through common ancestry, rather than as responses to con-temporary selection pressures. For example, it had long been assumed that bird species which use mating arenas (leks) were particularly sexually dimorphic. However, Jacob Höglund pointed out that across ten bird families in which lekking occurred, careful phylogenetic analysis did not reveal an evolutionary association between size dimorphism and lekking. Subsequently, an improved phylogenetic analysis by Ted Oakes did find an evolutionary association. The issue is still not fully resolved, but the warning is clear: phylogenetic relationships must be considered in any comparative study.

Sperm competition

In an influential series of papers during the 1970s, Roger Short suggested that differences in testis size among great apes might result from sperm competition: gorillas are four times the weight of chimpanzees, but chimpanzees have testes that are four times the weight of gorilla testes (see Chapter 1 by Roger Short). Since female gorillas mate with just one male per oestrus, that male needs only to produce enough sperm to ensure fertilization of the egg. In contrast, female chimpanzees mate with several males, thereby causing sperm competition. One way of increasing the chances of fathering the offspring is to have more tickets in the lottery: in other words, to produce more sperm and hence the larger testes. If Short was right, perhaps other primate species in which females mated promiscuously would also have males possessing large testes? The prediction turned out to be correct, and the results were not confounded by evolutionary history – in each of four monophyletic taxa with some species having sperm competition and others not, those with sperm competition have relatively large testes.

Short's idea has been taken further, with similar patterns having been found in other mammals and in birds. Comparative tests have been able to distinguish between sperm competition and sperm depletion as being causes of differences among species in testis size. For example, among birds, the largest testes occur in polyandrous species where females attempt to form pair bonds with more than one male per season, and hence the potential for sperm competition is particularly high (Figure 3.4). In these species the potential for sperm depletion is probably low, because the total number of females with which males mate is small. By contrast, in species using lek mating arenas, where individual males may copulate repeatedly, we might expect the greatest potential for sperm depletion, and yet these birds have the smallest testes (Figure 3.4). This is consistent with the fact that females may often mate with only a single male on leks, resulting in low sperm competition.

It is important to realize that the comparative patterns expected from sperm competition – a form of postmating sexual selection – are different from those expected from premating sexual selection. For example, there is more potential for male combat to be advantageous in polygynous primates than in monogamous ones but, among the former, only those species where females mate with multiple partners will be selected to have large testes. As a consequence we find the following pattern: the males of monogamous species have small testes and there is no sexual dimorphism in body size or canine size; the males in species where males fight for exclusive access to numerous females also have small testes, but there is sexual dimorphism in both body size and canine size; the males of group-living polyandrous species have large testes and there is sexual dimorphism in body size and canine size. Furthermore, there is now evidence that males from species where females typically mate with more than one male not only have larger testes but also larger ejaculates with more motile sperm per ejaculate. The next time a new species of primate is discovered, we should be able to deduce

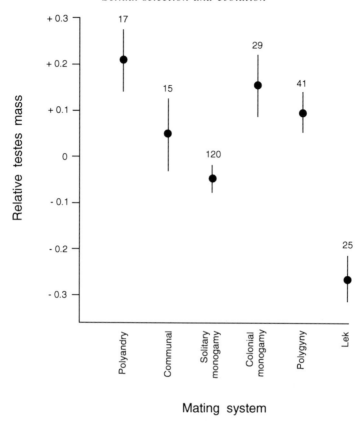

Figure 3.4. Testes weight relative to body weight in bird species in relation to social systems. Points are means (residuals controlling for body size in regressions), vertical lines are standard errors, and numbers are number of species (Birkhead and Møller, 1992).

its social behaviour by examining its testes and dimorphism in body and canine size.

Discussion

Whenever we see differences between the sexes in behaviour, morphology or physiology that exceed the basic necessities of reproduction, we should bear in mind the struggle which Darwin noted, namely 'the advantage which certain individuals have over other individuals of the same sex and species, in exclusive relation to reproduction'. We now have many examples from a variety of vertebrate taxa of relationships between the degree of sexual dimorphism and the extent of mate competition. These comparative studies, in conjunction with field and laboratory experiments, provide especially strong support for the importance of competition within the sexes, through direct combat and sperm competition. Comparative support for the role of mate choice has lagged behind, but we suspect that the evidence will improve as we learn more about the costs of the traits, as well as the benefits measured through genetic paternity analyses.

Although sexual selection has often been implicated in the more competitive sex evolving larger size, brighter colours or better armaments for combat, there are numerous examples of particular traits such as body size being less developed in the more competitive sex. For example, in most species of frogs, toads, snakes, insects, spiders and livebearing fishes, females are larger than males. Males are smaller in spite of a frequent advantage to large individuals in attracting mates and fighting with rivals. These species highlight the need to consider selection acting on *both* sexes rather than just using one to provide a baseline for comparison with the other. In these taxa, larger females produce more or larger offspring. Thus, natural selection for larger size in females probably exceeds natural and sexual selection for large size in males. Males can divert their energy more profitably into finding mates rather than into paying the costs of achieving and maintaining a large size, such as reproductive restraint early in life and high metabolic costs. Some species of anglerfish take this to the extreme, where tiny males become embedded permanently in the female's body, effectively parasitizing her while providing sperm. Therefore, 'sex differences' in any trait represent different evolutionary outcomes for each sex in response to different intensities of natural and sexual selection.

Dimorphisms in sexual traits have important ramifications for patterns of growth, development, mortality and sex ratios (see Chapter 15 by Tim Clutton-Brock). They also lead to important ecological differences between the sexes. Consider foraging behaviour: male guppies are less willing to forage in the presence of a predator than are females; male fiddler crabs feed at a slower pace than females; and, in some species of sandpipers, males and females feed in different locations.

Have these cases of ecological divergence occurred because of sex differences in sexual and natural selection, or because of some alternative? One possibility is that ecological divergence evolves because it minimizes competition between the sexes. This was a popular explanation for sex differences in traits such as the bill lengths of birds prior to the explosion of interest in sexual selection in the 1980s. We can use the comparative approach to test this argument. Ecological competition between the sexes will be strongest in species in which a monogamously pair-bonded male and female defend a territory together. Yet these are generally the species in which sexual dimorphisms are *least* pronounced (e.g. geese). This is contrary to expectation from the competition hypothesis, but in accordance with expectations from sexual selection. Ecological differences between males and females appear to be generally explainable as adaptations to differences in reproductive roles. In turn, these roles are established by natural and sexual selection.

Comparative studies have lent considerable support to Darwin's original insights into the evolutionary importance of reproductive competition. Explorations of asymmetries between males and females in sexual selection should lead to continued discoveries about the evolution of sex differences in physiology, ecology and life histories.

Acknowledgements

We thank Sarah Benton, Tim Benton, Isabelle Côté, Marion Petrie and Bill Sutherland for help and discussions, and Roger Short, Evan Balaban and the participants in the Comparative Physiology Conference for an exciting interdisciplinary meeting.

Further reading

Andersson, M. (1982) Female choice selects for extreme tail length in a widowbird. *Nature*, **299**, 818–20.

Balmford, A., Thomas, A. L. R. and Jones, I. (1993) Aerodynamics and the evolution of long tails in birds. *Nature*, **361**, 628–31.

Birkhead, T. R. and Møller, A. P. (1992) *Sperm Competition in Birds:Evolutionary Causes and Consequences.* Academic Press, London.

Clutton-Brock, T. H. and Vincent, A. C. J. (1991) Sexual selection and the potential reproductive rates of males and females. *Nature*, **351**, 58–60.

Clutton-Brock, T. H., Albon, S. D. and Harvey, P. H. (1980) Antlers, body size, and breeding group size in the Cervidae. *Nature*, **285**, 565–7.

Hamilton, W. D. and Zuk, M. (1982) Heritable true fitness and bright birds: a role for parasites? *Science*, **218**, 384–7.

Harcourt, A. H., Harvey, P. H., Larson, S. G. and Short, R. V. (1981) Testis weight, body weight and breeding system in primates. *Nature*, **293**, 55–7.

Harvey, P. H. and Bradbury, J. W. (1991) Sexual selection. In: *Behavioural Ecology: an Evolutionary Approach* (J. R. Krebs and N. B. Davies, eds.), pp. 203–23. Blackwell Scientific Publications, Oxford.

Jarman, P. J. (1988) On being thick-skinned: dermal shields in large mammalian herbivores. *Biological Journal of the Linnaean Society*, **36**, 169–91.

Møller, A. P. (1988) Female choice selects for male sexual tail ornaments in the monogamous swallow. *Nature*, **322**, 640–2.

Møller, A. P. (1990) Fluctuating asymmetry in male sexual ornaments may reliably reveal male quality. *Animal Behavior*, **40**, 1185–7.

Packer, C. (1983) Sexual dimorphism: the horns of African antelopes. *Science*, **221**, 1191–3.

Reynolds, J. D. and Gross, M. R. (1990) Costs and benefits of female mate choice: is there a lek paradox? *American Naturalist*, **136**, 230–43.

Reynolds, J. D. and Gross, M. R. (1992) Female mate preference enhances offspring growth and reproduction in a fish, *Poecilia reticulata. Proceedings of the Royal Society of London, series B*, **250**, 57–62.

Semler, D. E. (1971) Some aspects of adaptation in a polymorphism for breeding colours in the threespine stickleback (*Gasterosteus aculeatus*). *Journal of Zoology*, London, **165**, 291–302.

Short, R. V. (1979) Sexual selection and its component parts, somatic and genital selection, as illustrated by man and the great apes. *Advances in the Study of Behavior*, **9**, 131–58.

Thornhill, R. (1992) Female preference for the pheromone of males with low fluctuating asymmetry in the Japanese scorpionfly (*Panorpa japonica*). *Behavioral Ecology*, **3**, 277–83.

4

Sexual dimorphism in the Hymenoptera

J. WILLIAM STUBBLEFIELD and JON SEGER

Spectacular sex differences of many kinds occur abundantly among the wasps, bees and ants that make up the insect order Hymenoptera. In some cases these differences are so extreme that males and females of the same species have been classified in different genera for decades, until a chance observation of mating, or emergence from a single nest, establishes their identity. Even where the sexes are similar in morphology they lead very different lives. The hard-working females hunt for prey or other larval provisions, and in many taxa they carry these provisions back to a nest that they have constructed to protect their offspring. The males, by contrast, lead short lives (sometimes nasty and brutish), devoted to the single purpose of inseminating females. Countless variations on this theme have evolved during the long and successful history of the order, and other features of hymenopteran biology have allowed these sex differences of ecology to be translated into equally striking sex differences of behavior, morphology and physiology.

Hymenoptera provide excellent illustrations of the classical principles of sexual selection, and we believe that they also present opportunities to extend the study of sex differences in several directions. These opportunities are created by four basic characteristics of the group. First, the sexes tend to be more different from each other in Hymenoptera than they are in most other animals (see Figures 4.1, 4.4, 4.7, 4.9 and 4.10). Some possible reasons for this tendency will be mentioned below, but whatever the cause, it means that the 'signal' to be studied is relatively strong. Second, alternative phenotypes also occur *within* one sex or the other in many species (see Figures 4.2, 4.3 and 4.9). The most familiar example of intrasexual dimorphism is the difference between the female reproductive and worker castes in many social taxa, and in some ants there are further distinctions among two or more castes of workers. But there are also

71

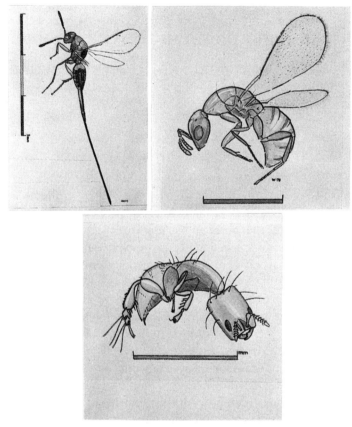

Figure 4.1. *Philotrypesis parca*, a parasitic fig wasp in the family Torymidae. The female (top left) uses her long ovipositor to place a few eggs into each of many different figs. Two kinds of male occur in this species. The winged form (top right) leaves its natal fig to search for mating opportunities elsewhere, while the wingless form (bottom) remains within the fig and fights with other wingless males for access to females that have not yet left the natal fig (scale in mm). (From Bouček, A., Watsham, A. and Wiebes, J. T. (1981) The fig wasp fauna of the receptacles of *Ficus thonningii* (Hymenoptera, Chalcidoidea). *Tijdschrift voor Entomologie*, **124**, 149–233.)

dramatic dimorphisms among males, both in solitary and social taxa, and distinct quantitative or qualitative differences between female morphs in various solitary taxa. Thus, within the order, and in some cases within a single species, it is possible to study alternative developmental pathways both within and between the sexes.

Third, the order is huge (perhaps as large as the beetle order

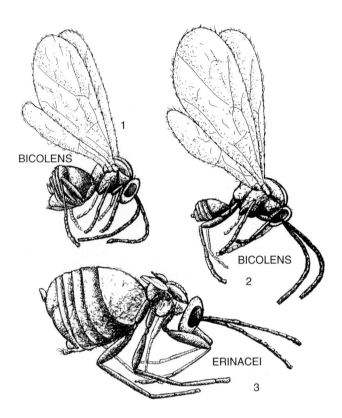

Figure 4.2. The gall wasp *Cynips erinacei*. Like many other members of the Cynipidae, this species alternates between sexual and asexual generations. Females (1) and males (2) of the sexual form mate, producing females of the asexual or 'agamic' generation (3), which do not mate but produce sexual males and females parthenogenically. (Redrawn from Kinsey, A. C. (1929) *The Gall Wasp Genus Cynips*. Indiana University Studies, vol. 16.)

Coleoptera), with many large families that are major groups in their own right. Because many kinds of sex differences (and alternative morphologies within one sex) have evolved repeatedly within the order, its abundance of related taxa can be used to generate detailed statistical descriptions of the relationships among morphological and other variables, and, as detailed phylogenies become available, comparative methods can be applied with unusual power. Fourth, many species can be reared in the laboratory or induced to accept artificial nesting substrates in the field. Thus, manipulations of many kinds are possible, and these can often be replicated more extensively than is feasible with vertebrates.

Figure 4.3. Major and minor workers of the Asian marauder ant *Pheidologeton diversus*. These nestmates (presumably sisters) differ in weight by a factor of 500, the largest known difference in adult size within any species of Hymenoptera (and probably any insect). Scanning electron micrograph by Mark W. Moffett/Mindon Pictures.

Although Hymenoptera are unequalled as a natural laboratory for the study of sex differences, we still have an incomplete and somewhat fragmented understanding of the distributions and correlates of such differences within the order, and of the proximate and ultimate mechanisms that give rise to them. For example, there is a rich tradition of behavioral study, but it has tended to focus on descriptions of mate finding and courtship (on the male side) and prey location and nesting (on the female side) in particular species, with relatively little attention to ecological and phylogenetic analysis. Morphological sex differences are described frequently in systematic works, but quantitative comparative studies are rare. And the developmental and physiological aspects of sexual dimorphism have hardly been touched.

As far as is known, all Hymenoptera have haplodiploid genetic systems, in which haploid males develop parthenogenically from unfertilized eggs, while diploid females develop from fertilized eggs. (Female parthenogenesis also occurs in a few groups, where it is derived from haplodiploidy.) Among the interesting consequences of this system is that mothers can determine precisely the sex of each offspring, by releasing (or not releasing) a stored sperm cell from the spermatheca, just prior to oviposition. Precise sex-ratio control has allowed the Hymenoptera to explore a universe of sex-allocation strategies that is not accessible to most other animals. Much effort has gone into unravelling the intricacies of hymenopteran sex allocation, but there has been little work on the implications of sex-ratio control for the evolution of adult sexual dimorphisms.

In the next section, we review briefly the evolution and diversity of Hymenoptera. Then we focus on some characteristic differences between the selective environments of the sexes and the ways in which these ecological differences have shaped male and female biologies. Next, we review the evolution of winglessness, which has occurred many times in both males and females, often in connection with unusual ecologies and mating systems. Males and females are often very different in size. In the final section we attempt to show why the Hymenoptera offer superb opportunities to study the evolution of size differences, both within and among species.

Who's who: a brief overview

The order Hymenoptera comprises a very large and diverse group of insects that play profoundly important roles in terrestrial ecosystems. More than 100 000 species have been described, but this is a small fraction of the species believed to exist; current estimates range from 250 000 to over a million species worldwide. We will never know with confidence how many species exist today (much less how many existed 100, or 1000, years ago), owing to the accelerating rate of habitat destruction around the globe.

Most Hymenoptera are carnivorous, preying on a great diversity of terrestrial arthropods including phytophagous insects, and, undoubtedly, their appetites keep the world greener than it would otherwise be. The order also includes phytophagous species; some attack leaves or stems, but a much larger number are invaluable pollinators of flowering plants. The evolution of flowers has been shaped in large part by the sensory physiology of bees, with their well-developed color vision and their fondness for pleasing fragrances and sugary fluids. These fortunate attributes have made the world a far prettier and sweeter place than it might otherwise have been.

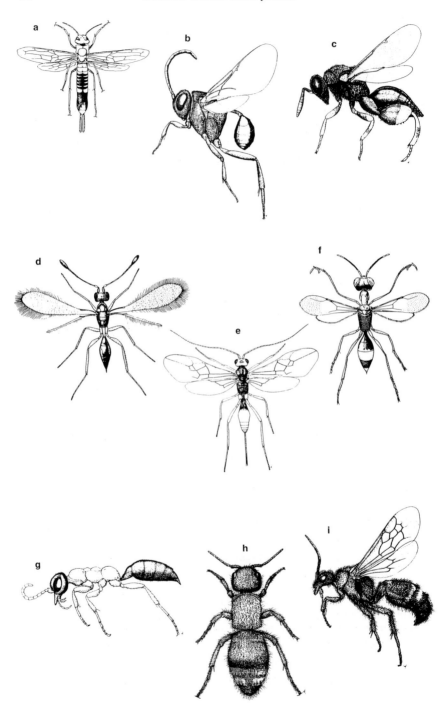

We are not aware that anyone has attempted to estimate the direct economic value of the regulation of insect numbers by carnivorous Hymenoptera and the pollination of wild and domesticated flowers by thousands of species of wild bees, but we are confident that the total would run to hundreds of billions, perhaps even trillions, of dollars in benefits received annually. Moreover, it is impossible to place a merely monetary value on the more general contributions that Hymenoptera make to the functioning of natural ecosystems (for example, the scavenging and soil-turning services of ants), but these activities are vital to our survival. Imagine a plague that eliminated only the order Hymenoptera. Uncontrolled outbreaks of destructive insects and the rapid demise of bee-dependent plants would totally transform terrestrial ecosystems all over the world. Life would survive, but it is doubtful that civilization would do so.

The order Hymenoptera includes the familiar wasps, ants and bees as well as many less familiar insects (Figure 4.4). Traditionally, the order is divided into two suborders, the Symphyta and the Apocrita. The suborder Symphyta, which includes the woodwasps, sawflies, and stem flies, is structurally the most primitive of extant groups (Figure 4.4a). The earliest known fossils are symphytans from the Triassic, around 200 million years ago. Almost all symphytans are phytophagous, but the family Orussidae parasitizes wood-boring symphytans. The larvae of many symphytans are external feeders on leaves or other plant parts and, thus, behave much like caterpillars, while others are internal plant feeders, living a concealed existence boring in leaves, stems or wood. Sexual and intra-sexual dimorphisms tend to be relatively modest in this group.

The suborder Apocrita is a monophyletic group derived from symphytan ancestry and characterized by a remarkable morphological innovation that is unique among insects, a new articulation between the first and second abdominal segments. Like most insects, symphytans have the abdomen broadly joined to the thorax (Figure 4.4a). In Apocrita, however, the apparent thorax is actually the true thorax

---

Figure 4.4. Representatives of several major branches of the Hymenoptera. All except (i) are females. (a) a symphytan, the pigeon horntail. (b–e) Various 'Parasitica'; (b) an evaniid or 'ensign wasp'; (c) a chalcid; (d) a mymarid, one of the smallest insects in the world; (e) a braconid. (f–i) Various aculeates; (f) a dryinid; (g) a wingless tiphiid; (h) the female and (i) the male of a mutillid wasp or 'velvet ant'. (a and f are from Evans and West Eberhard, © The University of Michigan (1970); the others are from Gauld and Bolton (1988) © The Natural History Museum, London.)

together with the first segment of the abdomen, which is separated from the true second abdominal segment by a narrow petiole, forming the 'wasp waist' (Figure 4.4b–i). This invention greatly increased the maneuverability of the ovipositor and accompanied a shift from phytophagy to parasitism of other arthropods. Apocrita include most members of Hymenoptera, and is therefore one of the most successful of all insect groups. The earliest known fossils date from the Jurassic with an extensive radiation during the Cretaceous.

Traditionally, Apocrita are divided into two major groups: 'Parasitica' and Aculeata. The former is an informal grouping including many thousands of species, most of which are parasitoids attacking other arthropods, but with some notable reversions to plant foods. Included here are the great superfamilies Chalcidoidea (see Figures 4.1 and 4.4c–d) and Ichneumonoidea (Figure 4.4e) with tens of thousands of species each, as well as a number of smaller groups such as Cynipoidea which produce the familiar galls on oaks (see Figure 4.2). A typical life history of a parasitican can be described as follows. Newly emerged adults mate, after which males soon die while inseminated females, each carrying a lifetime supply of sperm in a spermatheca, search for suitable host insects or spiders which will provide the sole food for their offspring. Single or multiple eggs are then laid on or in each host and these hatch into larvae which undergo several molts as they grow on a rich diet of host tissue. Eventually, they spin cocoons from which they emerge as adults to repeat the cycle. Since the host dies in the process, 'parasitic' Hymenoptera are usually referred to as parasitoids rather than parasites. If they feed at all, the adults generally have a very different diet from their larvae. Although the adults of some species are known to feed on the haemolymph of their hosts, most subsist entirely on sugary solutions such as the nectar of flowers or the 'honeydew' secretions of plant-sucking Homoptera such as aphids, scale insects and leaf hoppers.

Aculeata are an apparently monophyletic group derived from the 'Parasitica' and characterized by another dramatic morphological innovation, the sting. The females of all other Hymenoptera possess an ovipositor employed to place the egg on, near or within the larval food source. Glands associated with the ovipositor produce secretions that lubricate the passage of the egg and that also affect the host in various ways. Thus, the ovipositor serves both to place the egg and to introduce substances that modify host behavior, physiology or development. In Aculeata, however, the egg is released at the base of the ovipositor, which has become highly specialized for its one remaining role, that of injecting paralyzing or irritating venoms into prey or enemies.

The venoms of different aculeate taxa vary greatly in their composition and effects. In some species, the effects wear off quickly and the prey resumes its normal activities while the larva develops as a parasitoid. In species that construct nests, the venom induces permanent paralysis, but the prey remain physiologically alive for extended periods, until they are consumed by the developing larva. Although the sting evolved originally as an organ used to immobilize prey, in many social species it functions as a fearsome defensive weapon against vertebrates, as billions of people know from personal experience. The pharmacology of wasp and bee venoms is a vigorous field of research with important applications in medicine and neurobiology.

Since the aculeate sting is a modified ovipositor, only females have it. In some groups, however, males have evolved pseudo-stings. These are modifications of the male genitalia or associated sclerites into sharply pointed structures that can prick the skin of a vertebrate attacker. Pseudo-stings lack venom glands, but pseudo-stinging behavior often wins a male's release from predators which have prior experience of female aculeates. Freud would surely have been amused to learn of this evidence for 'sting envy' in aculeate males.

Many aculeates are parasitoids, with ways of life quite similar to those of their parasitican ancestors. Eggs are laid on the host wherever it is found, and there is no transport of prey or construction of a nest. This pattern is undoubtedly primitive for the aculeates, and it is found in a wide diversity of familes including Dryinidae (Figure 4.4f), Bethylidae (see Figure 4.10), Embolemiidae, Scoliidae, Tiphiidae (see Figures 4.4g and 4.10) and Mutillidae (Figure 4.4h–i), as well as some Chrysididae and Pompilidae. However, the aculeate radiation is dominated by taxa in which females transport larval food back to a central location: the nest. This breakthrough in domesticity has had major consequences for the evolution of sex differences, and it also set the stage for the evolution of sociality. The nest may be simply an existing cavity in which the egg and its food supply are sealed, or it may be constructed by the nesting female. Many species dig burrows in the ground; some make tunnels in wood; and still others build free-standing nests out of paper, mud, or tiny stones and mortar. Various glandular secretions may be used to glue together nesting materials and to waterproof the cells in which offspring will be reared.

The nest is the focal point of a female's life: the place to which she

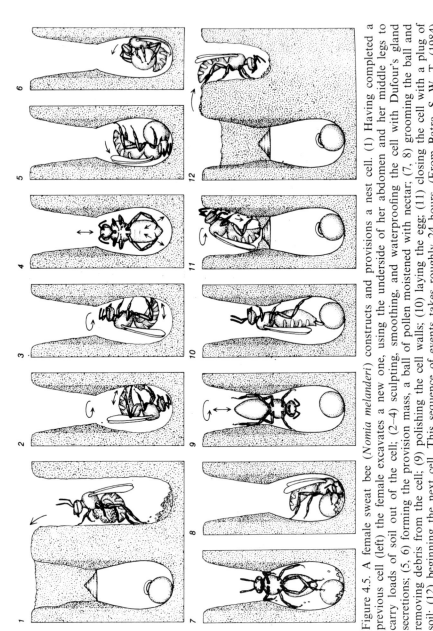

Figure 4.5. A female sweat bee (*Nomia melanderi*) constructs and provisions a nest cell. (1) Having completed a previous cell (left) the female excavates a new one, using the underside of her abdomen and her middle legs to carry loads of soil out of the cell; (2–4) sculpting, smoothing, and waterproofing the cell with Dufour's gland secretions; (5, 6) forming the provision mass, a ball of pollen moistened with nectar; (7, 8) grooming the ball and removing debris from the cell; (9) polishing the cell walls; (10) laying the egg; (11) closing the cell with a plug of soil; (12) beginning the next cell. This sequence of events takes roughly 24 hours. (From Batra, S. W. T. (1984) Solitary bees. *Scientific American*, **250**, 120–7. Copyright © (1984) by *Scientific American, Inc.* All rights reserved.)

returns repeatedly with larval food and with building materials. Remarkable powers of orientation and memory are required to find a nest entrance that may be no more than a tiny hole in the ground. In a few species, females maintain several nests simultaneously, at distances of tens of meters from each other. Most species rear several offspring per nest, each sealed in a separate cell. Some provide only a single prey item per offspring, but most provide several items. In most groups, larval food consists of paralyzed insects or spiders, but bees and some vespid wasps have switched to pollen and nectar; the bees are literally 'vegetarian wasps'. Larval provisions have proved to be an irresistible target for some aculeates who behave as cuckoos, laying their eggs in the nests of other species. The egg of the cuckoo species hatches into a larva that destroys the host larva and then consumes its food. The cuckoo way of life has evolved repeatedly and is found in most Chrysididae and in some spider wasps (Pompilidae), digger wasps (Sphecidae) and bees.

The nest-building aculeates differ from most other insects, and indeed from most other animals, in the very high levels of maternal care they provide for each offspring. Constructing and provisioning a rearing cell is typically a hard day's work, and may take even longer (Figure 4.5). As a consequence, species that pursue this way of life have mean and maximum lifetime fecundities that would be low even by avian or mammalian standards. In most solitary species, a female that produces 10 or 15 offspring is well above average, and in social species the productivity per worker is the same or even lower.

Most aculeates are solitary, with each female providing for her offspring entirely on her own. Even among nest-building species, the primitive and still typical condition is for each female to locate, to harvest and to transport larval provisions to a nest occupied by her alone. However, sociality has evolved repeatedly within several groups of wasps and bees and at least once among the ancestors of ants. The reasons for its relative commonness in aculeate Hymenoptera and its extreme rareness in all other animals are still not entirely clear, but the nest, the sting and haplodiploidy are all high on the list of probable contributing factors.

This brief sketch of hymenopteran evolution has identified some key features of the biology of the order that have strongly influenced the evolution of sex differences, often by placing males and females in what amount to different worlds. But it has certainly not done justice to the rich diversity of the group, or to its importance. The articles and books cited as further reading will give immediate help to anyone whose curiosity has been aroused.

The selective environments of females and males, and some of their effects

Natural selection can be viewed metaphorically as a contest among genes
or individuals, in which the object is to contribute as many genes as
possible to future generations. Even in species with haplodiploid genetics,
half of all genes in future generations will come from females and half
from males; females compete among themselves for their half, while males
compete for their half. Parental investment is an almost exclusively female
activity in many groups of animals, including Hymenoptera, and this has
important consequences for the nature of the evolutionary competition
within each sex. The reproductive success of females is limited mainly by
their ability to invest in offspring, while the reproductive success of males
is limited mainly by their ability to father the offspring produced by
females. In short, females compete for resources that can be converted
into offspring, while males compete for females.

Many female morphological adaptations are tools of the parental-
investment trade. For example, in many ground-nesting species, only the
females have a foretarsal rake which is employed to move soil during
construction of the burrow system. Similarly, many females have a
pygidial plate, a trowel-like modification of the apical segment of the
abdomen, which is used in nest construction. The females of some
ground-nesting wasps have greatly modified mouthparts that serve as
baskets for carrying sand out of the nest during construction. And female
leaf-cutter bees have massive, sharp-edged mandibles that are used to cut
precisely shaped pieces of leaves that are used to line the rearing cells in
their nests.

Some female wasps have remarkable adaptations used for prey capture
and transport. For example, the end of the female foreleg is modified in
most dryinids to form a grasping organ used to hold the prey while it is
being stung (Figure 4.4f). Nesting species use different techniques to carry
prey home. Some species simply grasp the prey in their mandibles and
drag them back to the nest, and others use their legs to cradle the prey
beneath the body while flying home (often from distances of hundreds of
meters). But many other species employ more exotic methods. Digger
wasps of the genus *Clypeadon* prey exclusively on worker ants which they
carry with the aid of an 'ant clamp', a modification of the apical segment
of the abdomen (Figure 4.6). In the digger-wasp genus *Cerceris*, females
often have prominent, species-specific projections on the front of the head.
These projections are thought to facilitate prey transport by fitting
precisely with the morphologies of the beetle prey and thereby helping

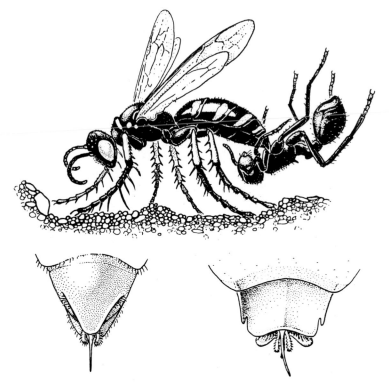

Figure 4.6. A solitary sphecid wasp, *Clypeadon laticinctus*, returning to her nest with a worker harvester ant (*Pogonomyrmex*) attached to the end of her abdomen. This feat is made possible by unusual modifications of the apical abdominal segment (lower right) that fit precisely into a space between the legs of the prey. For comparison, the apical abdominal segment of a related species that does not practice abdominal prey carriage (*Aphilanthops frigidus*) is shown on the left. (From Evans and West Eberhard, © The University of Michigan (1970).)

the female to maintain a firm grasp on it during flight. None of these specialized structures are present in males.

Bees provide pollen for their larvae and have evolved a variety of structures to facilitate pollen transport. Most female bees carry pollen on their hind legs, which are modified in various ways depending on the nature of the pollen. Megachilid bees carry pollen on the under surface of the abdomen, with the aid of long, highly specialized hairs. The machinery of pollen transport is always absent in males and in cuckoo bees of both sexes.

The reason why males do not develop these tools is presumably that they are costly and of little or no potential use to males. There are other animal taxa (such as birds) in which males cooperate with their mates in the defense and feeding of offspring. No such cases are yet known in the Hymenoptera, where the only significant form of male parental investment appears to be nest guarding. But even nest guarding is rare, and in the majority of cases where the guards are mates of the females whose nests they are protecting (as opposed to brothers or sons) the defense is probably as much against conspecific males as against predators or parasites.

One consequence of the fact that females are the high-investment sex is that they are expected to be much more selective in their choice of mates than are males, who have little to lose and who, therefore, can afford to be less discriminating. The relative lack of male discrimination has been exploited by certain orchids that induce males to copulate with structures that mimic females, both in appearance and in odor. Males duped into such pseudocopulation advance the orchid's reproduction at a (presumably) minor expense to their own.

As in other animals, many uniquely male characteristics can be understood as being adaptations that enhance a male's ability to find, to court or to defend females, even at the cost of a possible reduction in male survival. The males of many species appear to be subject to strong sexual selection, either through direct male–male competition (which tends to favor large size, conspicuous weaponry and pugnacious behavior) or through female choice (which tends to favor bright colors, ornaments and ritualized displays).

Current theories for the evolution of mating systems predict that the type and intensity of sexual selection will tend to be related in a logical way to the ecology of females, especially their spatial distribution. Many of these ideas can be tested in the Hymenoptera. To mate, a male must either go where the females are, or wait where they can be expected to appear. The males of most hymenopteran species search individually for receptive females, and engage in little or no male–male competition. Usually, males emerge before females and immediately begin searching at nesting areas, nectar sources, or other places where females are likely to be; in some species, they simply perch in promising areas and wait for the approach of females. As expected, searching and perching strategies are most common where females and their resources are dispersed widely.

Territorial defense is expected only where females or their resources are spatially clumped in a way that makes them defensible. In some bees, for example, males defend clumps of flowers against other males and attempt to mate with any females that come to harvest nectar or pollen. Another form of 'resource-defense polygyny' occurs in species where females emerge from nests that are clustered in small patches of suitable nesting substrate.

In the solitary wasp genera *Oxybelus* and *Trypoxylon*, a male often defends the nest of a provisioning female (both from other males and from parasitic flies) and mates with the female on most of her visits to the nest. This arrangement is highly unusual in at least two respects. First, it is unusual for hymenopteran females to mate more than once in their lives. And second, as was mentioned above, this is almost the only context in which male Hymenoptera provide any form of parental investment. However, there is a strong conflict of interest between the guarding male and his mate. Owing to haplodiploidy, the male may father daughters but not sons. In theory, this leads to a disagreement over the sex ratio of the female's offspring, and, as expected, in at least one species where this prediction has been tested, males tended to cease guarding nests at times when the female was likely to produce sons.

Male–male competition is expected to reach its extremes where females are so clumped that a single male might successfully defend many of them at once. This situation occurs in several groups of Hymenoptera, with spectacular consequences. Large size, special mechanisms for seizing females, and powerful mandibles have evolved repeatedly in groups where females are highly concentrated. For example, in bees of the genus *Nomia* the hindlegs of males are modified for holding on to the abdomens of females. These bees nest in dense aggregations and there are typically many suitors for each virgin female, so the ability to hold on to a mate is highly advantageous. The ultimate in male–male competition occurs in several unrelated taxa where males fight to the death for access to the females available within a restricted (often enclosed) mating arena or 'lek'. Such fighting males often have bizarre modifications including wingless-ness (as discussed in the next section) and enlarged, heavily sclerotized heads with powerful mandibles that can literally crush an opponent (Figure 4.1).

Just as male–male competition often leads to grotesque modifications and violence, female choice often leads to male structures and behaviors of unusual beauty. Well-known examples include the spectacular displays of some male birds such as peacocks and birds of paradise, but some male bees and wasps are also highly ornamented, with colorful markings, luxurious pubescence, or striking enlargements of various exoskeletal elements, especially segments of the legs (Figure 4.7) or antennae. In theory even arbitrary female preferences can drive the evolution of

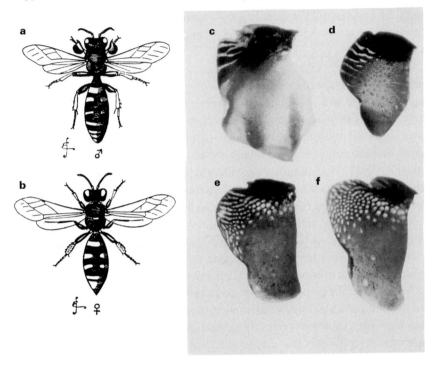

Figure 4.7. Tibial shields in males of the solitary sphecid wasp genus *Crabro*. In his book on sexual selection (*The Descent of Man, and Selection in Relation to Sex*, 1871), Darwin illustrated these plate-like enlargements of the foretibia in the British species *Crabro cribarius*, and discussed the idea that they might help the male to hold on to the female during copulation (a, b). Recent work has shown that instead they seem to be designed and used so as to provide the female with an informative 'light show'. The shields are thin, transluscent, and highly patterned (c–f). During courtship and copulation, the male places them over the female's compound eyes. The photographs on the right are interior views, showing the patterns of transmitted light that a female would see. The two upper panels are for *C. latipes* (left) and *C. tenuis* (right); the lower panels show two different individuals of *C. cribellifer*. (From Low and Wcislo, 1992.)

male characters, but it is also thought that female choice may involve characteristics that reveal something about the ecological 'quality' of a male's genotype. This seems especially likely to be true in Hymenoptera, where a male's genes pass immediately only into daughters. Unfortunately, very little is known about the mating systems of species with highly ornamented males, and we are not aware of any attempts to analyze the possible ecological relevance of such ornaments. The most thorough studies of Hymenopteran courtship concern small parasitoids that are not highly ornamented; in these species, males perform elaborate, ritualized behavioral displays that involve head bobbing, wing vibrations and other forms of 'dance'.

Several kinds of mass displays and leks occur in species where the only resource provided to females is sperm. Some male wasps establish territories on the tops of hills or ridges where females apparently come to choose among a number of potential mates. Some social Hymenoptera form immense mating swarms that consist of male and female reproductives from many colonies. Such swarms are especially dramatic in ants that have synchronized 'nuptial flights'; tens of thousands of reproductives (mostly males) may gather in a small area, with many males scrambling intensely in attempts to mate with each female that enters the swarm.

These systems seem to have elements of *both* male–male competition (physical combat) *and* female choice (purposeful travel to a concentration of males where comparisons can be made and acted on). And, more generally, it is not always obvious that a given structure or behavior is likely to have evolved more through female choice or through male–male competition. In the digger-wasp genus *Crabro*, for example, the male foretibiae of most species are greatly expanded into thin, flexible 'tibial shields' (Figure 4.7). Conspicuously patterned in contrasting yellow and black, tibial shields show major differences between species and minor differences among individuals. Darwin was familiar with tibial shields in a British species, and he argued that they serve as claspers to prevent separation during mating. However, more recent work has shown that they could not plausibly work as claspers, and that males hold them over the eyes of females during mating. This suggests that they indicate the male's species identity, or signal some attribute of his phenotype that females use in assessing male quality. Female choice is implicated as being the evolutionary mechanism, but the basis for the presumed preference remains obscure.

Mixtures of different male strategies may occur within populations of the same species, and are sometimes associated with striking (even discontinuous) variation in male morphology. Cases where small males employ a different mating strategy than do large ones are known in a number of wasps and bees. In the digger wasp *Bembecinus quinquespinosus*, for example, large males compete intensely to grab emerging females, and then fly off with them to mate without interference, while small males search for opportunities to mate with females that may have been missed by large males (Figure 4.8). More dramatic cases are known in which there are two distinctive male morphs that pursue different mating

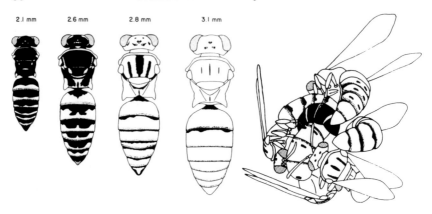

Figure 4.8. Male polymorphism in size, coloration, and mating strategy in the solitary sphecid wasp *Bembecinus quinquespinosus*. Males of this species vary enormously in size (left). Large males have predominantly light coloration and they search for newly emerging females by flying and walking over the open, sandy areas where nests occur. There is little shade in these areas, and temperatures near the ground can be very warm. Emerging females are often discovered by several large males who struggle with each other and with the female in an attempt to mate with her (right). Small males patrol nearby vegetation, presumably searching for females that escaped the notice of large males or that may be willing to re-mate. (From O'Neill and Evans, 1983.)

strategies. In some bees, ants and torymid wasps there is a flightless, fighting morph that engages in fierce combat over females at the site of emergence, and a fully winged morph that flies away to seek matings elsewhere (see Figures 4.1, 4.9 and 4.11).

Like dogs, most insects perceive a vividly fragrant world of chemical signals that go mostly unnoticed by visually and aurally specialized creatures such as ourselves. Chemical communication is particularly important in the sex lives of many insects (see Chapter 16 by Jean-François Ferveur and colleagues), including many Hymenoptera. In some wasps and bees, for example, males mark their territories with secretions that they spread on plants or other surfaces using clumps of hairs that have been modified to serve as brushes for this purpose. The resulting odor plumes advertise the territory (and its holder) both to females and to other males, and may indicate the male's specific identity and aspects of his phenotypic quality.

Chemical communication is often more important and more elaborate than this among female Hymenoptera, who have evolved an unparalleled diversity of exocrine glands. (Dozens are known, and, undoubtedly, more remain to be discovered.) As with other female adaptations, these glandular systems tend to be used in various aspects of parental

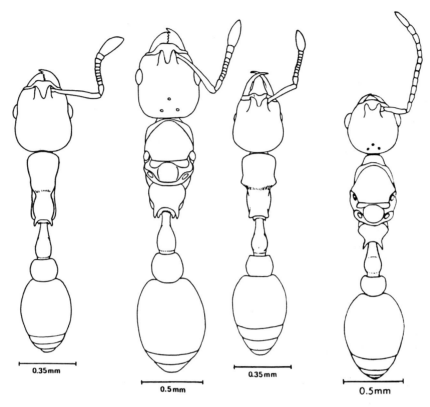

Figure 4.9. Winged and wingless males and females in the ant *Cardiocondyla wroughtonii*. From left to right: wingless worker, (initially) winged queen, wingless (ergatoid) fighting male, and winged dispersing male. Wingless males engage in lethal fights using their enlarged mandibles (see Stuart, R. J., Francouer, A. and Loiselle, R. (1987). Lethal fighting among dimorphic males of the ant, *Cardiocondyla wroughtonii*. *Naturwissenschaften*, **74**, 548–9; and Hölldobler and Wilson (1991), p. 186. Adapted from Kugler (1983). The males of *Cardiocondyla* with a description of the winged male of *Cardiocondyla wroughtonii*. *Israel Journal of Entomology*, **17**, 1–21).

investment. In social species, particular chemical signals are used to recruit nestmates to food sources, and others are used to arouse group defensive responses when the colony is threatened by attack. Many exocrine secretions serve purposes other than communication, such as nest construction, the suppression of bacterial and fungal infections within the nest, and others, but the line between 'communication' and these other functions is not always clear cut (as when the volatile components of a venom recruit workers to the site of an attack), and this line has probably been crossed repeatedly during the evolution of such systems. Understanding an organism is often a matter of learning to 'see' and to 'think' the way it does, which, in the case of insects (including Hymenoptera), often requires that we imagine ourselves in a world of tastes and smells unlike anything we experience directly.

To fly or not to fly

Winglessness (aptery) or near-winglessness (brachyptery) in one sex or the other has arisen many times in Apocrita, but it is very rare in Symphyta where only three species with apterous females are known. Two of these are tropical sawflies in which females guard eggs in confined locations. Many cases of aptery are known among the 'Parasitica' (including some species in several families of Cynipoidea, Chalcidoidea and Ichneumonoidea) and the Aculeata (including at least some species of Dryinidae, Embolemidae, Bethylidae, Sclerogibbidae, Formicidae, Tiphiidae, Mutillidae, Bradynobaenidae and Sphecidae).

Although either sex (rarely both) may be wingless, by far the most frequent pattern is female aptery (see Figures 4.2, 4.3, 4.4g, h, 4.9, 4.10). The typical syndrome is one in which females spend most of their lives in confined situations, such as burrows in the ground, where wings are worse than useless, while males spend much of their time flying in search of sugary food and females. For example, mutillid females parasitize the immature stages of ground-nesting wasps and bees, and spend much of

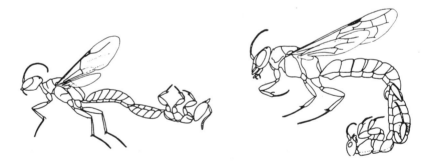

Figure 4.10. Phoretic copulation evolved independently in two different families of wasps. The pair on the left are bethylids (*Apenesia nitida*), and the pair on the right are tiphiids (*Dimorphothynnus haemorrhoidalis*). The female's position is venter-up in the bethylid pair, but venter-down (relative to the male) in the tiphiid pair, which requires an unusual rotation of the male's genitalia. Wingless females typical hunt for prey underground or in other confined spaces, and often (as here) have greatly reduced compound eyes; this bethylid female is completely blind. (From Evans, H. E. (1969) Phoretic copulation in Hymenoptera. *Entomological News*, **80**, 113–24.)

their time burrowing through the ground in search of hosts. Similarly, wingless female tiphiids parasitize underground beetle larvae. In these groups the females are obviously specialized for burrowing; for example, they tend to have short, stout, and powerful legs. In other cases, apterous females may be more gracile, but they also tend to live or to hunt in confined spaces.

Aptery improves a female's maneuverability in restricted places, and saves her the energetic cost of building wings and flight muscles. However, it also creates some problems. In particular, it restricts the area that she can search for adult food or larval provisions, and it may also restrict her ability to locate a mate. These potential disadvantages have been overcome in various ways. In ants, for example, only the reproductives have wings while the workers do not, and queens shed their wings after mating, thereby combining the advantages of having wings to locate mates and to disperse to new areas, with the advantages of unencumbered movement within the nest (Figure 4.9). Wingless females in the Australian tiphiid *Diamma* are sleek, metallic blue creatures with long legs that enable them to run at surprisingly high speeds.

In some tiphiids, mutillids and bethylids, males are larger than females and carry their mates about while joined *in copulo* (Figure 4.10). Such phoretic copulation has been studied in the thynnine tiphiids of Australia. Virgin females climb vegetation, assume a characteristic posture, and presumably release an advertising pheromone. An arriving male literally picks up the calling female, gives her a meal of nectar, and then drops her off at a site where she is likely to find the scarab larvae on which she will lay eggs. In some species, females are carried to flowers where they feed on nectar. In others, the male head is modified to form a 'nectar basket' used to carry an offering of food, and the female can sample the food before agreeing to indulge in aerial sex.

Wingless males seem to occur only in species that mate in a restricted area near the place of adult emergence (see Figures 4.1, 4.9 and 4.11). The most striking examples involve chalcidoid wasps associated with figs. A fig is an inflorescence 'turned inside out' to form a hollow structure in which hundreds of flowers line the inside surface. Figs are pollinated by tiny wasps in the family Agaonidae. The typical life history is one in which a mated female emerges from her natal fig and flies to another fruiting tree where she selects a fig and enters it through a tiny opening at the distal end. This involves negotiating a very narrow passageway, and, typically, females lose their wings in the process. Once inside, the one or more females entering a given fig oviposit in the tiny flowers lining the

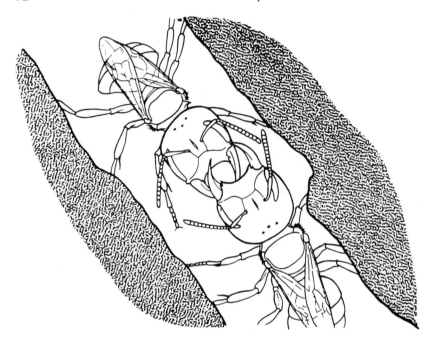

Figure 4.11. Flightless fighting males in a communally nesting sweat bee, *Lasio-glossum erythrurum*. In this species, as in other bees known to have fighting males, there is also a fully winged form that disperses from the natal nest, and the fighting males have disproportionately large heads and small, non-functional wings. (From Kukuk, P. F. and Schwarz, M. P. (1988) Macrocephalic male bees as functional reproductives and probable guards. *Pan-Pacific Entomologist,* **64,** 161–7. Reprinted by permission of Pacific Coast Entomological Society.)

interior cavity. Then they die having completed their lifetime reproduction within a single fig. The offspring develop into adults after feeding as larvae on tissues of the fig. The wingless males emerge prior to females and seek out virgin females before they emerge. A male gnaws a hole in the female's natal chamber, inserts his abdomen inside, and mates with the still imprisoned female, providing her with a lifetime supply of sperm. The female then leaves the fig, coated with pollen from the male flowers, and flies off to another tree to repeat the process. The wingless males spend their entire lives in their natal figs and all dispersal is by the winged females.

This life history favors a strongly female-biased sex-ratio. If only one foundress female entered the fig, then she should produce just enough sons to ensure that all her daughters are mated, because she maximizes her genetic contribution to future generations by maximizing the number of fertilized daughters that disperse to other figs. This contrasts strongly with the situation in large, randomly mating populations, where the evolutionary equilibrium is one in which equal totals of investment should go into each sex, as shown by R. A. Fisher in 1930. But, as first pointed out by W. D. Hamilton, where mating is local and only mated females disperse to establish new mating groups, males compete only for a limited subset of all females, and a given female can make a larger contribution to future generations by devoting more of her resources to female production than to male production. As expected, female-biased sex ratios are observed in fig wasps as well as in numerous other parasitoids with local mating and dispersal of mated females.

Several other parasitoids depend on figs for their reproduction without providing any pollination services for the fig. In particular, several genera in the family Torymidae parasitize the fig or the agaonid wasps that pollinate them. These parasitic forms have long ovipositors which they employ to introduce eggs into the fig from the outside. Typically, females oviposit in a number of figs, laying only a few eggs in each one. The females of parasitic species are always winged, but the males may be winged, wingless or a mixture of both. The level of male winglessness tends to be correlated with wasp abundance. Complete male aptery is favored where wings are rarely of any value. This occurs if populations are dense, so that males always find mates in their natal figs. Where populations are diffuse, males may find themselves without mates unless they can fly to other figs. As expected, common parasites tend to have wingless males, while rare ones tend to have winged males. In cases of intermediate abundance, both kinds of males have reliable expectations of finding mates and selection may favor a mixture of winged and wingless males.

Winglessness in parasitic fig wasps is associated with intense, often lethal male–male combat. Some males are modified bizarrely with large heads and powerful mandibles that can literally chop an opponent in two (Figure 4.1). Describing his observations in Brazil (see Blum and Blum), W. D. Hamilton remarked that the situation in the fig 'can only be likened in human terms to a darkened room full of jostling people among whom, or else lurking in cupboards and recesses which open on all sides, are a dozen or so maniacal homicides armed with knives'.

Not all fig wasps are so pugnacious, however, and the situation is quite different in the pollinating agaonids. The heads of male agaonids are not unusually large, and fighting is rare if it occurs at all. Instead, male–male competition takes the form of a scramble for access to virgin females in the galls where they develop. W. D. Hamilton pointed out that the intensity of fighting should be related to the average relatedness among males sharing a fig. In the agaonid case, only a very few females enter a given fig, and they each produce their entire brood within that fig. Consequently, males are often brothers and share many genes in common. In the torymid case, on the other hand, females visit numerous figs and lay only a few eggs in each one so that males are often unrelated, and male–male competition is correspondingly more intense.

Figs, and the community of wasps that depend on them, constitute a unique situation that is duplicated nowhere else in the living world. None-the-less, the ecological factors that influence the evolution of male winglessness and fighting in fig wasps have their parallels elsewhere with similar, if less extreme, consequences. For example, male aptery and lethal fighting occur in the ant genus *Cardiocondyla*. In most species of *Cardiocondyla* there is no nuptial flight, and mating takes place within the nests which are shared by several queens. Males in these species spend their entire lives within their natal nests where they fight intensely for access to virgin queens. A few species have both wingless (fighting) and winged (dispersing) males, which differ greatly in morphology (Figure 4.9). A similar situation occurs in a few bees in the genera *Lasioglossum* (Figure 4.11) and *Perdita*. These are communal nesters in which several females share a common nest but care only for their own offspring. As in some torymid fig wasps and some species of *Cardiocondyla*, the males are dimorphic. One morph is fully winged and substantially smaller than females, as is usual for bees. The other morph is much larger with reduced wings, enlarged head and powerful mandibles. The small morph leaves the nest and seeks females at flowers or other encounter sites, while the large morph remains within the nest and mates repeatedly with returning females. In keeping with what is now a familar pattern, the large males are highly pugnacious and fight to the death for possession of the nest and its female residents.

Male flightlessness in fig wasps, ants and bees is currently a subject of considerable interest among hymenopterists, because it illuminates a number of general issues in evolution. Dimorphic males are especially important owing to the unique opportunities they provide for examining fitness tradeoffs between alternative male strategies within a single population.

The economics of size dimorphism

Many fundamental aspects of an organism's ecology depend on how big it is. These include such things as metabolic rate, life span, fecundity and prey size, all of which tend to increase with size. Size can have very different consequences in the two sexes, leading to selection for size differences. Many studies involving a wide variety of different kinds of animals have shown that the degree of size dimorphism between the sexes tends to be correlated with the type of mating systems occurring in different species. In particular, males tend to be relatively large in taxa where they fight for control of females, or for control of limited resources needed by females. This makes sense, since the larger of two combatants is likely to have an advantage that may offset any disadvantages of large size, such as longer development time or greater nutritional needs. Males tend to be smaller than females in the Hymenoptera, but relatively large, combative males occur in some taxa, and there is much quantitative variation in size dimorphism even within groups where males are smaller than females. These overall patterns have long been appreciated, but there have been no large-scale comparative surveys of sexual size dimorphism and few attempts to explain why size differences tend to fall within certain characteristic ranges.

The nest-building aculeates differ from most other animals in two ways that make them uniquely valuable for testing general theoretical ideas about the evolution of size. First, the primary determinant of an individual's adult size is the amount of larval food provided by its mother. In other animals (including those with extended parental investment), most of the final difference in size between males and females is generated by growth *after* the offspring are released to fend for themselves. Thus, adult size differences are likely to reflect both early investment by the parents and later decisions by the offspring themselves. In most aculeates, however, adult size is determined almost entirely by the amount of food provided to each offspring by its mother, and the relationship between the size and number of offspring is expected to reflect the interests of the mother alone. Moreover, because larval provisions are such a major component of parental investment in aculeates, offspring size is probably a better index of investment in these insects than it is in most other animals.

Second, the ability to predetermine the sex of each offspring opens up a world of opportunities that are denied to most other animals. Some parasitoids, for example, produce female offspring on relatively large hosts and male offspring on relatively small hosts. This makes good sense if female offspring gain more from being large than do male offspring. Nest-building aculeates take this strategy a step farther, freely adjusting both the *number* of offspring of each sex they produce, and the amount of effort or *parental investment* they put into each male or female offspring,

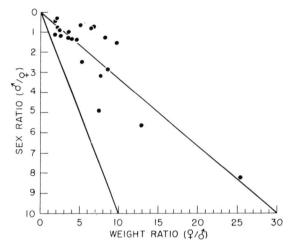

Figure 4.12. The sex ratio (males/females) as a function of the weight ratio (female/male) for 21 species of ants with one queen per colony. This famous figure illustrates the first test of the prediction that workers in some species of social Hymenoptera should tend to bias their colony's investment ratios toward female reproductives. The upper line shows the 3:1 ratio expected if the queen typically mates with just one male and the workers control the investment ratio; the lower line shows the 1:1 ratio expected if the queen mates many times or if she controls the ratio of investment. Note that in this sample of species, females are always at least twice as large as males, often five to 10 times as large, and in one case 25 times as large. (From Trivers, R. L. and Hare, H. (1976) Haplodiploidy and the evolution of the social insects. *Science*, **191**, 249–63. Copyright 1976 by the AAAS.)

so as to produce an appropriate overall sex ratio of investment. This intimate relationship between cost and number of offspring forms the basis of many tests of sex-ratio theory that have been performed in a variety of hymenopteran taxa. Owing to the great range of relative male and female sizes that can be found within many groups of aculeates (especially ants), these tests provide the strongest evidence yet obtained in support of R. A. Fisher's basic theoretical insight (as refined by W. D. Hamilton, Robert Trivers and others), that the target of selection should be the population-wide *ratio of investment*, not the numerical sex ratio *per se* (Figure 4.12).

Typically, sex-ratio studies treat male and female sizes as given, and regard the sex ratio as the only variable of interest. Although appropriate for the purposes of such studies, this may have tended to deflect attention from the equally interesting problem of size itself, as a dependent variable that is expected to reflect ecological and other factors. In contrast to animals lacking precise sex-ratio control, hymenopteran females can compensate efficiently for large differences between the costs of males and females by producing relatively *fewer* of the *more expensive* sex. Therefore, large cost differences can evolve without causing either the inefficiencies entailed in culling partially reared broods, or unbalanced (hence unstable) population-wide ratios of investment. Thus, for much the same reasons that they are the organisms of choice for the study of sex-ratio evolution, aculeate Hymenoptera are ideal organisms in which to ask fundamental questions about the economics of size in general and of sexual size dimorphism in particular.

Substantial size dimorphisms are the rule throughout the nest-building aculeates, but while females are usually larger than males, the magnitude of the difference varies considerably within and among taxa. What determines how much a mother should invest in each son or daughter? A general theoretical framework for thinking about offspring size was proposed by Christoper Smith and Stephen Fretwell, who pointed out that parents should attempt to adjust offspring size and number so as to maximize the fitness produced *per unit invested*. It is reasonable to assume that fitness increases with each additional unit invested, at least up to some point, but that the gain per unit invested declines for sufficiently high levels of investment. Under this very general assumption, the optimal investment from the parent's point of view is less, and often much less, than that needed to yield an offspring of maximal fitness (Figure 4.13). This amounts to the familiar economic distinction between getting 'the best' and getting 'the best value for the money'.

An important corollary is that the optimal investment per male or female offspring depends on the shape of the fitness gain curve within each sex. Since male and female Hymenoptera live very different lives, it is likely that their fitness gain curves also differ such that mothers are selected to make male and female offspring of different sizes. If the male gain curve begins to saturate (show declining fitness gains per unit invested) more quickly than the female curve, then each male should receive relatively less investment. Conversely, if the male curve saturates more slowly than the female curve, then males should receive relatively more investment. Thus, the question of why male aculeates are usually smaller than females amounts to asking why the male gain curve usually, but not always, saturates more quickly than the female curve. It seems likely that larger females tend to produce more eggs, live longer, and forage more efficiently than smaller females. For males, however, the advantages of large size may be more modest, since even a small male can produce many more sperm than a female will use in her lifetime. Only

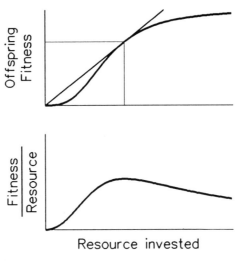

Figure 4.13. The Smith–Fretwell model of optimal offspring size. The optimal investment in an offspring depends on the shape of the fitness gain curve, which specifies offspring fitness as a function of the level of parental investment (upper panel). If the slope of the gain curve has an interior maximum (i.e. is steepest at intermediate levels of investment), then offspring fitness *per unit invested* will be greatest at some intermediate offspring size (lower panel). This size or level of investment in individual offspring is optimal from the parent's point of view because it gives the parent the greatest total offspring fitness under the constraint that total investment is limited. The optimal offspring size can be found graphically by drawing the line of greatest possible slope from the origin to a point on the gain curve; this slope is the 'rate of return' on investment at optimum. An important assumption of the model is that parents control both offspring number and offspring size with sufficient precision to make the size–number tradeoff meaningful. This model was first derived in 1974 by C. C. Smith and S. D. Fretwell (The optimal balance between size and number of offspring. *American Naturalist*, **108**, 499–506).

in species with intense male–male competition will large males tend to have substantially higher fitness than small males. To date, only a few studies have attempted the difficult task of directly assessing the relationship between size and reproductive success in nest-building aculeates, but the findings are generally consistent with these assumptions.

As part of an ongoing study of size structure in bee communities, we have made head-width and other morphological measurements on hundreds of species of bees and wasps. (Head width is a standard measure of size that is correlated strongly with dry weight, and one that can be determined easily from pinned museum specimens.) Although our primary focus is on female size, we have also measured males where available.

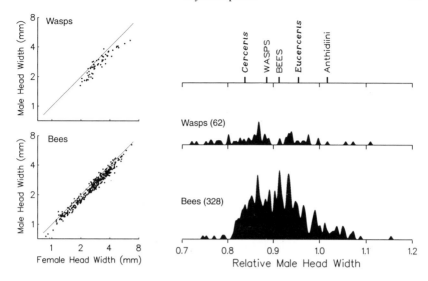

Figure 4.14. Size dimorphism in bees and sphecid wasps. Each point in the scattergrams (left) represents the mean male and female head widths for one species. The sample of wasps consists almost entirely of North American Philanthinae (mainly *Cerceris*, *Eucerceris* and *Philanthus*); the much larger sample of bees represents the faunas of New England and two localities in Wyoming and Utah. Note that the axes are scaled logarithmically, and that the 45-degree lines indicate equal male and female head widths. The distributions on the right show the same data as ratios of male to female head width (relative male size); means for several taxonomic groups of interest are shown at the top. Bees and wasps as a whole do not differ significantly from each other, but *Cerceris* (in which males tend to search individually for females) and *Eucerceris* (in which males are often territorial) differ significantly from each other and from the mean for all wasps, while Anthidiini (which tend to be highly territorial) differ significantly from the mean for bees.

Figure 4.14 shows scatterplots of mean male head width by mean female head width, for 328 species of bees and 62 species of sphecid wasps. On average, there is a remarkably constant proportionality of male and female sizes, independent of a species' absolute size, with male head widths typically about 0.9 times those of females. (If males and females were geometrically similar, this would imply an average relative male weight of 0.9^3 (0.73); in fact, males are usually more slender than females, and thus weigh even less than implied by their relative head widths.) However, there is much variation about this average relationship. Some of the variation is associated at generic and tribal levels with known tendencies of certain taxa to have particular kinds of mating systems (Figure 4.14). As expected, males tend to be larger than average (relative to females) in taxa, such as the anthidiine bees, where, typically, they defend territories; in some of these cases males are absolutely larger than females. Conversely, males tend to be relatively small in taxa where they are known to search for widely dispersed females. This suggests that extreme mating systems (or ones atypical of a given taxon) might be found easily by looking for extreme levels of size dimorphism (where males are relatively large or relatively small) within the taxon of interest.

Also, there is often much variation of size *within* the sexes, and our survey reveals a tendency for males to be more variable than females. There are several possible explanations for this pattern. One is that males, being haploid, express greater genetic variance than do diploid females; if the population contains genetic variation for traits that affect final size (given a fixed provision mass), then male size will be more variable, other things being equal. Another possibility is that male development is less well canalized than female development. Both hypotheses predict that the correlation between provision mass and adult size will be lower for males than for females, but we are not aware that this prediction has ever been tested. Yet another possibility is that the male optimum is usually defined less sharply than the female optimum, so that females allow themselves more latitude in determining the size of the provision mass when making a son than they do when making a daughter, but we know of no general reasons for thinking that this would often be true.

Finally, it is likely that in at least some species, females are selected to produce males of different sizes. There is good evidence from several wasp and bee species that males of different sizes tend to pursue different strategies (see Figure 4.8). What is not clear is whether any of the observed variation in male size is adaptive, and produced deliberately by mothers, or whether it is merely accidental, with the unfortunate smaller males adopting alternative strategies that 'make the best of a bad lot'. The dimorphic males considered above (wingless fighters and winged dispersers) provide a reasonably convincing case for the adaptive production of different kinds of males. The value of producing additional fighters that stay home to compete for a restricted set of females is likely to decline very rapidly after only a few are produced. If so, it would still pay to produce winged males if they have opportunities to find receptive females elsewhere, perhaps at sites that happen to lack sedentary fighters. Females might even prefer mating with dispersing males to avoid producing inbred offspring, where they are closely related to the fighting males at home.

Females can also be highly variable, and they too may sometimes exhibit multiple strategies associated with size. The most dramatic cases of intrasexual size variation occur among the workers of ant species with complex caste systems, where the largest and smallest workers may differ by one to two orders of magnitude in weight (see Figure 4.3). Such cases demonstrate that the hymenopteran developmental program is able to produce viable adults spanning a huge range of sizes, from a single genome, and that such variation may in fact evolve where ecological conditions call for it. The difference in size between reproductive male and female ants can also be enormous (see Figure 4.12) proving that aculeates of very different sizes can mate successfully.

These facts about ants raise an interesting question: Why is there not sometimes even more size dimorphism in bees and wasps? Why do males seldom weigh less than half as much as females? Why not a quarter, or an eighth? One possibility, which seems never to have been explored, is that females are selected to favor relatively large males because their size indicates both their mother's superior provisioning ability and their own metabolic efficiency in converting larval provisions into adult tissues. This bias might indeed prevent a slide toward really tiny males, but it would simultaneously create another dilemma: If females prefer large males produced by large, vigorous mothers, then why are males not sometimes two or three times the size of females? The formal answer is simply that various factors conspire to keep the male and female gain curves from diverging very far; in the end this merely restates the question in a different way. We need to measure the gain curves themselves in a variety of taxa and to learn what shapes them.

Conclusion

Generations of biologists have been drawn to the Hymenoptera by their great diversity and ecological importance, and by their endlessly fascinating behavior. The order is rich in sexual dimorphisms that are of interest in their own right, and that raise problems of fundamental significance for the biology of sex differences. Much of the relevant literature is scattered and known only to specialists, but increasing amounts of effort are being devoted to organizing existing knowledge according to principles of evolutionary ecology, and increasing numbers of field studies now focus on general theoretical issues. These developments have contributed to a growing awareness of the order's potential importance for research on sex differences.

Much remains to be done in quantifying sex differences, in describing general patterns of sexual dimorphisms, in finding the ecological correlates of such patterns, and in assessing how variation in sexually dimorphic character states affects male and female fitnesses. For example, size differences between the sexes depend in theory on the fitness gain curves within each sex, but there have been very few attempts to estimate these curves in nature. The gain curves are expected to depend (in part) on mating systems, but the mating systems of most species still await description. Likewise, some groups exhibit strikingly dimorphic male structures that undoubtedly play roles in courtship, display or male–male competition, but we know very little about how these structures are actually used in most species, and even less about their fitness consequences.

The developmental and physiological bases of sexual dimorphism in the Hymenoptera remain largely unknown, although there is increasing interest in the mechanisms of caste differentiation in social species. Both quantitative and qualitative differences in larval nutrition play important roles in caste differentiation, and larval nutrition might well play a similar role in solitary species that have multiple morphs within one sex. Such morphs always differ both in shape and size. Allometric changes in shape as a function of size can be studied using natural variation, but more could be learned by altering larval nutrition experimentally. Such studies would allow us to explore the set of nutritionally accessible morphologies, and would surely deepen our understanding of the ecological consequences of varying the level of investment in each offspring.

Major advances in our understanding of hymenopteran phylogeny are occurring at a rapid pace, owing largely to the application of powerful new techniques for inferring phylogenetic relationships, and to the availability of new sources of phylogenetically informative characteristics such as DNA sequences. As evolutionary relationships within the group become known with more certainty and in greater detail, it will become increasingly feasible to test historical and adaptive hypotheses using the sophisticated comparative methods now being developed. Hymenoptera provide unequalled opportunities to apply these methods to a broad range of questions about the evolutionary forces that shape differences between the sexes.

Acknowledgements

We thank E. Balaban and V. J. Rowntree for comments on the manuscript.

Further reading

Blum, M. S. and Blum, N. A. (eds.) (1979) *Sexual Selection and Reproductive Competition in Insects.* Academic Press, New York.

Danforth, B. N. (1991) The morphology and behavior of dimorphic males in *Perdita portalis* (Hymenoptera: Andrenidae). *Behavioral Ecology and Sociobiology*, **29**, 235–47.

Evans, H. E. and O'Neill, K. M. (1990) *The Natural History and Behavior of North American Beewolves.* Cornell University Press, Ithaca.

Evans, H. E. and West Eberhard, M. J. (1970) *The Wasps.* University of Michigan Press, Ann Arbor.

Gauld, I. and Bolton, B. (eds.) (1988) *The Hymenoptera.* First published by Oxford University Press in association with the Natural History Museum, London.

Hölldobler, B. and Wilson, E. O. (1990) *The Ants.* Springer-Verlag.

Krebs, J. R. and Davies, N. B. (1993) *An Introduction to Behavioural Ecology* (third edition). Blackwell Scientific Publications, Oxford.

Kukuk, P. F. and Schwarz, M. P. (1988) Macrocephalic male bees as functional reproductives and probable guards. *Pan-Pacific Entomologist*, **64**, 131–7.

Low, B. S. and Wcislo, W. T. (1992) Male foretibial plates and mating in *Crabro cribellifer* (Packard) (Hymenoptera: Sphecidae), with a survey of expanded male forelegs in Apoidea. *Annals of the Entomological Society of America*, **85**, 219–23.

O'Neill, K. M. and Evans, H. E. (1983) Alternative male mating tactics in *Bembecinus quinquespinosus* (Hymenoptera: Sphecidae): correlations with size and color variation. *Behavioral Ecology and Sociobiology*, **14**, 39–46.

O'Toole, C. and Raw, A. (1991) *Bees of the World.* Facts on File, New York.

Ross, K. G. and Matthews, R. W. (eds.) (1991) *The Social Biology of Wasps.* Cornell University Press, Ithaca.

Seger, J. (1991) Cooperation and conflict in social insects. In *Behavioural Ecology, An Evolutionary Approach* (third edition). Eds. Krebs, J. R. and Davies, N. B. Blackwell Scientific Publications, Oxford.

Thornhill, R. and Alcock, J. (1983) *The Evolution of Insect Mating Systems.* Harvard University Press, Cambridge.

Trivers, R. (1985) *Social Evolution.* Benjamin/Cummings, Menlo Park.

Waage, J. and Greathead, D. (eds.) (1986) *Insect Parasitoids.* Academic Press, London.

Wilson, E. O. (1971) *The Insect Societies.* Harvard University Press, Cambridge.

5

Sex change in fishes – how and why?

DOUGLAS Y. SHAPIRO

Many shallow-water marine fishes are sequential hermaphrodites – individuals mature and function first as one sex and later in life they change and function as the opposite sex. The two forms of sequential hermaphroditism are protogyny and protandry. Individuals of protogynous species mature and reproduce first as females and later change into males, while members of protandric species are males first and later become females.

In this chapter, I summarize important aspects of the behavioral mechanisms initiating sex change in fishes and review what is known currently about the genetics and physiology of sex change. Then, evidence is described that relates adult sex-change to initial sexual differentiation both in fishes and vertebrates generally and suggests that in spite of the direction of adult sex-change in protandric hermaphrodites there may be an underlying primacy of female development controlling sexual differentiation and adult sex-change. Finally, I discuss some approaches to how and why sex change might have evolved.

General patterns and processes of sex change

Sex change represents a dramatic event in the life of the individual. For example, when a female reverses sex its gonads beome completely reorganized. In most species, the ovary is round in microscopic cross-section and consists of oogonia and oocytes of varying sizes arranged in figure-like lamellae of tissue that project into a central cavity (Figure 5.1). In some species, the ovary also contains a few tiny islets of testicular cells nestling among the oocytes or a small region of testicular tissue on the periphery of the ovary. In all cases, when the oocytes ripen, just prior to spawning, they ovulate into the central cavity and pass through this cavity

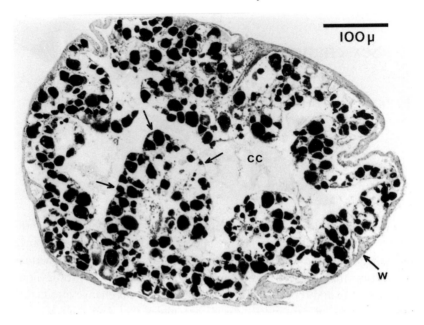

Figure 5.1. A typical immature ovary in a protogynous species, *Pseudanthias squamipinnis*, with a central cavity (cc) and an ovarian wall (w). Arrows delineate a lamellar projection into the central cavity. (From Shapiro, D. Y. (1981). Size, maturation, and social control of sex reversal in the coral reef fish *Anthias squamipinnis* (Peters). *Journal of Zoology*, **193**, 105–28.)

into an oviduct from which the eggs are released during mating into the outside world.

When sex change begins, the ovarian tissue within the gonad degenerates. Testicular tissue develops from the peripheral wall of the gonad (Figure 5.2), or from within the pre-existing islets or separate regions of spermatogenic cells in the ovary, and progressively overgrows the entire gonad (Figures 5.3 and 5.4). The final result is a solid testis, which may contain a remnant of the previous ovarian cavity as the sole indication that the testis had passed through an earlier phase as an ovary (Figure 5.5).

Many sex-changing species are sexually dimorphic in color or fin shape. As the individual alters gender, its external coloration changes. In the protogynous fish, *Pseudanthias squamipinnis*, females are a uniform, yellow-orange body color with clear, pale-yellow fins and a lovely purple stripe running from the eye to the base of the pectoral fin (Figure 5.6). In contrast, a deep red-purple decorates the head, caudal peduncle, posterior region of the dorsal fin and parts of the other fins of males (Figure 5.7).

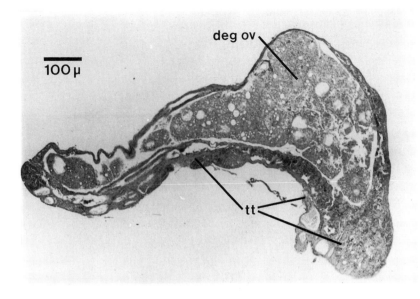

Figure 5.2. An early stage of protogynous sex change in *Pseudanthias squamipinnis*; testicular tissue (tt) develops from the periphery and ovarian tissue degenerates (deg ov). (From Shapiro (1981), see Figure 5.1.)

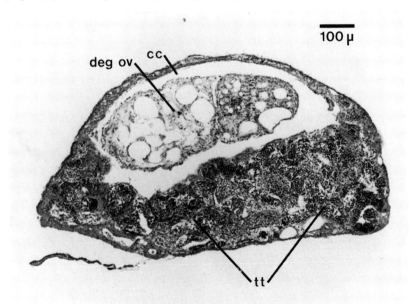

Figure 5.3. An early intermediate stage of protogynous sex change in *Pseudanthias squamipinnis* (symbols as for Figures 5.1 and 5.2).

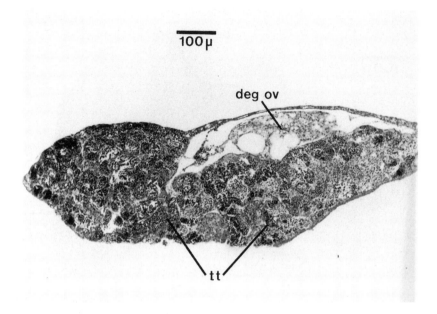

Figure 5.4. A late stage of protogynous sex change in *Pseudanthias squamipinnis* (symbols as in Figures 5.1 and 5.2). (From Shapiro (1981), see Figure 5.1.)

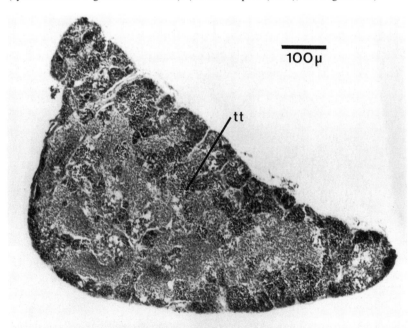

Figure 5.5. A typical testis of a protogynous species, *Pseudanthias squamipinnis*, containing compact testicular tissue (tt).

Figure 5.6. A female *Pseudanthias squamipinnis* with yellow-orange body and clear, pale-yellow fins (From Shapiro, D. Y. (1979). Social behavior, group structure, and the control of sex reversal in hermaphroditic fish. *Advances in the Study of Behavior*, **10**, 43–102.)

Figure 5.7. A male *Pseudanthias squamipinnis* with red-purple head, dorsal, anal and caudal fins, and a red spot on the upper part of the pectoral fin. (From Shapiro (1979), see Figure 5.6.)

Figure 5.8. The earliest recognizable color change in the sex reversal of *Pseudanthias squamipinnis* is dark pigment on the extreme tips of the pelvic fins (arrow). (From Shapiro, D. Y. (1981). The sequence of coloration changes during sex reversal in the tropical marine fish *Anthias squamipinnis* (Peters). *Bulletin of Marine Science*, **31**, 383–98.)

A red-purple spot also appears prominently on the male's pectoral fin. When a female changes sex, its body color gradually alters through a defined and regular sequence of events. The earliest sign of sex change is the appearance of melanistic pigment on the extreme tip of the pelvic fins (Figure 5.8). Over subsequent days, this pigment increases in intensity and spreads anteriorly up the fin. At the same time, black pigment begins to appear on the dorsal and caudal fins. Several thin, horizontal black lines become visible on the pectoral fin. Within a few days, these pectoral lines thicken and flow together to form the characteristic spot on the upper part of the fin. In several weeks the male color pattern has developed fully.

As the sex-changing female undergoes gonadal restructuring and colorational reattirement, it also changes its behavior. In all species, the new male begins to court and to spawn with females. In some species, non-reproductive behaviors, such as aggressive movements and social displays, either change in frequency or appear anew. For example, prior to changing sex, *P. squamipinnis* females seldom perform a characteristically

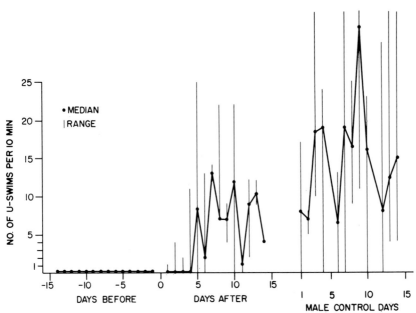

Figure 5.9. Rate of performance of U-swims, a characteristically male behavior, by individual *Pseudanthias squamipinnis* before and after the onset of sex change (on day zero), with typical male performance rates on the right. (From Shapiro, D. Y. (1981). Behavioural changes of protogynous sex reversal in a coral reef fish in the laboratory. *Animal Behaviour*, **29**, 1185–98.)

male display called a U-swim. Within four days of the onset of sex change, however, the newly emerging male begins to direct this behavior toward females and the frequency of the display increases thereafter (Figure 5.9).

There is solid evidence in a number of species that male coloration and behavior are secondary sex characteristics dependent upon androgen secretion. Thus, sex change entails not only anatomical reorganization of the gonad and the appearance of new color and behavior, but also a complete shift in the physiological control of hormone secretion. Since, in vertebrates generally, gonadal hormone secretion is modulated by a complex set of interactions involving the hypothalamus, pituitary and gonad, sex change depends upon alterations at many levels of internal physiology and results in striking changes in external appearance and behavior.

Mechanisms of sex change

In the early years of studying sex-changing fishes, demographic data suggested that protogynous and protandric species had different, but

Somatic sexual dimorphisms

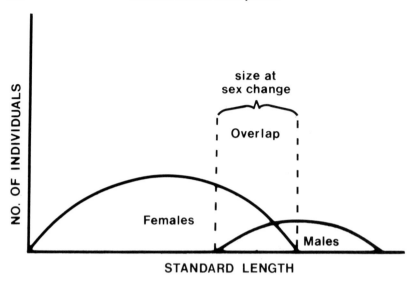

Figure 5.10. Individuals in a population were once thought to change sex in a single size range within the range of overlap between the sizes of females and the sizes of males. A frequency distribution of individuals from within the overlap range was taken to represent the size of sex change in the population (From Shapiro, D. Y. (1984). Sex reversal and sociodemographic processes in coral reef fishes. In *Fish Reproduction: Strategies and Tactics*, pp. 103–18. Eds. G. W. Potts and R. J. Wooton. Academic Press, London.)

characteristic, population structures. In protogynous fishes, all small individuals were female, all large individuals were male, and there was a relatively narrow range of overlap between the sexes, representing the size at which individuals changed sex (Figure 5.10). In protandric fishes, small individuals were male and large ones were female. Consequently, it was thought likely that sex change occurred at approximately a uniform size throughout the population and it was proposed that this size was determined genetically. This line of thought led to the first hypothesis about how sex change might be controlled. In protogynous species, when females attained a genetically controlled critical size, internal physiology would alter, resulting in sex change. In a reverse manner, this developmental hypothesis would apply to protandric sex-change.

Now we know that sex change is controlled in many species by behavioral or demographic alterations within the fish's social system. There is little direct evidence that individuals initiate sex change simply by attaining a uniform size, age or stage of development. In fact, in many

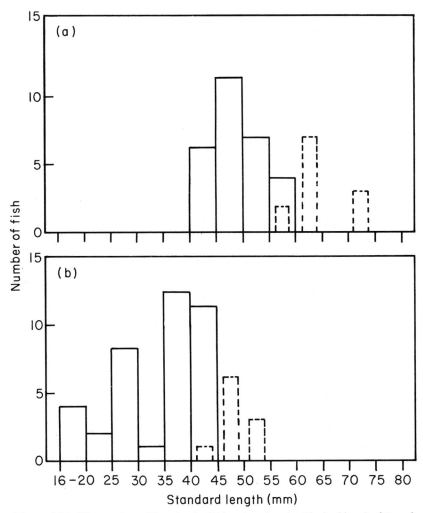

Figure 5.11. The number of females (solid bars) and males (dashed bars) of *Pseud-anthias squamipinnis* of different sizes at two sites (a, b) 600 m apart on a fringing reef at Aldabra Island, Indian Ocean. Individuals at site (a) were significantly larger than those at site (b) when they changed sex. (From Shapiro (1981), see Figure 5.1.)

species, sex change can occur at virtually any adult size. For example, the size at which males appeared in social groups of *P. squamipinnis* varied significantly between two sites 600 m apart on the same reef (Figure 5.11) and often differed among neighboring social groups only a few meters apart. Intrapopulation variation in the size at which individuals change sex is known for a number of species, resulting in a population structure consisting of many sets of overlapping male–female size frequency distributions, each set specific to one social unit or subpopulation. In other words, there is no critical size that applies equally to all individuals in the population and we must seriously doubt that the onset of sex change is a size- or age-dependent developmental phenomenon.

Sex change is known to be controlled behaviorally in at least a dozen species. In protogynous hermaphrodites, removal of a male from the social system stimulates a female to change sex. In protandric species, loss of a female induces a male to change sex. Apart from these simple observations, most details about how loss of one sex induces a member of the opposite sex to change gender concern protogyny.

Three aspects of the mechanism controlling protogyny are of primary importance. First, the mechanism is unlikely to depend on females having an endogenous tendency to change sex that is inhibited by some aspect of male presence or behavior. If this hypothesis were true, then we would never expect to find a female living without a male, because, in the absence of a male, the female's endogenous tendency would push it into sex change. Yet, all-female social groups have been found in two protogynous fishes, *Dascyllus aruanus* and *P. squamipinnis*.

In *P. squamipinnis*, all-female groups contain a mix of juvenile and adult females that do not mate within their groups and do not leave their groups to mate elsewhere, even when mating is active in neighboring bisexual groups. In studies in the wild, sex change was significantly less likely to occur in all-female groups over a two-month period than in unimale, bisexual groups following the removal of a male. When females did change sex in all-female groups, the onset of change was delayed significantly beyond that seen after male removal in bisexual groups. Subsequent laboratory studies confirmed these results under controlled conditions and established that females in all-female groups were physiologically capable of changing sex. When a male was added to each of a number of all-female groups in which no female had changed sex during an initial 60-day period, and then the male was removed two weeks later, a large female immediately began to change sex in each group.

These results imply that behavioral induction of sex change is more complicated than just the removal of male inhibitory cues from a female. Male absence, as in all-female groups, differs in its effect on sex change from male removal in groups in which a male had previously been present. It would appear that, firstly, a female must be exposed to a male and then the male must disappear in order for the female to be induced to change sex. Since most individuals in most populations would be exposed to males from the moment they settled onto the reef as small juveniles, this condition would normally be easily satisfied.

Secondly, sex change depends on interactions a female has both with males and with other females. In several species, separation of a single female from a male fails to induce the female to change sex. If several females remain together, however, one of them will change following male removal. The principle established by such observations is that a minimum number of females must reside together with a male in order for male removal to induce one of them to change sex. The minimum number of females is two in some wrasses and four or five in *P. squamipinnis*. The principle applies in reverse to protandric species. At least one juvenile (often called a subadult male in the literature even though its gonad contains both ovarian and testicular tissue) and one adult male must live together with a female for loss of the female to stimulate the male to change sex.

Thirdly, it is possible to find single measures of the behavioral interactions a female has with males and other females that explain most observations concerning sex change. For example, in *P. squamipinnis*, members of social groups receive six types of behavioral acts from both sexes. Each of the six discrete behaviors is expressed as a proportion of the total number of acts received by a fish and together these proportions describe a measure of behaviors received, called a 'profile' (Figure 5.12). In a sequence of earlier studies, this profile had a different, characteristic shape for each of six adult females living in laboratory groups with one male. When the male's contribution to each female's profile was removed, the profile of the largest female in each group changed dramatically, the profile of the second largest female changed less, and the profile of all smaller individuals changed virtually not at all (Figure 5.13).

The largest female's profile of behavior changed dramatically on removal of the male fish and, consequently, this female changed sex. So, it may follow that it is the change in the profile of behavior that controls the onset of sex change. Likewise, in all-female groups where the profile of behavior remained constant, females did not change sex. Finally, a minimum number of females are required to live together before this type of sex change can occur, since characteristic profiles that are capable of change can be developed only in relation to other fishes.

Thus, quite possibly, the central nervous system (CNS) of each female *P. squamipinnis* monitors the relative proportion of each of six discrete behaviors received both from males and other females cooccupying the

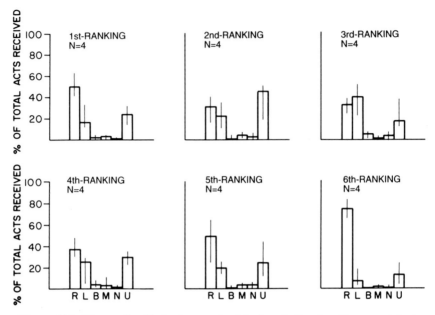

Figure 5.12. The profile of behaviors received for female *Pseudanthias squamipinnis*, where individual fish are ranked in order of decreasing dominance (from 1st, most dominant to 6th, least dominant), in laboratory social groups each containing one male and six adult females. The height of each bar represents the median proportion of the total number of behavioral acts received by an individual from all other members of the group. The thin vertical line is the range. The six different behaviors are identified along the horizontal axis: R, rush; L, lateral display; B, bent approach; M, mouth-to-mouth fight; N, nose bump; U, U-swim. (From Shapiro, D. Y. (1981). Intragroup behavioural changes and the initiation of sex reversal in a coral reef fish in the laboratory. *Animal Behaviour*, **29**, 1199–212.)

social group. When the internal representation of that profile of behavior alters beyond some critical level, the CNS is stimulated to release physiological signals that generate the internal and external alterations we recognize as being a sex change.

The details of behavioral control are likely to vary from species to species. For example, in the Caribbean goby *Coryphopterus glaucofraenum*, females communicate their presence to one another with chemical cues; and in the Hawaiian wrasse, *Thalassoma duperrey*, where there are two types of males (one resembling females in color), the minimum number of 'females' may be composed entirely of females or may include similarly colored males.

Nevertheless, in all species for which there is evidence, the behavioral

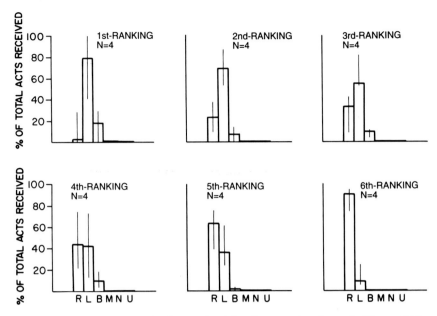

Figure 5.13. Profiles of behavior received for the same females as in Figure 5.12, except that the contribution of the male in the group has been excluded from the profile. This profile is what one would expect immediately following the removal of the male from the group. Labels as in Figure 5.12. (From Shapiro (1981), see Figure 5.12.)

mechanism is such that individuals change sex only under specific demographic or behavioral conditions. When these conditions occur, an individual changes sex regardless of its absolute size or age, even though the 'choice' of which individual to change may depend on the size of the individual relative to that of other fish in the social unit. Thus, we can easily see that the variation in size at which individuals change sex among social units and subpopulations is a direct result of the behavioral mechanisms controlling sex change in conjunction with spatial variation in demographic and behavioral conditions.

Genetics and physiology of sex change

In most fishes, the chromosomes are too small to distinguish sex chromosomes from autosomes morphologically. Breeding studies of a small number of temperate, freshwater fishes with sex-linked morphological markers indicate that sex-related genes are spread over a number of autosomes. In three laboratories, adult sex-change has been related to changes in the concentration of H-Y antigen.

H-Y antigen is a cell-surface, membrane component that is present in mammals on all cells of the body of males, and is absent in females. It is also sex-specific in other vertebrates (except that in birds and some reptiles, where the female is heterogametic, it is the female that possesses H-Y antigen and the male that lacks it). The synthesis of H-Y antigen is controlled by one or a small number of genes. At one time, it was thought that H-Y antigen might be the initial product of the sex-determining gene in vertebrates. While this idea is no longer accepted, H-Y antigen is at least associated with sexual development in many vertebrates.

In *P. squamipinnis* and two protogynous wrasses, H-Y antigen is present in higher concentrations in sex-changed males than in females. This finding suggests that the behavioral cues in the social realm that trigger sex change also influence H-Y genes to increase the synthesis of the antigen. However, a search for sex differences in the quantity and genomic distribution of satellite Bkm DNA, a segment of DNA that has been found to be sex-specific in many animals, revealed no consistent differences. Thus, we remain largely in the dark concerning genetic involvement during sex change.

Physiologically, injections of exogenous androgens have produced complete gonadal sex-change successfully in a few cases. In two species, gonadal sex-change has been produced fully by injecting females with mammalian luteinizing hormone (LH). In nature, 11-ketotestosterone increases as female-to-male sex-change progresses in several species, but there is little clear evidence indicating where in the causal chain of physiological events this increase occurs.

Recent cues come from studies on the number of cells in the hypothalamus producing luteinizing hormone releasing hormone (LHRH) that controls the release of LH from the pituitary. This work was done on a protogynous Caribbean fish, the bluehead wrasse. In this species, there are two developmental pathways for the production of males. If an individual differentiates a testis as a juvenile, it matures into a 'primary' male that is colored exactly like a female. If an individual differentiates an ovary as a juvenile, it matures as a female that may later change sex and become a 'secondary' or 'terminal' male whose color is strikingly different from that of females. Protogynous species with two types of males are called 'diandric'.

In the bluehead wrasse, the number of LHRH cells was significantly lower in females than in terminal males (Figure 5.14). Injections of

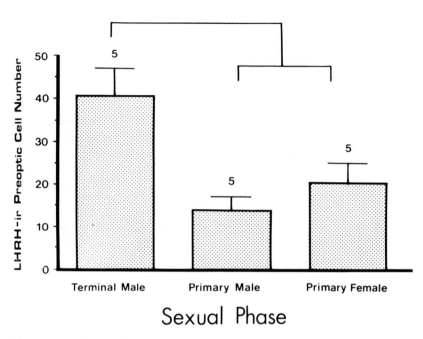

Figure 5.14. The number of LHRH-ir (luteinizing hormone releasing hormone, immunoreactive) cells in the preoptic tectum region of the hypothalamus of females and both types of males (primary and terminal, see text) in the protogynous bluehead wrasse. The height of the bar is the mean and the thin vertical line is the standard error of the mean. Numbers above the bar are sample sizes. The asterisk represents a significant difference between the means of the terminal male versus the primary male and female. (From Grober and Bass (1991) S. Karger AG, Basel.)

females with 11-ketotestosterone raised the number of LHRH cells in the hypothalamus to levels matching those of terminal males (Figure 5.15). The injections also changed the female's color to that of a terminal male. This result suggests that, early in the sex-change process, a surge of 11-ketotestosterone elevates the number of LHRH cells in the hypothalamus, and this alteration may reorganize the function of the entire hypothalamic-pituitary-gonadal axis. The individual would cease to function as a female and thereafter would function as a male.

Unfortunately for this interpretation, the number of LHRH cells is also low in primary males and injections of 11-ketotestosterone elevate their cell numbers. Since primary males have large testes and release large

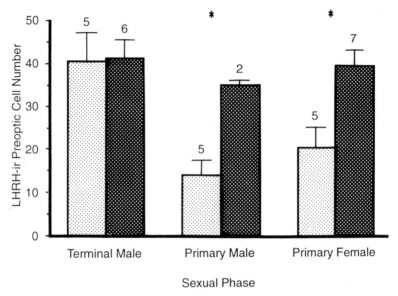

Figure 5.15. Number of preoptic LHRH-ir (luteinizing hormone releasing hormone, immunoreactive) cells in female and male bluehead wrasse, who were either treated with 11-ketotestosterone or who were controls. Symbols and abbreviations as in Figure 5.14. Asterisks represent significant differences between control and treated groups. (Grober, M. S., Jackson, I. M. D. and Bass, A. H. Gonadal steroids affect LHRH preoptic cell number in a sex/role changing fish. *Journal of Neurobiology*, **22**, 734–41. Copyright © 1991. Reprinted by permission of John Wiley & Sons, Inc.)

numbers of sperm during matings, the increase in LHRH cells following androgen administration is not specific to sex change itself.

Primary males differ behaviorally from terminal males: the latter spawn in pairs with individual females, whereas the former spawn in groups. Both males and females rush up in the water column together and release sperm and eggs at the top of the spawning rush. The fertilized eggs drift away in the current. In contrast, primary males congregate at particular locations; when a female approaches, between five and 40 males rush up in the water with the single female and all release their gametes simultaneously. The LHRH cell number seems to be related more closely to the development of alternative male reproductive behaviors (that of secondary versus primary males) than to sex change itself.

Sex change and sexual differentiation

In general, the association found between the presence of H-Y antigen and adult sex-change in fishes is similar to that of initial sexual differentiation in vertebrates, suggesting that these processes may share features in common. Indeed, from a broad ontogenetic perspective, adult sex-change is just the final stage of sexual differentiation.

In mammals, sexual differentiation follows a simple rule: individuals develop as females unless a specific, masculinizing mechanism intervenes. For many years it has been thought that sexual development in fishes could not be explained by the mammalian rule because of the existence of diandric species with two types of males and because of protandric hermaphroditism. However, recent evidence turns this possibility around and makes it viable.

In the diandric, bluehead wrasse, the gonads of all individuals have been found to pass through an early ovarian phase. For individuals fated to become females, the gonads continue their development as ovaries. When the fish matures, the ovary releases ripe oocytes and the individual spawns as a female. Later, under appropriate demographic conditions, the female may change sex and the gonads redifferentiate as a secondary testis. On the other hand, the gonads of juveniles fated to become primary males spend only a short period as ovaries before testicular tissue overgrows them and organizes them into testes.

The morphological structure of primary and secondary testes are sufficiently similar that they provide no support for a hypothesis that the two pathways for testicular differentiation (primary versus secondary) are qualitatively different, once testicular development begins. In other words, the main difference between primary and secondary testicular development probably lies in the time at which a masculinizing mechanism switches on. If it turns on early in juvenile life, the individual becomes a primary male; if it turns on after sexual maturity, the individual becomes a secondary male. Since both developmental sequences involve initial differentiation of an ovary, the processes can be described, at least morphologically, as following a female pathway unless a masculinizing mechanism intervenes.

Protandric hermaphroditism would appear to form a more formidable barrier to the application of the mammalian rule. If individuals develop first as males, and later change into females, this sequence could not be described as following an underlying female developmental pathway. Fortunately, recent evidence suggests a reinterpretation of the meaning of protandric sex-change within a broader ontogenetic framework.

Anemonefishes are protandric hermaphrodites that live in small social groups. Only the two largest fish reproduce. The largest is a mature, reproductively active female. The second largest is a reproductive male. Smaller individuals in the social group do not spawn as long as the two largest individuals remain in the group. However, if the female is removed, the male changes sex and becomes a female and the third largest individual in the group begins to reproduce as a male.

In a recent study of the anemonefish, *Amphiprion melanopus*, the proportion of the gonad devoted to ovarian, as opposed to testicular, tissue was plotted as a function of the position of the individual in the size hierarchy within the social group. The gonad of the smallest juvenile was approximately half ovarian and half testicular. As the position of the individual in the size hierarchy increased, the proportion of the gonad that was ovarian increased (Table 5.1). The gonad of the largest juvenile (the third largest fish in the group) was predominantly ovarian and the gonad of the largest fish, the active female, was entirely ovarian. The only individual in the group that did not fall into this regular developmental sequence was the second largest fish, the active male, whose gonad was predominantly testicular.

Thus, in spite of the fact that the direction of adult sex-change was from male to female, these results suggested that all individuals followed an underlying female developmental sequence unless a masculinizing mechanism intervened. The mechanism could be switched on behaviorally in a large juvenile, following the loss of the female and the adult sex-change of the male. The juvenile would then move out of the female developmental path and into a male sequence. It would remain in that sequence, i.e. would remain a male, until the masculinizing mechanism switched off (triggered by loss of the female), at which point the individual would revert to the underlying female sequence and its gonad would complete its development as an ovary.

This hypothesis was tested by simultaneously removing both the female and the male from a number of social groups of another protandric anemonefish, *A. ocellaris*, and investigating whether the gonad of the largest juvenile completed its development directly as an ovary or passed through an intervening testicular phase. The evidence was conclusive that the gonad did not pass through an intervening testicular phase. In other words, all juveniles were following a female developmental sequence.

Table 5.1 *Gonadal development in a social group of* Amphiprion melanopus

Size rank	Sex	% Ov
1	F	100
2	M	25
3	J	93
4	J	87
5	J	77
6	J	70
7	J	50

The proportion of the gonad devoted to ovarian tissue (% Ov) is listed as a function of the position of the individual (F, reproductively active female; M, reproductively active male; J, juvenile) in the size hierarchy within the group, with rank 1 being the largest and rank 7 being the smallest

Source: Adapted from Shapiro (1992).

In actuality, in order for the mammalian rule to apply to fishes, we would have to know that the physiological control of sex differentiation is such that individuals would differentiate ovaries until a physiologically male mechanism switched on. Not enough is known about the physiology of sex differentiation and sex change in fishes to be confident that the mammalian rule applies. We can say only that the descriptive, morphological evidence raises the possibility, for the first time, that fishes and mammals follow the same underlying paradigm.

How and why sex change might have evolved

During the period in which it was thought that protogynous and protandric fishes had simple population structures, with all small members of the population being one sex and all large members being the opposite sex, an elegant hypothesis was proposed to explain the selective advantage of sex change. The hypothesis was called the 'size advantage model' and it conformed nicely to what was known about population structure at the time.

The model assumed that the reproductive value of an individual varied with its body size and that the curve relating reproductive value to size differed between the sexes. For example, in one published scenario, the model assumed that reproductive value increased linearly with size for females. This is a reasonable assumption, because female fecundity generally increases with female body size in most fishes. The scenario also assumed a social system in which small males were relatively disadvantaged reproductively. Either small males could not compete successfully for access to females, or females just preferred to mate with large males. As male size increased, the male became increasingly competitive, and its reproductive value rose and eventually surpassed that of females (Figure 5.16). The result would be that male reproductive value is less than that of females at small body sizes and greater than that of females at large body sizes. An individual living in such a social system would do best reproductively over its lifetime by reproducing as a female when small and as a male when large.

For individuals served by this set of reproductive value functions, natural selection would favor those that changed sex at the size represented by the point of intersection of the male and female curves. All individuals in the population could be expected to change sex at or near that size. In other words, the model predicts that we should find a

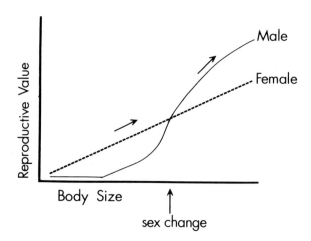

Figure 5.16. The relation between reproductive value and body size in males and females in one example of the size advantage model. Arrows along the surface of the curves represent the life-history track that maximizes lifetime reproductive value (female when small and male when large). The vertical arrow identifies the size at which natural selection should favor sex change. (Based on Warner, R. R. (1984). Mating behavior and hermaphroditism in coral reef fishes. *American Scientist*, **72**, 128–36.)

unimodal frequency distribution for the size at which individuals change sex in a population.

We know now, however, that population structure of hermaphroditic fishes is not simple. Often there is not a single unimodal frequency distribution for the size of sex change, but rather there are many such distributions. Each distribution describes a separate social unit or subpopulation. Furthermore, sex change is not generally triggered by the attainment of a critical size, but by the occurrence of particular behavioral and demographic conditions. Whenever those conditions occur, the individual changes sex regardless of its absolute body size. On the surface, the size advantage model would appear to be disproved by this evidence.

However, the size advantage model might still serve as an adequate explanation if each social unit or subpopulation were served by a separate set of reproductive value curves. The only additional condition that would need to apply is that each social unit or subpopulation should be self-recruiting, i.e. that offspring produced within each unit should remain there, or that offspring should recruit into a unit served by a set of reproductive value curves identical to those applying to their parents. Either of these conditions would allow natural selection to operate over successive generations to produce a different size for sex change within different social units.

With one or two exceptions, the life history of all known shallow-water marine fishes fails to conform to these two conditions. Instead of having self-recruiting young, adults disseminate their offspring widely by having pelagic eggs and/or larvae. During the pelagic phase, offspring spend weeks or months floating in the open ocean, probably mixing with offspring from a large number of social units, subpopulations, separate reefs and islands scattered over thousands of square kilometers. The chance that offspring will settle into an adult social unit or subpopulation served by the same set of reproductive value curves that serves their parents is so remote as to be negligible, at least for species in which there is wide variation among social units in the size at which sex change occurs.

What seems to be required to explain the selective advantage of sex change and the observed variability in size at sex change is a different kind of model in which reproductive value is assumed to vary with behavioral and demographic conditions rather than with body size. One approach to the construction of such a condition-dependent model would be to compare the costs and benefits of sex change under various conditions. The costs of changing from female to male probably include: an energetic investment in altering gonadal structure, reorganizing the hypothalamic-pituitary-gonadal axis, and changing external color and behavior; an increased mortality risk while the sex-changing individual looks and behaves differently from normal males and females so that it falls easy prey to predators; and a loss of reproductive time during the period in which sex change is completed and the individual is reintegrating reproductively into the social system. The benefit of changing from female to male is that, potentially, the new male can mate with many females.

Let us assume, as an example, that the costs and benefits can be measured in some common fitness unit and that the cost of female-to-male sex-change is independent of the adult sex-ratio in the social unit, but that the benefits increase linearly with sex ratio (Figure 5.17). In a social system in which males mate with several females, as sex ratio increases each male will mate on average with a linearly increasing number of females. Below the point of intersection of these curves, the costs of changing sex outweigh the benefits and we would expect natural selection not to favor sex change. Beyond the point of intersection, however, the benefits outweigh the costs and we would expect a mechanism to evolve that would allow individuals to change sex only when adult sex-ratio exceeded the value at the intersection point.

The critical sex-ratio, beyond which sex change is favored in this example, might correspond to the minimum number of females known

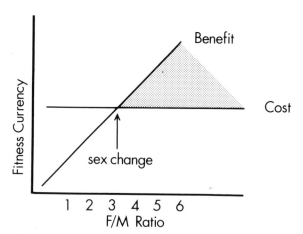

Figure 5.17. Fitness currency as a function of adult female-to-male sex ratio (F/M ratio) in subpopulations or social units in one example of a condition-dependent model for sex change. Shading represents the part of the curve where benefits of changing sex exceed costs. The arrow indicates the point on the curve where sex change becomes selectively advantageous.

to be needed in order for male removal to induce a female to change sex. In other words, a model of this type can be seen immediately to explain potentially at least some of the demographic and behavioral conditions known to be required in generating sex change. Here is another example of the potential power of this type of model: the loss of a male from within a social unit would immediately raise the female-to-male sex-ratio within the unit. If the resulting sex-ratio exceeded the critical ratio that makes sex change reproductively advantageous, then the model would explain why natural selection has favored a mechanism initiating sex change in response to male loss.

We might even use the model to help to explain why females do not change sex for relatively long periods in all-female groups. A reproductive advantage of changing into a male occurs whenever the male can mate with sufficient females to counterbalance the costs of changing sex. However, a male must presumably have to mate with those females over an extended period of time, i.e. on more than one or a few days, to reap substantial reproductive benefit. In *Pseudanthias*, all-female groups are recently established groups that formed during an earlier recruitment period by settlement of small juveniles at a new location on the reef. As juveniles mature and are joined by later recruits, the all-female group assumes its characteristic composition. Most such newly formed groups do not survive longer than six months on the reef. Perhaps the delay in changing sex in such groups allows a female to assess the probability that the group will persist. Only if the group proves its survivability will it be likely that a potentially new male will have access to more than the minimum number of females for sufficiently long to ensure that benefits outweigh costs.

A sex-ratio model is not the only type of condition-dependent model that one might construct to explain selection for behaviorally controlled sex-change. Other models can be found in the literature, based on the ratio of numbers of large and small individuals or on other factors. The important point is that some type of condition-dependent model is needed if we wish to explain, without violating observed empirical facts, the wide variability observed in the size at which individuals change sex in the population.

Conclusion

Behavioral control of adult sex-change provides a mechanism for initiating sex reversal only when demographic and behavioral conditions alter in particular ways. If future studies reveal that these alterations are the only conditions under which the reproductive value of being one sex is surpassed by the reproductive value of being the other sex, then we may conclude that natural selection has favored the evolution of condition-dependent control of sex change from the former to the latter sex. Before we reach the point of definitively drawing such a straightforward conclusion, many additional studies will be required to document the conditions under which sex change does and does not occur and the reproductive value associated with each of these conditions. One approach for revealing the full range of such conditions is to enquire analytically about the operational details of the mechanism controlling sex change.

Once natural selection has produced the behavioral control of sex change, the stage is set for the emergence of additional life-history characteristics secondary to that control mechanism. For example, it has been argued that an individual in a protogynous species, in which all juveniles are female, might increase its overall reproductive value by settling initially into a social group that is widely spaced from neighboring groups. There are three points to this argument. Firstly, more widely spaced groups will attract more juveniles than less widely spaced groups (assuming that, just before settlement, juveniles are spread homogeneously over the area occupied by adult groups). Secondly, the more juveniles entering a group, the more rapidly will the adult female-to-male sex-ratio of the group increase and the more frequently will an adult female change sex. Thirdly, the sooner that a female changes sex, the greater the proportion of its life that will be spent reproducing as a male with several or many females. Logically, this argument suggests that the mechanism controlling sex change behaviorally, in conjunction with early demography of these fishes, has led to the evolutionary emergence of a particular type of settlement behavior by juveniles that produces evenly spaced social groups on the reef.

Another example concerns the longterm stability of social groups. In some species of protogynous fishes, social groups may contain virtually any number of individuals, from few to thousands. Large group size provides a numerical buffer against wide swings in female-to-male sex-ratio. In such groups, it may take a long time before increases in female-to-male sex-ratio favor females changing sex. Small groups lack such a numerical buffer; the addition of relatively few new females can sway the sex ratio within small groups much more than addition of the same number would sway sex ratio within large groups. Thus, recruitment and subsequent growth of juveniles is more likely to produce conditions that are favorable to sex change in small rather than in large groups. A female could enhance her probability of changing sex by entering or establishing a small group rather than by remaining indefinitely in a large group. If this argument is true, then we should expect movement by females between large and small groups or subsets of female group members leaving a large group and moving elsewhere to establish a new, small group. In other words, we might expect to observe group fission. In one species of protogynous fish, group fission has now been documented during a longterm study in the wild. In this case, then, behavioral control of sex change has apparently influenced the evolutionary development of a particular type of adult behavior.

The point of these examples is to illustrate the evolutionary power that behavioral control of sex change may exert over a wide range of individual and social characteristics in these hermaphroditic species. Consequently, sex-changing fishes represent wonderful subjects for investigating a diversity of topics, from environmental influences on sexual differentiation, through the evolutionary development of particular types of behavior to the selective advantage of specific social systems under particular types of demographic conditions.

Further reading

Charnov, E. L. (1982) *The Theory of Sex Allocation.* Princeton University Press, Princeton.

Fricke, H. and Fricke, S. (1977) Monogamy and sex change by aggressive dominance in coral reef fish. *Nature,* **266**, 830–2.

Grober, M. S. and Bass, A. H. (1991) Neuronal correlates of sex/role change in labrid fishes: LHRH-like immunoreactivity. *Brain, Behavior and Evolution,* **38**, 302–12.

Pechan, P., Shapiro, D. Y. and Tracey, M. (1986) Increased H-Y antigen levels associated with behaviorally-induced, female-to-male sex reversal in a coral reef fish. *Differentiation,* **31**, 106–10.

Policansky, D. (1982) Sex change in plants and animals. *Annual Review of Ecology and Systematics,* **13**, 471–95.

Robertson, D. R. (1972) Social control of sex-reversal in a coral-reef fish. *Science,* **177**, 1007–9.

Ross, R. M. (1990) The evolution of sex-change mechanisms in fishes. *Environmental Biology of Fishes,* **29**, 81–93.

Shapiro, D. Y. (1987) Differentiation and evolution of sex change in fishes. *Bioscience,* **37**, 490–7.

Shapiro, D. Y. (1988) Behavioral influences on gene structure and other new ideas concerning sex change in fishes. *Environmental Biology of Fishes,* **23**, 283–97.

Shapiro, D. Y. (1992) Plasticity of gonadal development and protandry in fishes. *Journal of Experimental Zoology* **261**, 194–203.

Wachtel, S., Demas, S., Tiersch, T., Pechan, P. and Shapiro, D. Y. (1991) Bkm satellite DNA and ZFY in the coral reef fish, *Anthias squamipinnis*. *Genome*, **34**, 612–17.

Warner, R. R. (1975) The adaptive significance of sequential hermaphroditism in animals. *American Naturalist*, **109**, 61–82.

6

Teeth, horns and antlers: the weapons of sex

G. A. LINCOLN

To dominate is to possess priority of access to the necessities of life and reproduction. This is not a circular definition; it is a statement of a strong correlation observed in nature. With rare exceptions, the aggressively superior animal displaces the subordinate from food, from mates, and from nest sites. It only remains to be established that this power actually raises the genetic fitness of the animals possessing it. On this point the evidence is completely clear.

E. O. Wilson (1975) p. 287.

In nature, there are two quite different situations where conflict between individuals favours the evolution of weaponry. The first is the conflict between predator and prey, where the predator uses specialized weapons such as teeth and claws to kill its prey, while the prey uses another range of weapons in defence. The second is the more general conflict between individuals within the same species; here the weapons are used to determine dominance, which permits access to food, space and mates. In both situations of conflict, the outcome can have a major influence on survival and reproductive success. Furthermore, since these are two-sided conflicts involving aggressor and aggressed, they constitute the essential components of an arms race, which encourages an escalation in the development of methods of offence and defence. Predictably, selection in the long term will favour the evolution of larger and more effective weapons, as long as the functional benefits outweigh the overall costs of developing and carrying such an armoury.

A brief survey across the different species of mammals reveals many familiar examples of specialized weaponry. These include: the conspicuous tusks of the walrus, elephant, bush pig and hippopotamus; the elaborate spiral or curved horns of antelope, sheep, goats and cattle;

131

Figure 6.1. Mammals with different types of weaponry – *spot the odd one out*: (a) gemsbok, *Oryx gazella*, Kalahari Gemsbok National Park, South Africa; (b) hippopotamus, *Hippopotamus amphibius*, South Africa; (c) reindeer, *Rangifer tarandus*, Cairngorms, Scotland; (d) giraffe, *Giraffa camelopardalis*, Etosha National Park, Namibia. *Answer*: gemsbok – these are females and the others are males. (Photographs (a), (b) and (d) are courtesy of The Anthony Bannister Photo Library, Lanseria, South Africa; photograph (c) is by the author.)

the skin-covered pedicles of giraffe and the complex multi-branched antlers of deer (Figure 6.1). Interestingly, in most, if not all these cases, it appears to be competitiveness between individuals within the species (intraspecific selection) which best accounts for the evolution of the weaponry. In particular, it is the competition between males for social dominance, harems or territory which favours the development of weapons. Thus, males have bigger teeth, horns and antlers than females, and males of polygynous species, where competition for mates is most intense, have the biggest weaponry of all.

In the case of females, the development of large teeth, horns or antlers is a notable feature in some species but not in others. The functional significance of such equipment in females may relate to its value in defence from predators, particularly the defence of the offspring, and to its value as weaponry in intraspecific competition when females compete with females (or possibly males) for food, space and shelter. In most social species there is a clearly defined dominance hierarchy within female groups, and dominance allows access to the best feeding and resting places with its obvious benefits. The best examples of females with well-developed weaponry are in species which live in large, mixed-sex groups where competition for food may be especially intense.

One example is the gemsbok, a large species of antelope from South Africa, whose females have very impressive horns (Figure 6.1). This is a social species which lives in semi-arid areas where the food supply varies according to the time of the rainy season and can be very patchy in its distribution. The adaptive significance of the weaponry in females appears to relate to the use of the horns in competition over the clumped herbage; better feeding presumably results in more viable offspring. The reindeer (alias caribou) is another species whose females have weapons; in this case, competition for food is especially intense in the snow in winter.

While intraspecific competition appears to be the major determinant in the evolution of weaponry, it is clear that different selective forces operate in males compared to females. For males, it is competition for mates which is the key to success, and success is measured by the ability to leave more offspring. For females, it is competition for resources, plus the importance of defence from predators, which markedly influences the lifetime reproductive success. These different types of competition result in a different type of weapon in males and females, and thus sexual dimorphism (Figure 6.2). This is illustrated nicely in the case of cattle (Figure 6.3). Sexual dimorphism in the size and shape, or presence/absence of weaponry, is a noticeable feature in most mammals.

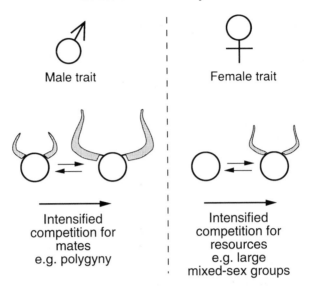

Figure 6.2. Diagram to summarize how intraspecific competition favours the evolution of weaponry; selection operates differently in males and females which results in sexual dimorphism in the size and shape, or presence/absence of horns or antlers.

Focus on antlers

Antlers in deer provide an excellent example of weaponry which has evolved through intrasexual selection. Of the 40 species of deer living today, 36 species develop antlers, and in all but one species it is exclusively a male characteristic. The antler grows from the frontal bone of the skull, and consists of a permanent basal pedicle and an upper part (commonly referred to as 'the antler') which is cast and regenerated annually. In the species where antlers are a male characteristic, the growth of the pedicle commences at puberty, stimulated by secretion of testosterone from the maturing testes. The cycle of casting and regrowth of the antler in the adult is then regulated by the seasonal cycle in the secretion of testosterone which occurs as the testes wax and wane in activity associated with the mating season. In this situation, the hard antler phase is synchronized with the period of maximum sexual activity, at which time the antlers are used as weapons and display organs when males compete for females in the rut.

The reindeer/caribou is the one species in which the females also develop antlers. In this case, the antlers begin to develop before puberty in both sexes, thus the growth of antlers *per se* is not a secondary sexual

Figure 6.3. Photograph of Highland cattle, *Bos taurus*, in Scotland to illustrate sexual dimorphism in the shape and size of the horns. The bull (right) has thick horns projecting low from the skull which are adapted to take the full weight of the animal in male–male fighting. The cow (left) has thinner, longer, more upright horns which are adapted to prod rivals to gain access to preferred feeding or to protect offspring.

characteristic, unlike in the other species. However, males have much larger antlers than females; thus size is a secondary sexual characteristic, and the sex hormone cycle still acts to synchronize the hard antler phase to a specific time of year for females as well as for males (see later).

Antlers take over from tusks

The deer share a common ancestry with other ruminants including camels, giraffes, pronghorn antelopes, goats, cattle, sheep and antelopes. Some 40 million years ago, the common ancestor of all these groups was a diminutive animal not unlike the mouse deer of today. These animals had large canine teeth but no horn-like organs. During the evolution of the deer, the antlers have replaced canines as the primary weapon. Some insight into this evolutionary progression can be gained from a comparison of the living species of deer (Figure 6.4). There are clear differences between species in the form of the weaponry and

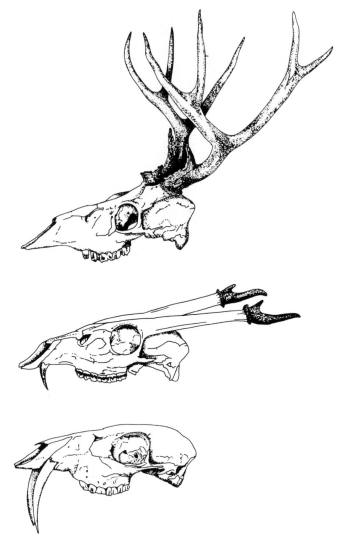

Figure 6.4. Skulls from three different species of deer illustrating the inverse relationship between the development of the canine teeth and antlers: (top) black-tailed deer, *Odocoileus hemionis*; (middle) Indian muntjac, *Muntiacus muntjak*; (bottom) Chinese water deer, *Hydropotes inermis*. (Specimens from the British Natural History Museum and drawings by Ted Pinner.)

fighting behaviour which illustrate that weaponry and fighting behaviour must coevolve.

The male Chinese water deer resembles the ancestral stage, since this species has no antlers, but the males have enlarged canine teeth developed as tusks up to 7 cm in length. These are sufficiently long that they protrude below the muzzle. This species is territorial and the male uses its sharp tusks as slashing weapons when fighting with other males. The Indian muntjac illustrates another evolutionary stage, with the males possessing both large canine teeth and small antlers which are supported at the end of long pedicles. In this case, the males are also territorial but have a more complex fighting behaviour utilizing both forms of weaponry. Here, the antlers tend to be used to ward off opponents and to prevent the use of the canines in aggressive encounters.

The larger species of deer, including the black-tailed deer, reindeer, moose and red deer, illustrate the final stage in the evolutionary sequence. These are the most social species living in groups and sometimes in large herds in winter. The males are polygynous and highly competitive in the rut. They have large, multi-branched antlers, while the canine teeth are small or totally absent. Thus the antlers are the primary means by which the males compete for dominance. In some species the vestiges of a canine ancestry are still apparent. For example, red deer stags have a pigmented area on the muzzle of the lower jaw which appears as an image of a tusk, and part of their aggressive repertoire includes a chin-up display during which the upper lip is raised to expose the small canine teeth. The canines are never used in actual physical fighting even though they are notably larger in males than in females.

Antler size, body size and mating group size

A detailed comparison between species has revealed that there is a positive allometric relationship between antler size and body size in deer. Thus, for the larger species, the antlers are larger relative to body size, compared with the smaller species. The size of the antlers of the extinct giant fallow deer, or 'Irish elk', fits on a line extrapolated from the living species (Figure 6.5). Of more relevance to the current topic is the observation, initially made by Tim Clutton-Brock (see Chapter 15), that species of deer with a highly polygynous mating system have bigger antlers relative to body size compared with species that form small mating groups (Figure 6.5). This is consistent with the idea that increased competition between males, which occurs with increasing polygyny, favours the evolution of larger and more complex weapons. The selective pressure associated with polygyny also results in an increased body size, musculature and aggressive behaviour, all attributes that contribute to fighting success.

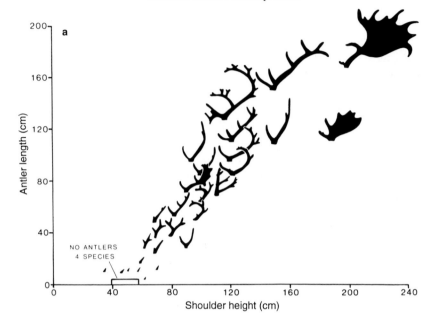

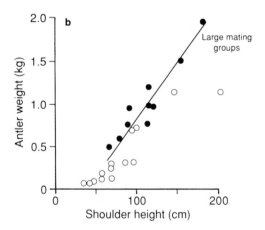

Figure 6.5. Relationship between antler size and body size in different species of deer: (a) Shape and size of antlers related to body size for all 40 living species of deer plus the extinct giant Irish elk, *Megaloceros giganteus* (top right); (b) Weight of antlers related to body size for 24 of the living species of deer separated according to the degree of polygyny (open circles, small mating groups; filled circles, large mating groups). Note, the polygynous species have relatively larger antlers. (Data from Lincoln, G. A. (1992) Biology of antlers. *Journal of Zoology*, London, **226**, 517–28; and Kitchener (1991).)

The red deer, wapiti, barasingha, fallow deer and reindeer are typical of the large social species which have elaborate antlers and fight overtly during a brief rutting season. Selection for larger weaponry explains, at least in part, the positive allometry with body size. This is because large deer tend to be grazers living in large social groups and this allows for a polygynous mating system. Thus, ecology influences social behaviour, which ultimately influences sexual dimorphism through sexual selection. Based on these generalizations, it is possible to predict aspects of social behaviour from the degree of sexual dimorphism in weaponry. Therefore, it is most probable that the giant Irish elk, with its spectacular male weaponry, lived in large social groups and had a short rutting season; at this time the biggest stags with the best weaponry would have monopolized large harems much like red deer do today. The large antlers would have been a key determinant of reproductive success.

Annual replacement of the antlers and fighting behaviour

In all species of deer in which the males develop antlers, the antlers are cast and regrown annually. This cycle is regulated by the seasonal change in the activity of the testes. In red deer, for example, the old antlers are cast in the spring when the secretion of testosterone from the testes declines to a minimum (Figure 6.6). The casting is followed immediately by the regrowth of new antlers which develop by a process akin to wound healing. The regeneration process involves the production of new skin, hair follicles, bone, blood vessels and nerves and is a unique feature amongst mammals. The regrowth occurs from apical growing points and can be extremely rapid, even by embryonic standards. The main beam of the growing antler may increase in length by about 1 cm a day in the larger species and the new antlers are complete within three months.

The growing phase is terminated by the increase in the circulating levels of testosterone derived from the testes as they regain functional activity prior to the mating season. The increase in androgen titres causes the final calcification of the bone of the antler followed by the shedding of the overlying velvet-like skin. As a consequence the antler tissue dies, forming the insensitive hard antler which is the specialized weapon used in the rut. The dead antler then remains attached to the living pedicle as long as the testes continue to secrete testosterone; in this situation the hormone acts to prevent the rejection process which would normally remove the dead structure. This rejection finally occurs in spring when the secretion of testosterone declines to a minimum. Castration at any stage during the

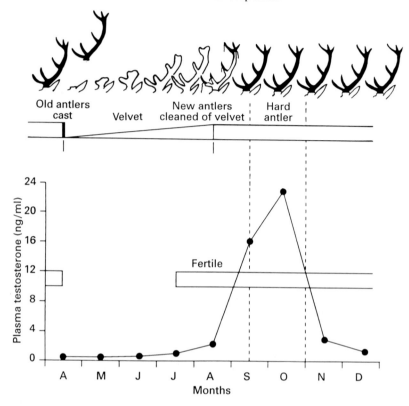

Figure 6.6. Seasonal cycle in casting, growth and cleaning of the antlers related to the secretion of testosterone and the period of fertility in adult male red deer, *Cervus elaphus*. Testosterone acts to induce the hard antler state and to induce the increase in sexual and aggressive behaviour associated with the rut; thus, stags have functional fighting weapons when they are fertile and most competitive. (Data from Lincoln, G. A. (1992) (see Figure 6.5).)

hard antler period results in premature casting of the antlers within two to four weeks. Such castrated males redevelop new antlers, which are smaller than normal and with a reduced number of branches; these antlers remain permanently in velvet, since there is no seasonal cycle in testosterone to induce their periodic maturation.

The seasonal reproductive cycle in the stag is regulated by the gonadotrophic hormones (LH and FSH, luteinizing hormone and follicle-stimulating hormone respectively), secreted from the anterior pituitary gland, which dictate the spermatogenic and androgenic activity of the testes. The release of the gonadotrophic hormones is governed by the brain through the secretion of a decapeptide hormone (GnRH) produced by the hypothalamus. The annual cycle in the activity of the hypothalamo-pituitary-gonadal axis is generated intrinsically within the central nervous system, but is entrained by seasonal environmental cues, in particular the annual cycle in daylength. This highly sensitive mechanism allows for very precise synchrony of the reproductive cycle with time of year. This is evident in the antler cycle where individuals can be very predictable in the timing of casting and cleaning of the antlers. It is not unusual for an adult stag to cast its antlers on the same day in successive years. The physiological system, whereby daylight length (acting via the brain) regulates gonadal activity, and testosterone from the testes induces the development of secondary sexual characteristics, mating and aggressive behaviour, ensures the temporal coordination of all these seasonal events. This means that males have functional hard antlers during the rut when they are fertile and most competitive.

Stags fight frequently during the rut. For example, Fiona Guinness and her colleagues, working on the Isle of Rhum situated off the west coast of Scotland, have shown that mature stags are involved in a serious fight on average every five days, and injuries are common. The animals use roaring behaviour to broadcast their presence and to intimidate other stags. A challenge between two well-matched rivals involves a ritualized display. The two animals parade side by side a few yards apart (Figure 6.7). If neither animal withdraws, a fight will develop as the opponents turn to face each other. The antlers are interlocked and animals viciously attempt to force each other backwards and off balance. The sharp points of the antlers act as dangerous weapons, while the curved tines form a shield to block the opponent. The outcome of a fight is usually decisive and the vanquished animal is chased from the scene. Virtually all matings are achieved by the stags which can exclude rivals, and can monopolize a harem during the main period of the rut when the hinds are coming into oestrus and ovulating. Thus, fighting success is translated into sexual success.

Outside the rut, stags live together in loose associations in bachelor groups. The antlers are still used to exert dominance, and there is a clearly defined social hierarchy amongst the males. The individuals learn to recognize each other and then observe the dominance relationships without contest. This situation changes suddenly in spring when the stags begin to cast their antlers (Figure 6.8). It is usually the older, high-ranking animals which cast their antlers first. Almost immediately, the newly polled animals are challenged by a succession of subordinates which still retain their antlers. The cast stags are now at a clear disadvantage, and

Figure 6.7. Drawing of an aggressive encounter between two mature male red deer, *Cervus elaphus*, competing for possession of a harem during the autumn rut. (a) During the initial ritualized phase, the animals parade side-by-side in a contest of intimidation – the antlers provide a visual index of age and dominance. (b) During the next phase, the contest escalates into a fight and the antlers are used as both an offensive weapon and defensive guard. (Drawings by Ted Pinner from photographs by the late Julius Behnke.)

are soon relegated to the bottom of the social hierarchy. At this time they often seek seclusion to avoid contact with the other stags.

Over the following six weeks or so, all the males lose their antlers, and the social relationships are re-established, although not quite as before. The antlerless stags now use their feet in aggressive encounters, more like females. This fighting technique is apparently less decisive than the use of the hard antlers, since the hierarchy amongst the bachelor males is no longer linear and there are some poorly resolved dominance relationships.

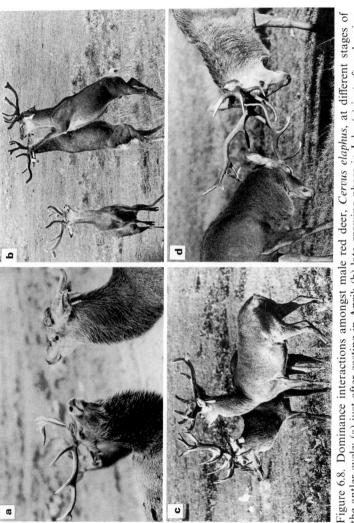

Figure 6.8. Dominance interactions amongst male red deer, *Cervus elaphus*, at different stages of the antler cycle: (a) just after casting in April; (b) late growing phase in July; (c) just after cleaning the velvet in August; and (d) during the hard antler phase in December. The front feet are used to exert dominance only during the period when the antlers are regrowing during the spring and summer. (Data from Lincoln, G. A. (1971) 'The Reproductive Physiology and Behaviour of Red Deer Stag', Ph.D. Thesis, University of Cambridge.)

Finally, once the new antlers have finished growing and have come into hard horn, the situation changes again. This time, the older stags are the first to shed the velvet and reacquire the use of their new antlers. Within a few days the animals take advantage of their new weaponry to establish absolute dominance (Figure 6.8).

It seems likely that the stags learn the dimensions of their antlers during the growing period by rubbing the richly innervated tips of the growing antlers against the body, and by touching objects with them. Once the antlers are cleaned, the animals indulge frequently in sparring contests which are presumably another learning exercise. Sparring is essentially play fighting, and often involves young males which interlock antlers and wrestle gently against each other. Sometimes a young animal will approach a much larger animal and use submissive vocalizations to initiate a sparring contest. Usually, it is the subordinate which terminates the contest when the tussle gets too vigorous.

In red deer the antlers are retained for many months after the rut; however, in reindeer, white-tailed deer and moose, the casting occurs earlier and the males are without antlers throughout the winter. This is related to an abrupt 'switch off' of testicular activity associated with a more restricted mating season. This early casting has the advantage that the males do not carry their heavy weaponry when there is no competition over females; however, the animals are at a potential disadvantage in competition for food with younger animals which retain their antlers for a longer period, and, in the case of reindeer, with females which have antlers in winter. However, the large body size of the adult males may allow dominance to be maintained even without the benefit of antlers.

Manipulating the size of antlers experimentally

There have been a number of studies to investigate the role of antlers which have involved experimentally altering the size of the hard antler when the bone is dead. In one of our experiments the main portion of the antler above the brow points was removed from a high-ranking red deer stag in the wild on the Isle of Rhum (Figure 6.9). This was carried out in September, shortly before the rutting season, and involved dart-immobilization of the animal to saw off part of the antlers. The effect was dramatic. When the animal rejoined his bachelor group, after recovering from the immobilizing drug, his dominance was challenged immediately. The first approach was made by a stag which was next down in the social hierarchy. The result was a brief and vicious fight which was won

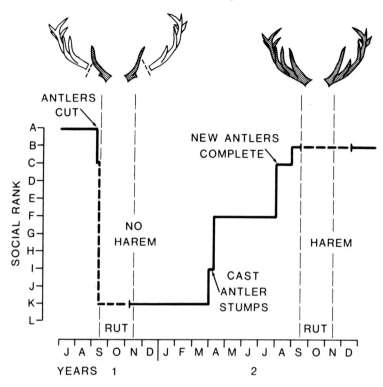

Figure 6.9. Diagram to illustrate the effect of removing the antler beam and top tines from a dominant male red deer, *Cervus elaphus*, living in the wild on the Isle of Rhum. This caused a rapid fall in social dominance and failure to secure a harem in the rut, since the animal was challenged repeatedly by other stags and was unable to fight successfully with the antler stumps. The animal recovered dominance and rutting success the following year, after it had redeveloped a new set of antlers. (Data from Lincoln, G. A. (1972) The role of antlers in the behaviour of red deer. *Journal of Experimental Zoology,* **182,** 233–50.)

by the experimental animal in spite of his partially amputated antlers. Interestingly, in this encounter the experimental animal presented his antlers initially just as if they were still complete. He attempted to engage the tips of his imaginary antlers with the tips of the antlers of the rival, and for an instant there was no contact. A moment later the animals crashed together and the experimental animal used his superior weight to win the encounter.

This was not the end of the story. After the first fight, a series of further challenges were made by other subordinate stags within the group and, after a series of defeats, the experimental animal was relegated to near the bottom of the social hierarchy. During the ensuing rutting season in the following month, the animal was excluded almost totally from the company of hinds by other stags, and never formed a stable harem as in previous years. His lowly status continued throughout the winter, and lasted until the antlers were cast in the spring. Then, once he redeveloped a normal set of hard antlers, he regained his dominance and was again successful in the following rut.

Although this was an experimental intervention, loss of parts of antlers can occur naturally when they are broken during fighting. The loss of a single tine from an antler appears to be of little consequence, but the loss of a major part of one antler can impair fighting ability severely. The conclusion from these observations was that there are two aspects to the function of antlers; one relates to their function as a physical weapon, and the other relates to their function as a visual symbol of the animal's age, size and dominance. There is no doubt that the basic shape of the antler has evolved through its function as a physical weapon. The interesting question is, to what extent has the shape been modified by selection to fulfil its visual function?

Antlers as status symbols as well as a weapon

Andrew Kitchener, currently working at the Royal Scottish Museum in Edinburgh, has made a detailed study of the structural and functional relationships of antlers. This has involved the measurement of the density of bone at different places on the antler and a calculation of the strength of the bone. The results show that the antler is strengthened mechanically by the deposition of extra bone around the main beam and tines, where extra stresses would be predicted when antlers are used in fighting. In all species studied, the actual strength was many times greater than that necessary merely to support the weight of the antlers. This applied equally to an analysis of the antlers of the extinct giant Irish elk which revealed that the antlers of this species were some 67 times stronger than required for support alone. The obvious conclusion is that antlers are designed to fulfil their role as fighting weapons, and even the Irish elk was likely to have used its formidable antlers for physical conflict, not just for ornament.

There appears to be no species whose antlers are not effective fighting weapons. However, this does not preclude adaptations related to the role of the antlers in display. There are some features of the antlers in the larger species of deer which are not readily explained in terms of their function as a physical weapon. For example, in male reindeer, red deer, wapiti and barasingha the tops of the antlers are formed into multiple branches which are not normally used when the animals engage in fighting. However, these features do add markedly to the visual impact of the antlers. In fallow deer, moose and the Irish elk the upper tines of the antlers are linked together by bone to form a palmate structure which is again very visually apparent from the side or front, and acts to accentuate the movements of the head. In the Pére David's deer, the antlers have many additional small bumps and branches which collect mud and vegetation readily, and this can enhance their appearance. It is a common observation that rutting males of the larger species of deer carry vegetation on their antlers and appear to use this to intimidate other males.

Based on these features, it appears that the basic weapon has been refined to fulfil additional functions: in particular to provide a visually impressive structure which varies in size related to body size and age. Thus, the antlers act as a visual index of fighting ability, and the appearance of the opponent's antlers can be used along with the other signals in deciding whether to challenge or to withdraw. This is advantageous to an individual, since it avoids some of the physical costs associated with actual fighting.

Stags with no antlers

In the highlands of Scotland, a few red deer stags fail to develop antlers. They are called hummels, a corruption of the word 'humble' (hornless or awnless), and represent about 0.3% of the population. These animals are apparently normal in all other respects; they develop male secondary sexual characters like the thickened neck, and show rutting behaviour at the normal time in the autumn. But there is a widely held belief amongst deer stalkers that hummels are able to dominate stags with normal antlers. The story is that, since hummels are spared the nutritional burden of growing antlers each year, they grow larger bodies than antlered stags, providing them with a relative advantage over these rivals. The legendary success of the hummel was described by Frank Fraser Darling in his pioneer study of red deer in the Scottish highlands:

A hard dunt in the ribs from the polled head of a hummel seems to upset his opponent more than a sharp jab from the points of an antler. Were it possible to take a count of services by each stag in a large population, I think it would be found that hummel stags would have covered individually a larger number of hinds than each of their antlered fellows.

(F. F. Darling, *A Herd of Red Deer*. 1937, p. 158, Oxford University Press)

This claim raises a number of obvious questions in view of what has been stated in this chapter. Can it be true that antlers are of no special advantage in fighting? Also, why are hummels so rare if they are so successful in the rut? Part of this story is explicable in terms of the past attitudes of the stalkers and landowners. The requirement has always been for stags with large, multi-pointed antlers which provide good hunting trophies to mount on the wall. Thus, a stag with no antlers was the epitomy of a 'bad' stag, and so a price was always on its head. Occasionally, these animals had been seen to dominate antlered stags and to hold a harem of hinds, and this added to their evil reputation.

The assumption that the hummels are actually larger and superior to normal stags appears to be a misunderstanding. When a direct comparison is made, hummels are found to be in the normal range for body weight (Figure 6.10). The confusion appears to arise from the marked effect of age on body size and dominance. Since all stags continue to grow larger and heavier throughout most of their lives, older animals will normally dominate younger animals. It is very likely that an old hummel may dominate a younger antlered rival especially if it has learned to cope with the antler deficit, but this does not indicate that hummels are in any way superior.

With the help of Roger Short and John Fletcher, I have studied four different hummels in captivity. These animals were all captured from the wild in different parts of Scotland (Braemar, Jura, Sutherland and Balmoral, Figure 6.10). Contrary to their reputation, they were all relatively submissive in behaviour when kept with other stags (Figure 6.11). There was no indication that the hummels were superior in fighting ability, although they were able to dominate some younger antlered stags. By mating the hummels with normal hinds we showed that they are, indeed, fertile. Some 45 male offspring were produced from such matings, and, significantly they all developed normal antlers. Even when a hummel was back-crossed with his daughters, this still failed to produce any antlerless male offspring. So what is the cause of the hummel condition?

This was partly revealed by a simple experiment. All the captive hummels had small, rudimentary antler pedicles on their foreheads which were barely visible above the scalp, and of course no antlers. The experiment consisted of removing a small piece of tissue from the tip of one of the rudimentary pedicles from the Braemar hummel when he was

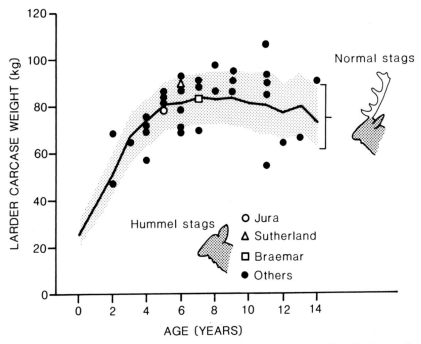

Figure 6.10. The relationship between body weight and age for 34 hummel red deer (naturally polled stag, *Cervus elaphus*) from the highlands of Scotland compared to data for stags with normal antlers (mean ±SEM, stippling) showing that hummels are not consistently larger than normal stags. (G. A. Lincoln, unpublished results.)

about 12 years old. The effect was remarkable. As the wound healed over the damaged pedicle, antler tissue began to develop for the first time. During the next two months, a complete antler was produced. This was then cleaned of velvet before the rut, and subsequently cast and regrown just like a normal antler. The rudimentary pedicle on the other side remained unchanged, acting as the perfect control.

This experiment was confirmed in other hummels. It was also found that the antlers produced by the treated hummels in succeeding years continued to increase in size and the degree of branching, and there was a concomitant enlargement of the pedicle each year as occurs in normal stags. This shows that it is the localized response within the pedicle to the process of antler growth and casting which induces the change in the mass of the pedicle and thus dictates the age-dependent increase in size of the antler.

The studies with the hummels show clearly that these animals are

Figure 6.11. Photograph illustrating the submissive behaviour of a hummel (naturally polled stag, *Cervus elaphus*) towards a normal stag with antlers. The hummel was mated with hinds in captivity but none of the male off-spring inherited the polled condition. (Data from Lincoln, G. A., Guinness, F. and Fletcher, J. (1973). History of a hummel – part 3. Sons with antlers. *Deer*, **3**, 26–31.)

normal except for a defect in the initial formation of the pedicles. By traumatizing the immature pedicle, it was possible to use the wound healing response to trigger the development of antler tissue. Once this had been produced, the sex hormones took over to dictate the antler cycle. So, why did the pedicles fail to develop in these animals? With no evidence of a genetic effect, we have concluded that this is probably a result of malnutrition early in life, which inhibits testosterone production and stunts the growth of the pedicles at puberty. The pedicles then become unable to complete their growth even though the animal subsequently becomes sexually mature. It seems probable that there is a genetic predisposition to this condition but this still needs to be tested by rearing the offspring of a hummel on restricted feeding to mimick the situation which occurs in the wild on the barren moorlands of Scotland.

Deciduous nature of antlers

Deer antlers are cast and regrown on a regular basis. Many reasons have been proposed to explain the deciduous nature of antlers. These include a mechanism for replacement in case of breakage, or a means of ensuring that the size of the antlers can change with body size as a result of increasing age. Bearing in mind that the antler, in its functional state, is a dead structure, both alternatives are logical. The way the casting cycle permits the adjustment of the size of the antlers with age is illustrated in Figure 6.12. This is in contrast to the situation in other ungulates which develop living keratinized horns as fighting weapons, where growth occurs incrementally (Figure 6.12).

The deciduous nature of antlers can be traced back through the fossil record to the Miocene some 20 million years ago. The early ancestors of deer are thought to have had permanent bony protuberances on the skull which were covered with skin, with an outward appearance rather like the pedicles of a giraffe. It was from these that the more complex antlers of the modern deer evolved. The wound healing mechanism which results in the regeneration of the antlers may have evolved at an early stage. This would have allowed the replacement of tissue damaged through fighting. Since the progenitors evolved in a temperate climate, it can be assumed that they were seasonal breeders, just like the living species. The coincidence of seasonal damage to the antlers due to fighting in the mating season, and the cycle in the secretion of sex hormones, would have favoured the evolution of the mechanism whereby the sex hormones took control of the calcification in the extremities of the ancestral antlers, thus producing a more effective fighting weapon. This is not unexpected, since sex hormones have an important effect on bone growth elsewhere in the body. The final product was a structure consisting of two parts; the lower living portion which was attached permanently to the skull (the pedicle) and the upper dead portion which was deciduous (the antler). The rising levels of sex hormones caused the death of the tissue of the antler in the mating season, thus providing an effective weapon, and the wound healing mechanism replaced the antler once the sex hormones declined.

Once this stage had been reached it opened the way for the evolution of large, complex antlers. Since the antlers were replaced annually, it was possible for successive sets of antlers to increase in size. In many respects

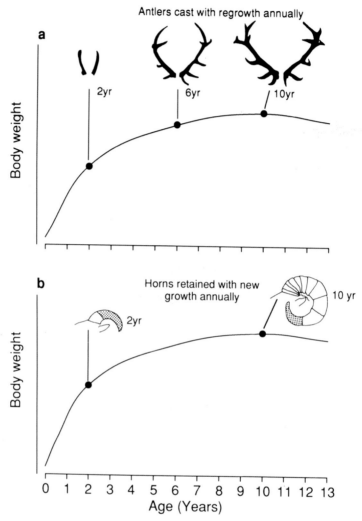

Figure 6.12. Diagram showing the way the size of the weaponry changes with increasing age: (a) red deer, *Cervus elaphus*; (b) mouflon sheep, *Ovis musimons*. In red deer, the hard horn antlers are formed of dead bone and the annual replacement allows the increase in size. In mouflon sheep, the horns are living throughout (the keratinized sheath provides protection) and growth occurs in annual increments from the base.

the ontogeny of the antlers in each life cycle recapitulates their phylogeny. The initial single spiked antlers produced by the young deer resemble the ancestral situation, and the subsequent antlers, which are more complex in shape and a product of wound healing, reflect this evolutionary history.

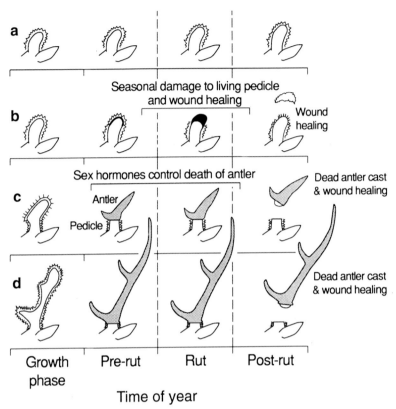

Figure 6.13. Hypothetical stages in the evolution of the antlers in male deer: (a) permanent pedicles; (b) permanent pedicles with seasonal damage and repair by wound healing; (c) simple deciduous antlers with the seasonal cycle of death and casting controlled by the seasonal cycle in sex hormones; and (d) complex deciduous antlers as found in most extant deer. (G. A. Lincoln, unpublished results.)

The hypothetical stages involved in the evolution of deciduous antlers is illustrated in Figure 6.13.

Based on these observations, the regular replacement of the antlers is an obligatory part of their physiology. This has also been illustrated in experiments in which the annual antler cycle was blocked by the administration of sex hormones using subcutaneous implants. In this situation the hard antlers may be retained for more than two years. However, while the continuous presence of elevated levels of sex hormones from an implant effectively blocks the casting of the old antlers, it does not halt the die-back process at the junction between the dead antler and

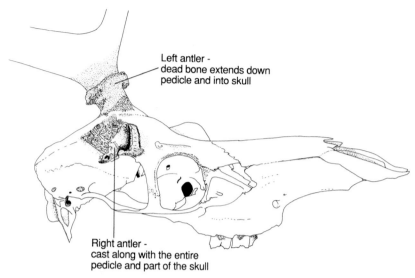

Left antler -
dead bone extends down
pedicle and into skull

Right antler -
cast along with the entire
pedicle and part of the skull

Figure 6.14. Skull of a male red deer, *Cervus elaphus*, which was treated with a subcutaneous implant of oestradiol-17β which prevented the casting of the hard horn antlers for more than two years. During this time, the junction between the dead antler and the living pedicle moved down the pedicle and into the skull. When casting of the antlers finally occurred due to the exhaustion of the hormone implant, the entire pedicle and part of the skull was cast along with the antler and was fatal. (Skull provided by T. J. Fletcher and the drawing produced by Ted Pinner.)

the living pedicle. This continues to progress slowly down the pedicle and, after two to three years, it extends into the skull and is potentially fatal (Figure 6.14). In the normal animal, the regular casting of the dead antlers maintains the live–dead junction within the pedicle. The ancestral deer had very long pedicles like the living Indian muntjac (Figure 6.4), and the long pedicles may have ensured that the dead and dying tissue of the antlers was kept well away from the skull.

Initially, the deer evolved in temperate climates, but there are now many species living in equatorial regions where the seasonal changes in the environment are less extreme. These include the rusa, sambar, axis, hog deer and muntjac in South-East Asia and the white-tailed deer, pampas deer, huemul, pudu and various species of brocket deer in Central and South America. In all cases the males cast and regrow their antlers on a regular basis, even though the cycles of the individuals are not necessarily synchronized with each other or to a particular season as in the species from temperate and cold climates. The persistence of this cyclicity in the antlers may reflect the evolutionary history: having evolved the deciduous antler with its unique dead tissue phase, this specialized weapon could not be abandoned readily even for animals living in a less seasonal environment.

Reindeer as the exception

The reindeer (alias caribou) differ from all other species of deer, because females develop antlers as well as males. The antlers of females are much smaller than those of males, but are cast and regrown annually. Reindeer are adapted to living in tundra at very high latitudes, where the seasonal changes in the environment are extreme. They live in large herds in the winter, and feed on lichens by digging down through the snow. It seems likely that it is the intensified competition over the feeding sites which has favoured the evolution of the weaponry in females, as mentioned earlier. The intriguing question is how the growth of antlers is regulated in female reindeer, in view of the key role of testosterone in the regulation of antler growth in males of other species of deer.

One notable feature is that in both female and male reindeer the antlers begin to grow soon after birth, and not at the time of puberty (Figure 6.15). Furthermore, the removal of the gonads from reindeer calves within 1–2 weeks of birth does not block the subsequent growth of antlers in either sex, and gonadectomized animals still cast and regrow their antlers each year. These observations indicate that the growth and replacement of the antlers is not totally dependent on the secretion of gonadal hormones as in other species. However, there is good evidence that the sex hormones do act to synchronize the antler cycle to the seasonal reproductive cycle in reindeer of both sexes.

Of particular interest is the result of a recent study conducted in collaboration with Nick Tyler at the University of Tromso in Norway, which showed that the removal of the ovaries from adult female reindeer in October when in hard antler, results in the premature casting of the antlers. This is a similar response to that induced by castration in males of other species. This finding demonstrated that the ovaries in reindeer secrete a hormone, possibly oestradiol-17β, which acts like testosterone in the male to induce the hard antler state.

Another interesting feature in reindeer is that the initial growth of the antlers is not preceded by the development of pedicles covered in scalp hair as in other deer species. Instead, the antler tissue with its characteristic velvet-like hair, grows directly from the frontal bones of the skull in the

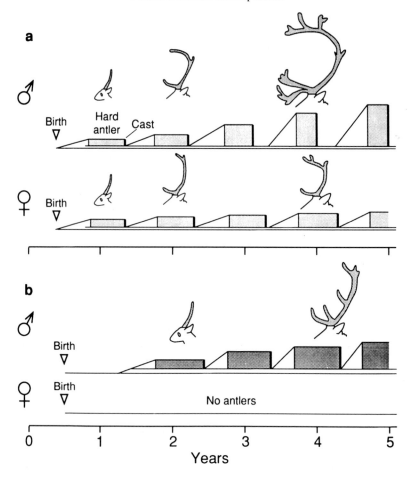

Figure 6.15. Diagram to illustrate the differences in the ontogeny of antler development in (a) reindeer, *Rangifer tarandus*, in which both sexes develop antlers starting soon after birth, and (b) red deer, *Cervus elaphus*, in which only the males develop antlers, commencing at puberty. Horizontal bars represent the antler cycle (velvet, angled; hard antler, shaded) and height of bars indicates relative size of antlers. In reindeer, the growth of antlers can occur in the absence of the gonads, but the secretion of sex hormones from the gonads normally acts to synchronize the antler cycle to the seasonal reproductive cycle in both sexes; thus both males and females have functional hard horn antlers when they are most competitive in the rut (males) or in the winter (females).

young animal, and there are never long pedicles, even in the adult, in either sex. This may indicate that the development of antlers before puberty in both sexes has been achieved by omitting the pedicle stage, which is normally androgen-dependent, and by producing antler tissue directly from the antler primordia. Even in other species of deer, the growth of the antler tissue itself is not androgen-dependent. This special situation in reindeer may represent the selective pressure for precocious antler development, where it is an advantage to produce weapons in prepubertal animals to compete for food in winter. The genetic change which allows for early antler development may have arisen in males, and this then permitted the development of antlers in females. The reindeer is a relative newcomer in evolutionary terms, having diverged only some 2 million years ago, so the precocious growth of antlers is the latest twist in the antler saga.

Benefits versus costs

Antlers are often described as extraordinary, bizarre or extravagant, and the casting of the old antlers appears to be wasteful. This is especially the case in the larger species of deer where many kilograms of bone are laid down to form the antlers only to be cast off within a few months. In addition the antlers are used frequently when males compete for dominance and injuries through fighting are common. Male deer of most, if not all species, have a shorter average lifespan than females and their premature death is often related to the extravagances of competition in the rut and the injuries through fighting. Based on these aspects, it is tempting to emphasize the negative side of the evolution of weaponry and fighting behaviour. Perhaps the costs outweigh the benefits – are large antlers maladaptive?

The simple answer must be 'no'! There are clear examples in this chapter which illustrate that the possession of antlers has a marked effect on competitiveness. In situations of competition with other animals with antlers, the possession of equally efficient fighting weaponry is an absolute requirement for success. The stronger the competition, the more essential is the possession of weaponry. This is most obvious in the case of polygynous species where males compete with each other for access to females. Here, failure to dominate other males can mean that none of the genetic characteristics of an unsuccessful individual are passed to the next generation. Conversely, success can mean that traits which enhance competitiveness are propagated. This may involve a large investment in the paraphernalia of combat, such as antlers, and all the effort and risks associated with achieving dominance, but for reproductive success there is no alternative.

Acknowledgements

I am grateful to the many people who replied to my requests for material in the preparation of this article including Dr I. Boyd, Dr F. H. Fay, Dr R. L. Gentry, Professor R. R. Hofmann, Dr A. A. Macdonald, Dr E. H. Miller, Dr R. Reeves and Dr N. Tyler. The photographs in Figure 6.1 were kindly provided by The Anthony Bannister Photo Library, Lanseria, South Africa, and the artwork in the other illustrations was produced by Ted Pinner and Tom McFetters of the MRC Reproductive Biology Unit, Edinburgh, UK. I am also indebted to Professor R. V. Short for his constructive comments on the manuscipt.

Further reading

Bubenik, G. A. and Bubenik, A. B. (1990) *Horns, Pronghorns and Antlers.* Springer-Verlag, New York.

Clutton-Brock, T. H. (1982) The function of antlers. *Behaviour,* **79**, 108–25.

Cockburn, A. (1991) *An Introduction to Evolutionary Ecology.* Blackwell Scientific Publications, Oxford.

Cronin, H. (1991) *The Ant and the Peacock.* Cambridge University Press. Cambridge.

Geist, V. (1966) The evolution of horn-like organs. *Behaviour,* **27**, 175–214.

Goss, R. J. (1983) *Deer Antlers: Regeneration, Function and Evolution.* Academic Press, New York.

Kitchener, A. C. (1991) The evolution and mechanical design of horns and antlers. In *Biomechanics and Evolution,* pp. 229–53. Ed. J. M. V. Rayner. Cambridge University Press, Cambridge.

Lincoln, G. A. (1984) Antlers and their regeneration – a study using hummels, hinds and haviers. *Proceedings of the Royal Society of Edinburgh,* **82B**, 243–59.

Putman, R. (1988) *The Natural History of Deer.* Christopher Helm, London.

Wilson, E. O. (1975) *Sociobiology: The New Synthesis.* Harvard University Press, Cambridge, Massachusetts.

7

The evolution of sexual size dimorphism in primates

ROBERT D. MARTIN, LESLEY A. WILLNER
and ANDREA DETTLING

In primate species, adult males and females can be strikingly different in a variety of physical features and any such differences between the sexes are labelled collectively '*sexual dimorphism*'. A simple and obvious form of sexual dimorphism can be found in overall body size. For example, adult males are approximately twice as heavy as adult females in baboons (*Papio* species) and gorillas. In adult humans, males are on average about 20% heavier than females. Sexual dimorphism can also occur in dimensions of the teeth and an obvious example here is provided by the canine teeth of many primates. In the extreme case of the gelada baboon (*Theropithecus gelada*), the upper canines of the adult male are $3\frac{1}{2}$ times as long as those of the adult female. More subtle sexual dimorphism can be found in the other teeth (incisors, premolars and molars). Sexual dimorphism can also be present in skeletal features. Although this is most obvious in the shape of the pelvis (in direct association with the birth canal of females), there are more subtle differences in the rest of the skeleton. For instance, male primates generally tend to have more robust skeletons than females. Males and females can also differ in their external appearance, for example in the colour and patterning of the fur, and, in some species, one sex possesses specific externally visible structures, such as the mane of the hamadryas baboon (*Papio hamadryas*) and the cheek flanges of the fully adult male orang-utan (*Pongo pygmaeus*). Sexual differences can, of course, also be found in aspects of physiology and behaviour (see Chapter 11 by Evan Balaban, Chapter 12 by Manfred Gahr and Chapter 13 by John Wingfield), but the discussion in this chapter will be limited to morphology.

One important point about sexual dimorphism is that its manifestations in different aspects of the phenotype are independent of one another, at least to some extent. For example, sexual dimorphism in adult body size and in the size of the canine teeth is virtually lacking among lemurs and gibbons, yet some species of the genus *Lemur* (e.g. *Lemur macaco*) and some gibbons (e.g. crested gibbons, *Hylobates concolor*) show striking colour differences between the sexes. Although dimorphism in adult body size and dimorphism in canine size show some correlation with one another, they also show a large degree of independent variation. Different kinds of dimorphism can occur even within a given system such as the dentition or in the post-cranial skeleton. Given such variability between species in the expression of different kinds of sexual dimorphism, it is likely that several interacting factors are responsible for physical differences between the sexes, and, therefore, a simple explanatory basis for the phenomenon is ruled out.

Discussion of the evolution of sexual dimorphism in primates requires a survey of its occurrence among living species, followed by an investigation of the presence or absence of sexual dimorphism in fossil primates, insofar as the relevant details are both fossilizable and correctly identifiable in practice. As a basis for both undertakings, one can take an outline phylogenetic tree of the primates, showing the likely relationships among the six major groups of living primates: (1) lemurs, (2) lorises, (3) tarsiers, (4) New World monkeys, (5) Old World monkeys, and (6) apes and humans (Figure 7.1). The identification of sexual dimorphism in fossil primates is, of course, fraught with difficulty because the same criteria (e.g. overall body size and body dimensions) may be used both to distinguish species and to separate males from females within species. Therefore, the interpretation of sexual dimorphism in the primate fossil record deserves special attention. This is particularly true with respect to interpretations of sexual dimorphism in human evolution, since there has been a long history of confusion between species differences and sexual differences.

A central problem for any assessment of the evolution of sexual dimorphism in primates concerns the *direction* taken by evolution. In the first place, such assessment requires some assumption about the ancestral condition from which evolution has taken place. Secondly, there is a fundamental difficulty in interpreting sexual dimorphism (e.g. in body size), since we are confronted with a difference between males and females with no obvious indication of the direction taken by evolution. Since many approaches to the topic have been based on analyses of correlations between relevant variables, it is important to emphasize that mere correlation should not be equated with causation. To identify causes in the evolution of sexual dimorphism, we must go far beyond the level of simple correlational analysis.

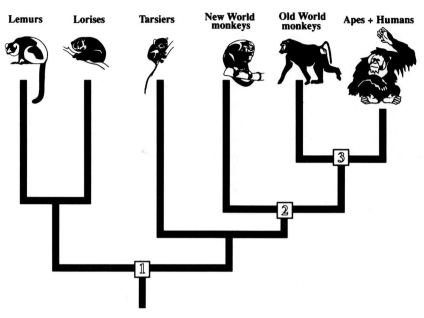

Figure 7.1. Outline phylogenetic tree showing inferred relationships among the six major groups of living primates (after Martin, R. D. (1990) *Primate Origins and Evolution.* Chapman and Hall, London). With respect to the evolution of sexual dimorphism, three successive ancestral stocks are of particular interest: (1) the stock giving rise to all living primates; (2) the stock giving rise to the simian primates (monkeys, apes and humans); and (3) the stock giving rise to Old World simian primates. Modern prosimians uniformly lack sexual dimorphism and it is likely that dimorphism was lacking or only weakly developed in stock 1. Modern New World monkeys show, at the most, moderate sexual dimorphism only, and weakly developed dimorphism is likely for stock 2. Pronounced sexual dimorphism is relatively common among Old World simians, so quite marked dimorphism may well have been present in stock 3. (N.B. The fossil simians from the Oligocene Fayum deposits are generally thought to be quite close to stock 3.)

Because there are numerous spectacular examples of sexual dimorphism among simian primates, there have been many attempts to explain this phenomenon. One of the clearest relationships that has emerged repeatedly is that sexual dimorphism in both body size and canine size is typically lacking among species that live in monogamous family groups and is generally confined to polygynous species living in either harem groups or multi-male groups. As a general rule among mammals and birds, sexual dimorphism in body weight is lacking among monogamous species, so primates conform to the general pattern in this respect. In view of this widespread relationship, there has been a natural tendency to seek a unitary explanation, and the primary focus has been on selection acting on the male. One common explanation, dating back to Darwin's seminal discussion of sexual selection in 1871 (*The Descent of Man, and Selection in Relation to Sex*), is the '*sexual competition hypothesis*', according to which males are commonly larger than females in polygynous species because there is greater competition for breeding access to females (see Chapter 3 by John Reynolds and Paul Harvey). A different interpretation, the '*anti-predator hypothesis*', is that males of species living in relatively large polygynous groups are commonly bigger than females, because there is greater selection for protection against predators. It is, of course, also possible to argue that both influences act in concert.

Figure 7.2. Two hypotheses accounting for the evolution of sexual dimorphism in adult body size. In principle, such dimorphism could arise either through increase in body size of the male (hypothesis A) or through reduction in body size of the female (hypothesis B). Of course, it is also possible that *both* kinds of change may occur in concert.

In fact, in themselves, both of these widely favoured hypotheses reflect a curious form of sexual dimorphism on the part of investigators. One possibility that has received surprisingly little attention is that selection may act on the female to generate dimorphism. It is commonly assumed that the female sex represents some kind of baseline against which the male condition can be assessed, and this leads almost inevitably to the assumption that selection always acts on the male to generate sexual dimorphism (see Figure 7.2, hypothesis A). Logically, however, it is perfectly possible that sexual dimorphism in adult body size, for example, can be produced by selection acting to reduce female body size (see Figure 7.2, hypothesis B) and this possibility should also be tested. Overall, it is likely that the evolution of sexual dimorphism has involved selection pressures affecting both sexes and the possibility of diverging responses of males and females should be borne firmly in mind whenever an explanation is attempted.

Sexual size dimorphism in primates
Patterns of size dimorphism

Perhaps the most basic form of sexual dimorphism in primates occurs as a difference in adult body weight between males and females. The mandrill (*Mandrillus sphinx*) reportedly holds the record among primates, with an adult male body weight more than $2\frac{1}{2}$ times greater than that of adult females. The simplest way to express this form of dimorphism is to take a ratio between average adult body weights of males and females (Figure 7.3). As can be seen, males generally tend to be heavier than females in primates, although there are a few cases in which females are heavier than males. Humans occupy an intermediate position in the histogram.

In fact, sexual dimorphism in body size is not found in all primate groups. If we examine the degrees of sexual dimorphism found in three main groups of living primates (Figure 7.4), it emerges that there is little or no dimorphism in prosimian primates (lemurs, lorises and tarsiers) and that sexual dimorphism in body size is restricted essentially to simian primates (monkeys, apes and humans). Further, it can be seen that sexual dimorphism in body weight is relatively limited among New World monkeys and that extreme forms of sexual dimorphism are limited to Old World simians (monkeys and apes). With respect to this latter point, it should be noted that a correlation has been found between terrestrial habits and sexual dimorphism in both body weight and canine size among

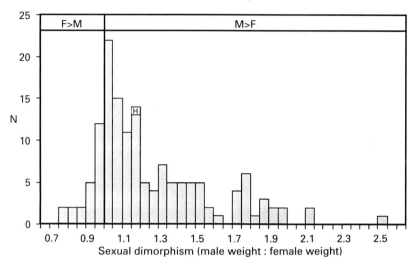

Figure 7.3. The distribution of male to female ratios for adult body weight among living primates ($n = 145$ species). (Body weights of wild specimens have been used as far as possible; weights from captive specimens have been used for less than 10% of the species shown.) In the great majority of cases, males are heavier than females and the distribution is skewed, with a relatively small number of species showing extreme sexual dimorphism. Humans (H) occupy an intermediate position. The highest degree of dimorphism (adult male to female weight ratio of just over 2.5) has been reported for the mandrill, *Mandrillus sphinx*.

primates. As terrestrial habits are rare among primates other than Old World simians, this could be a factor influencing the prevalence of sexual dimorphism among the latter.

The difference between prosimians, New World monkeys and Old World simians could potentially be explained on the grounds that the degree of sexual dimorphism shows a tendency to increase with increasing body size of the species concerned (see below). The modal body weight of simian primates (5 kg) is approximately 10 times greater than the modal body weight of prosimian primates (500 g). Further, the New World simians are on average markedly smaller than Old World simians. Therefore, it might be expected that size dimorphism in simians should be more marked than in prosimians and that the most extreme examples of such dimorphism should be found among Old World simians. In the first place, however, the largest-bodied extant lemurs have body weights that overlap with those of dimorphic Old World monkeys, yet these prosimians show no sexual dimorphism in body size. In addition, several large-bodied species are known among the subfossil lemurs of Madagascar, and the largest (*Megaladapis edwardsi*) had a body size comparable to

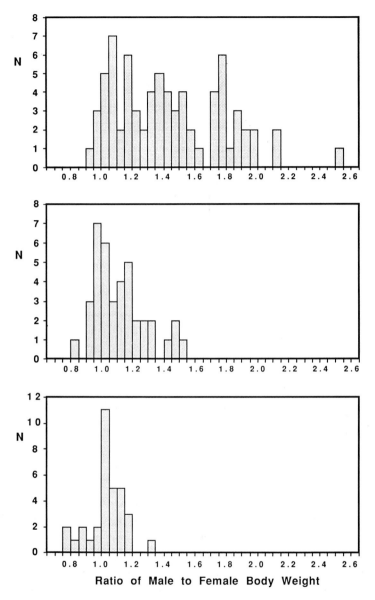

Figure 7.4. Degrees of sexual dimorphism in body weight, based on ratios for adult males and females, in three different groups of primates: (top) Old World simians (*n* = 73 species); (middle) New World monkeys (*n* = 39 species); and (bottom) prosimians (*n* = 33 species). It can be seen that sexual dimorphism is virtually lacking among prosimian primates and is only weakly developed among New World monkeys. Marked sexual dimorphism is restricted to Old World monkeys and apes, which show a wide range of male to female body weight ratios.

that of a female gorilla. But none of these large-bodied subfossil lemurs seems to show any indication of sexual dimorphism in body size. Hence, the absence of sexual dimorphism in adult body weight among both extant and subfossil prosimian primates presents a genuine contrast to the common occurrence of such dimorphism among simian primates (especially Old World species). Accordingly, any explanation that is advanced to account for the occurrence of sexual dimorphism in simian primates should at the same time indicate why this phenomenon is generally lacking among prosimians and is only weakly evident among New World monkeys.

As has already been indicated, the ratio of adult male to female body weight must be considered in the light of a general trend that has been reported for size dimorphism in animals. The degree of dimorphism in overall body size in mammals tends to increase with increasing size of the species concerned. As a general rule, larger species tend to show a greater degree of sexual dimorphism than smaller species, a pattern that is also found with birds (Rensch's rule). While an explanation for this finding remains elusive, this trend clearly applies to primates, as is obvious from a plot of adult male body weight against adult female body weight for an almost complete sample of 49 primate genera (Figure 7.5). An extremely tight relationship is seen in which body weight (measured in grams) of males (W_m) increases relative to that of females (W_f) according to the following scaling formula:

$$W_m = 0.54 \cdot W_f^{1.11} \ (r = 0.995).$$

If the ratio between adult male and female body weight remained constant with increasing overall size of the species, a simple proportional (isometric) relationship would apply and the value of the exponent in the equation would be 1.00, rather than the 1.11 actually found. Although the exponent value in the equation exceeds unity only slightly, the 95% confidence limits (1.08–1.14) are tight enough to rule out simple proportional scaling. For a female body weight of 1 kg, male body weight is 1.15 kg; for female body weight of 10 kg, male body weight is 14.8 kg; for female body weight of 100 kg, male body weight is 192 kg.

It is necessary to take this pattern of scaling into account when comparing primate species of different body sizes. This can be done by calculating the expected value for male body weight for each species using the above equation and then by determining the residual value (the logarithmic difference between the actual and expected values). As can be seen from a histogram of residual values for primate species (Figure 7.6),

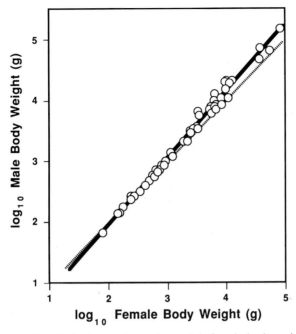

Figure 7.5. Adult male body weight against adult female body weight (both in grams; logarithmic scales), taking 49 pairs of average values for primate genera. This plot is based on average values for primate genera, rather than raw values for individual species, in order to reduce any bias due to over-representation for species-rich genera. The major axis (heavy line) has a slope of 1.11. This is significantly greater than a slope value of 1, which would correspond to simple proportional (isometric) scaling. The points would lie on the dotted line if adult male and female weights were equal for all species.

one outcome of this refinement is that humans turn out to have a strikingly low residual value in comparison to other primates. In other words, because humans are relatively large-bodied in comparison to other primates, the degree of sexual dimorphism in adult body weight is actually very low in comparison to expectation. A further implication of this pattern is that, on grounds of body size alone, sexual dimorphism was probably not pronounced in the earliest primates, which were relatively small-bodied.

Development of size dimorphism

As a general principle in evolutionary biology, consideration of development (ontogeny) can yield a valuable additional perspective, and the

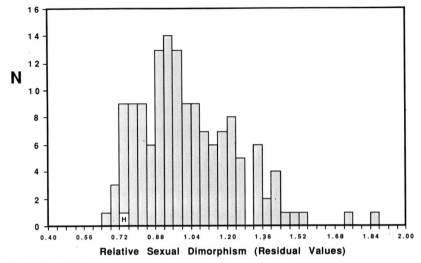

Figure 7.6. The distribution of residual values for sexual dimorphism in adult body weight among living primates (*n* = 145 species; sample as for Figure 7.3). When Rensch's rule is taken into account, humans (H) prove to have a very low degree of sexual dimorphism for their body weight.

phenomenon of sexual dimorphism is no exception. Logically, there are two mechanisms by which a difference in size between males and females at sexual maturity could be achieved during development. Such a difference could arise either by a difference in the *rate* of growth, with males growing faster and hence maturing at the same time as females, or by a difference in the *duration* of growth, with females maturing earlier than males (Figure 7.7). There is, of course, also the intermediate possibility that sexual dimorphism in adult body size could be achieved by differences between males and females in both duration and rate of growth. It is certainly true that males and females of primate species, including humans, generally differ little, if at all, in birth weight and that they follow very similar growth trajectories early in life. The attainment of sexual dimorphism in adult body weight is principally a result of later divergence between male and female trajectories, usually beginning close to the time of puberty. If larger adult male body size is achieved by an increased rate of growth, this means that males require access to greater resources for faster growth. Alternatively, if larger male body size is achieved by a more extended period of growth, this is highly likely to result in males being older than females when they first breed ('sexual bimaturism'). For males, sexual maturity and first breeding are often not

Hypothesis A:

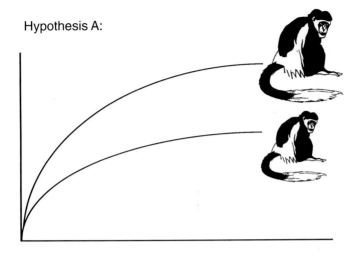

Hypothesis B:

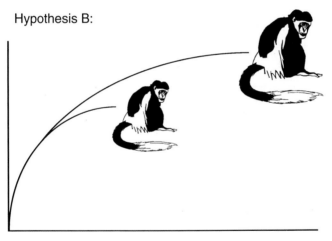

Figure 7.7. Two hypotheses accounting for the development of sexual dimorphism in body size for adult males and females. In principle, such dimorphism could arise either through a more rapid *rate* of growth by the male (hypothesis A) or through a greater *duration* of growth by the male (hypothesis B). Of course, it is also possible that *both* effects may occur in combination.

synonymous. Even in humans, boys (for example) become sexually mature at the age of 12 or 13 years, i.e. they are producing sperm, but this is a far cry from first breeding.

The predicted effect of greater duration of growth on the age of first

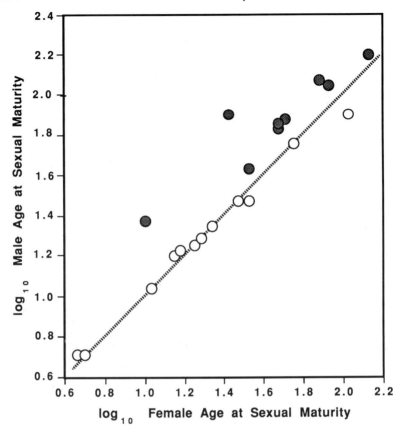

Figure 7.8. The age at sexual maturity for males plotted against females (both in months; logarithmic scales) for a sample of 21 primate species. White points indicate monomorphic species (adult male:female body weight < 1.15); shaded points indicate sexually dimorphic species (adult male:female body weight > 1.15). The dotted line indicates identity between male and female weights. (Data from Willner, L. (1989) 'Sexual dimorphism in primates'. Ph.D. thesis, University of London.)

breeding can be tested by plotting age of sexual maturity of males against that of females for a sample of primate species (Figure 7.8). As can be seen, males and females of monomorphic species (conveniently defined as having a male to female body weight ratio of less than 1.15) tend to reach sexual maturity at approximately the same age, whereas in sexually dimorphic species (male:female body weight > 1.15) males typically become sexually mature later than females. In other words, sexual dimorphism is typically associated with sexual bimaturism in primates. The fact that males usually breed later than females in species with sexual dimorphism in body size must therefore be taken into account in any explanation of such dimorphism. Any benefits obtained from larger body size in males must, in effect, offset the 'cost' of delayed reproduction. Conversely, females in sexually dimorphic species may gain an advantage through earlier breeding because of their relatively smaller body size.

The direct link between sexual dimorphism and later attainment of sexual maturity by males could suggest an explanation for the absence of sexual dimorphism among lemurs, if not for its absence among prosimian primates generally. The extant lemurs of Madagascar are generally characterized by strictly seasonal patterns of breeding and it may be that, because any delay of breeding by males would have to be in multiples of one year, sexual bimaturism cannot evolve easily. Although clear-cut seasonal breeding patterns are found with a minority of simian primate species, it is rare to find strictly seasonal patterns comparable to those found among lemurs. Many simian primates breed throughout the year.

In a detailed analysis of growth patterns in 31 sexually dimorphic simian primate species, Steven Leigh examined the relative contributions to sexual dimorphism in adult body size of differences between males and females in rate and duration of growth. The age of attainment of an identifiable plateau in adult body weight, rather than the age of first breeding, was taken as the criterion for physical maturity. Taking this age of maturity, considerable variation was found among simian species in the relative contributions of differences in rate and duration of growth between males and females. In one case (the talapoin monkey, *Miopithecus talapoin*), sexual dimorphism was achieved exclusively through a difference in duration of growth between males and females (see Figure 7.7, hypothesis B), whereas in four cases (the blue monkey, *Cercopithecus mitis*; the Sulawesi macaque, *Macaca nigra*; the drill, *Mandrillus leucophaeus*; and the douc langur, *Pygathrix nemaeus*), sexual dimorphism was achieved exclusively through a difference in rate of growth (see Figure 7.7, hypothesis A). Accordingly, full adult body weight was achieved later by males than females in 27 of the 31 primate species examined, and, across the sample, differences in rate and duration of growth contributed approximately equally to observed dimorphism in body weight. In most cases, therefore, sexual dimorphism in body weight is associated with achievement of full body weight being later for males than for females ('*growth bimaturism*'). Given the marked variation among species in the patterns of divergence between males and females in growth trajectories, Leigh concluded that explanations for sexual dimorphism in body weight should encompass selection acting on both sexes.

It is, in fact, apparent from the data provided by Leigh that full adult body weight may not be attained until some time after the age of first breeding. According to species, both males and females may continue to increase in body weight after the age of first breeding and this means that the concept of 'bimaturism' is actually quite complex. If the age of sexual maturity is taken as the criterion for adulthood, it emerges that the difference in body weight between males and females at the time of first breeding is due primarily to a difference in the duration of growth, with any difference in the rate of growth playing no more than a subsidiary role. In sexually dimorphic species, females typically breed earlier than males, even if the termination of female growth occurs as late as in males. An extreme example is provided by the blue monkey (*Cercopithecus mitis*) in which the ratio of adult male to female body weight is 1.72. In this species, females breed markedly earlier than males (at just over 4 years, as opposed to 6 years of age), but, in fact, they take somewhat longer than males to reach a plateau in adult body weight.

Causes of size dimorphism

In most cases, explanations for sexual dimorphism in body weight among primates have focused on the two hypotheses mentioned in the introduction: sexual competition and anti-predator defence. There are, however, a number of problems with both hypotheses. In the first place, there is one conspicuous exception to the general rule that monogamous species lack sexual dimorphism in body weight. In de Brazza's guenon (*Cercopithecus neglectus*) the adult male to female body weight ratio is high (1.72), but one field study of this species indicated a monogamous pattern. Although it is possible that this single known exception among primates may be explained as a result of phylogenetic inertia (retention of a condition from some previous ancestral stage adapted to different conditions), such a departure from the rule indicates a need for caution.

A more fundamental problem is that, although sexual dimorphism in body size is indeed confined essentially to polygynous primate species, by no means all polygynous species are dimorphic (Figure 7.9). Polygynous simian species show a range of degrees of sexual dimorphism, overlapping at the lower end of their distribution with monogamous primates. Further, as has been noted already, sexual dimorphism is uniformly lacking among prosimian species, many of which have polygynous social systems (Figure 7.9). Therefore, it is necessary to explain how selection could have favoured marked dimorphism in some (but not all) polygynous

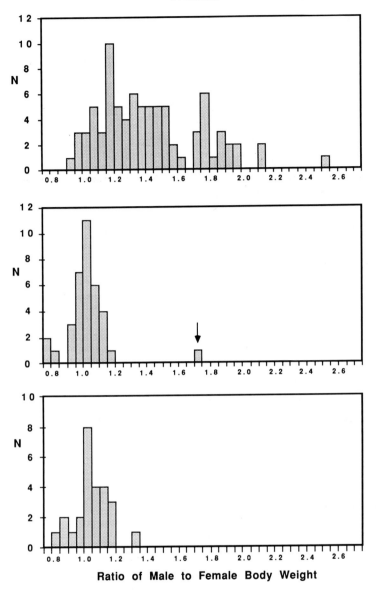

Figure 7.9. Degrees of sexual dimorphism in body weight, based on ratios for adult males and females, in three different categories of primates: (top) polygynous simians (*n* = 83 species), (middle) monogamous primates (*n* = 36 species) and (bottom) polygynous prosimians (*n* = 26 species). Sexual dimorphism is essentially confined to polygynous simians, although it is also present in de Brazza's guenon (*Cercopithecus neglectus*; arrowed). In fact, polygynous simians show a wide range of male to female body weight ratios and some of them are essentially monomorphic.

simian species while not affecting polygynous prosimian species in the same way. A potential explanation for the wide spread of values for degrees of sexual dimorphism in polygynous simian species shown in Figure 7.9 is that the degree of sexual dimorphism varies according to the level of competition between males in a social group. One possible test of this explanation is to examine the relationship between the degree of sexual dimorphism in body weight and the socionomic sex-ratio (the ratio between numbers of adult females and adult males in a social group). When the relationship between body weight dimorphism and socionomic sex-ratio was examined for a sample of 42 primate species, it seemed at first sight that there was the expected positive correlation. However, this initial analysis was flawed, since it included 13 monogamous species and 29 polygynous species. As has been shown above, monogamous primate species are typically monomorphic in body weight. They are also smaller on average than polygynous primate species, so they will tend to bias a best-fit line in a plot of sexual dimorphism against socionomic sex-ratio. When the analysis was repeated after excluding the mono-gamous species, no significant correlation was found between dimorphism in body weight and socionomic sex-ratio. Accordingly, there has, as yet, been no convincing demonstration of a link between sexual dimorphism in body weight and the level of competition among males for access to females.

Sexual dimorphism in primate dentition
Patterns of canine dimorphism

One of the most pronounced forms of sexual dimorphism found among primates occurs in the size of the canine teeth. In many species, the canines of males are both stouter and longer than those of females. For practical purposes, a convenient measure of the size of the canines is provided by the basal area, since this is relatively unaffected by wear or breakage of the projecting tip of the tooth and is hence also more easily applicable to fossil primates. As with body weight, a direct ratio of male to female canine size, using basal area, can be taken as being a simple indicator of sexual dimorphism. Once again, a histogram of ratio values (Figure 7.10) provides a skewed distribution, with a few species showing very high values. As has been found with body weight, sexual dimorphism in canine dimensions is virtually lacking among prosimian primates. In comparison to other primates, humans show a relatively low degree of sexual dimorphism in the canine teeth, although the difference is nevertheless

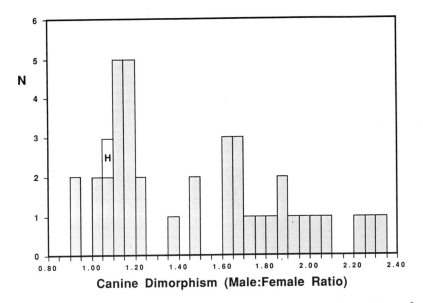

Canine Dimorphism (Male:Female Ratio)

Figure 7.10. The distribution of male to female ratios for basal area (in mm^2) of the upper canine tooth among living primates ($n = 40$ species). In all but two cases, males have stouter canines than females and the distribution is skewed, with a relatively small number of species showing extreme sexual dimorphism. Humans (H) have a very low degree of sexual dimorphism in their canine teeth. The highest degree of dimorphism (ratio of adult male to female canine size is approximately 2.4) is found in the pigtailed macaque, *Macaca nemestrina*. (Non-human primate data from Harvey *et al.* (1978). Human figure from Goose, D. H. (1963) Dental measurement: an assessment of its value in anthropological studies. In *Dental Anthropology*, pp. 125–48. Ed. D. R. Brothwell. Pergamon Press, Oxford.)

great enough for it to be used as an accessory criterion for sexing fragmentary human remains.

A plot of male canine size against female canine size (Figure 7.11) shows a pattern very similar to that found with dimorphism in body size; the degree of sexual dimorphism in canines tends to increase with increasing canine size of the female. A very regular relationship is seen in which the basal canine area (measured in mm^2) of males (C_m) increases relative to that of females (C_f) according to the following scaling formula:

$$C_m = 0.85 \cdot C_f^{1.17} \ (r = 0.975).$$

This scaling relationship, in which the exponent value is even greater than for the equation relating male weight to female weight (1.17; 95% confidence limits: 1.11–1.23), shows that a similar principle also governs

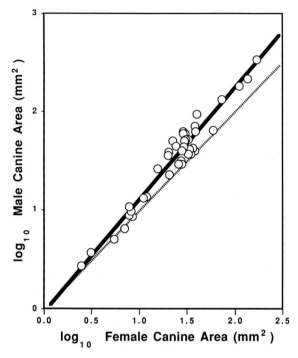

Figure 7.11. The basal area of the upper canine for males plotted against females (both in mm^2; logarithmic scales), for a sample of 40 primate species (data as for Figure 7.9). The major axis (heavy line) has a slope of 1.17. This is significantly greater than a slope value of 1, which would correspond to simple proportional (isometric) scaling. The points would lie on the dotted line if male and female canine sizes were equal for all species.

sexual dimorphism in canine size. Here, too, some kind of explanation is required if sexual dimorphism is to be understood fully.

Further, it emerges once again that dimorphism in canine size is virtually lacking among primates that live in monogamous groups, whereas polygynous primates show an overall trend towards sexual dimorphism in the canines. Indeed, if canine length is measured, the separation between these two categories is found to be even more pronounced than for sexual dimorphism in adult body size.

As with dimorphism in body size, our understanding of sexual dimorphism in the canine teeth among primates would undoubtedly benefit from a detailed study of developmental aspects. It is, for instance, likely that the very large canine teeth found in some primates require a considerable period of growth and that bimaturism is involved once again.

As yet, however, there has been no comparative study of developmental schedules for the canine teeth in male and female primates.

Causes of canine dimorphism

Following the general tendency to seek selection pressures acting on the male, explanations for sexual dimorphism in canine size have also focused on sexual competition between males and on protection against predators. Once again, however, there are a number of problems with such inter- pretations. For instance, the lack of sexual dimorphism in canine teeth among prosimian primates is puzzling, particularly because there are several diurnally active lemur species that are comparable in body size to monkeys and also live in social groups. Why have sexual competition and/or the threat of predation not led to the development of sexually dimorphic canines in these lemur species? As with sexual dimorphism in body weight, de Brazza's guenon (*Cercopithecus neglectus*) is again an exception, because marked dimorphism in canine teeth is present, despite the fact that this species has been reported in one study to live in monogamous groups under natural conditions.

A somewhat surprising finding is that the degree of sexual dimorphism in canine size is not correlated tightly with the degree of sexual dimor- phism in adult body weight (Figure 7.12). In fact, although the correlation between these two forms of sexual dimorphism is significant, it is not very strong ($r = 0.77$) and the coefficient of determination ($r^2 = 0.59$) indicates that less than 60% of sexual dimorphism in canines is accounted for by sexual dimorphism in body weight, or vice versa. Indeed, if the analysis is restricted to polygynous species, thus removing the cluster of points at the lower end of the line in Figure 7.12 (representing monogamous species with little sexual dimorphism in body weight or canine teeth), the figure falls to below 50%. This result is surprising because, other things being equal, it is only to be expected that in any species showing significant sexual dimorphism in body weight the canine teeth of a large-bodied male should be larger than those of a small-bodied female. Therefore, the moderate correlation between dimorphism in canine size and dimorphism in body weight indicates that these two forms of sexual dimorphism can vary independently. Although it could be argued that a given selection pressure might favour alternative responses in the evolution of sexual dimorphism in different primate species, sometimes favouring dimorphism in body weight and sometimes favouring dimorphism in canine size, it is necessary to explain why these different responses should occur. In this

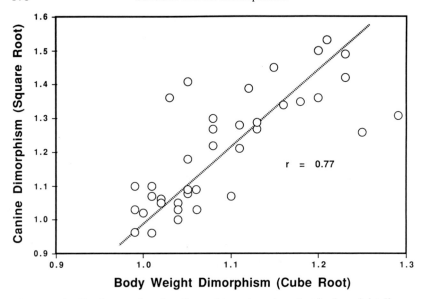

Figure 7.12. The degree of canine dimorphism plotted against body weight dimorphism for a sample of 38 primate species. The cube root of the ratio between adult male and female body weight (see Figure 7.3) and the square root of the ratio between male and female basal canine area (see Figure 7.12) have been taken to linearize the data, hence permitting direct comparison. The best-fit line (dotted) is the major axis.

respect, current explanations for sexual dimorphism in primates are particularly inadequate. Paul Harvey and Peter Bennett have examined sexual dimorphism in body size and canine size in relation to the relative size of the testes in different primate species. The latter measure provides a fairly direct indicator of the degree of sexual competition among males for females, and the testes are found to be relatively larger in species living in multi-male groups than in those living in uni-male groups (harem or monogamous family groups) where direct mating competition between males is limited or absent (see Chapter 1 by Roger Short). One might expect that any measure of sexual dimorphism that is linked to sexual competition between males should be most apparent in multi-male species. In fact, however, sexual dimorphism in relative canine size is more-or-less comparable for primates living in multi-male and harem groups, while sexual dimorphism in body size is actually more pronounced for primates living in harem groups than for those living in multi-male groups.

Another study took a different tack by examining the relationship between canine size, jaw size and the extent to which the jaws can be opened (gape). It was shown that canine size in males, but not in females, is clearly related to constraints imposed by the size and shape of the jaws. The canines of males are generally close to the size limit permitted by gape, whereas female canines are usually below this limit. In fact, the size relationships of canines in male simian primates parallel those found in mammalian carnivores, while the canine teeth of females generally lack such close similarity. This suggests that the canine teeth of male simians can be regarded as being weapons that are directly comparable to those of carnivores, and, accordingly, it is justifiable to infer an offensive role for them. Thus, there is a more convincing basis for linking sexual dimorphism in canine size with male fighting ability, whether this be related to sexual competition among males or to the potential for warding off predators, than is the case for sexual dimorphism in body weight. It is therefore possible that the relatively weak correlation between degree of sexual dimorphism in canine size and degree of sexual dimorphism in body size reflects the fact that the latter is connected less directly with male competition than the former.

Sexual dimorphism in the primate fossil record

Any reconstruction of the evolution of sexual dimorphism in primate morphology should ideally take into account relevant fossil evidence. To some extent at least, this should permit us to test hypotheses concerning the emergence of sexual dimorphism during primate evolution. For instance, it should (in principle) be possible to test the inferences that sexual dimorphism first arose during the evolution of the simian primates and that it then became relatively pronounced during the early evolution of the Old World simian primates (Figure 7.1). The presence or absence of sexual dimorphism in direct fossil relatives of humans (hominids) is of special interest, since it may provide clues to patterns of social organization in the hominid lineage.

Patterns of sexual dimorphism

As might be expected from the relatively common occurrence of marked sexual dimorphism among modern Old World monkeys and apes, cases of apparent sexual dimorphism have been reported for a variety of their fossil relatives. There have been reports of sexual dimorphism in both canine size and overall skull size for a variety of fossil Old World monkeys such as *Mesopithecus*, a presumed relative of leaf-eating monkeys in the subfamily Colobinae, and *Theropithecus*, a direct relative of modern gelada baboons in the subfamily Cercopithecinae. In fossil monkeys that clearly are related quite closely to modern species, as with Pleistocene *Theropithecus*, inference of sexual dimorphism can be based on a direct parallel and is relatively convincing. There are also numerous reports of sexual dimorphism in fossil apes, for example *Proconsul*, *Sivapithecus* and *Laccopithecus*. The latter report is, in fact, somewhat surprising, since *Laccopithecus* has been interpreted as being a direct fossil relative of modern gibbons (Hylobatidae), which are monogamous and uniformly lack sexual dimorphism. In fact, sexual dimorphism has also been reported for some of the earliest known Old World simians from the early Oligocene Fayum deposits in Egypt, such as *Aegyptopithecus*. In all cases, a combination of differences in tooth size (especially canine size) and in overall jaw dimensions has provided the basis for inference of sexual dimorphism. In all, this evidence confirms the inference (see Figure 7.1, stock 3) that relatively pronounced sexual dimorphism appeared early in the evolution of Old World simian primates.

Given the fact that sexual dimorphism is relatively limited among modern New World monkeys, it is not surprising to find that little evidence of dimorphism has been reported for fossil relatives of these monkeys. Fossil evidence of New World monkeys is, admittedly, rather limited, but it would nevertheless seem that sexual dimorphism was not pronounced in the earlier members of this sister-group of the Old World simians. This confirms the view that sexual dimorphism has never been pronounced among New World monkeys and that sexual dimorphism would have been developed only weakly, at most, in the ancestral stock of simians (see Figure 7.1, stock 2).

As has been noted above, sexual dimorphism in body size or canine size is virtually lacking among modern prosimians and it is unlikely that dimorphism was present in the small-bodied ancestral primates (see Figure 7.1, stock 1). Therefore, it is somewhat unexpected to find that evidence for sexual dimorphism has been reported for Eocene lemuroids of the family Adapidae, both for the European genera *Adapis* and *Leptadapis* and for the North American genus *Notharctus*. These lemuroids represent a relatively primitive grade of primate evolution, and some authors regard them as being direct relatives of modern lemurs and lorises.

Difficulties with sex determination of fossil specimens

Identification of sexual dimorphism in the fossil record is problematic, partly because the evidence is usually fragmentary and partly because it is exceedingly difficult to define reliable criteria that will distinguish differences between sexes from differences between species. A cautionary tale is provided by the supposed evidence for sexual dimorphism in *Adapis* and *Leptadapis*. Originally, it was believed that skulls attributed to these two genera represented just two species of radically different body sizes, *Adapis parisiensis* and *Leptadapis magnus*. On the basis of canine tooth size, Philip Gingerich suggested that it was possible to distinguish males and females in each of these two species, and a plot of cranial width against cranial length also seemed to divide each species into two clusters of smaller and larger individuals (presumed to be females and males). Although numerous well-preserved skulls are available for *Adapis* and *Leptadapis*, there is a major drawback with this material, because it was not excavated under controlled conditions, so that stratigraphic context of individual specimens is unknown.

A recent re-analysis of the skulls of *Adapis* and *Leptadapis* compared variation in skull dimensions with the level of variation found in modern primate species (including both monomorphic prosimian species and dimorphic simian species). This study showed quite clearly that the level of variation in skulls of *Adapis* or *Leptadapis* greatly exceeds that found in any single living primate species, even when strongly dimorphic species are taken for comparison. Indeed, in some cases the degree of variation in skull dimensions found in *Adapis* and *Leptadapis* approaches that found at the generic level in modern primates. These findings have been confirmed by a more penetrating study of muzzle shape in these fossil primates. Therefore, there is little doubt that the skulls of *Adapis* represent several different species (perhaps five or even more), and that there are also at least three separate species of *Leptadapis* represented by known skulls. This does not necessarily mean that sexual dimorphism has now been shown to be lacking in *Adapis* and *Leptadapis*, but any assessment of sexual differences will require detailed study of the individual species that can now be identified.

The evidence for sexual dimorphism in *Notharctus* is better than that for *Adapis* and *Leptadapis* because the fossil specimens concerned come from a well-documented stratigraphic context. Dental evidence indicates that specimens identified as belonging to single species of *Notharctus venticolus* can be divided into two groups that differ only moderately in dimensions of the cheek teeth (premolars and molars) but quite markedly in canine dimensions. These two groups have been identified as being females and males, respectively. This material from *Notharctus* now represents the only possible evidence for sexual dimorphism in fossil prosimians. John Alexander has recently found convincing evidence of sexual dimorphism in the skulls of *Notharctus* and of the related genus *Smilodectes*.

If a given sample of teeth is derived from a species with marked sexual dimorphism, in principle this should be revealed by the presence of two separate peaks (bimodality) in a histogram of values for a relevant dimension. Quite often, the coefficient of variation (the ratio between the standard deviation and the mean for a given dimension) has been used as an indicator of sexual dimorphism, following the argument that this coefficient should be higher for a bimodal distribution than for a unimodal one. In practice, however, the coefficient of variation does not consistently provide a reliable indicator of sexual dimorphism. For instance, it has been shown that in certain primate species with pronounced sexual dimorphism (e.g. the orang-utan, *Pongo pygmaeus*, and the colobus monkey, *Colobus polykomos*) the coefficient of variation for the length of the lower first molar tooth is not markedly greater for the combined sexes than for males alone.

The issues concerned in distinguishing males from females on this basis can be illustrated with idealized test data. As a working assumption, it can be taken that the values for a given tooth measurement are distributed normally in males and females, respectively, and that the pattern seen in combined data will depend on the separation between the mean values for the two sexes and on the degree of variation (as measured by the standard deviation, SD) for the two samples. For simplicity, we can consider cases in which the sample sizes and standard deviations are identical for males and females. (It should be noted here that the range of values covered by the mean plus/minus 2SD will represent approximately 95% of the data, while the range of values covered by the mean plus/minus 3SD will represent approximately 99% of the data for each sex. Accordingly, if the male and female means are separated by a total of 6SD, the distributions will be virtually non-overlapping.) A range of different possibilities is illustrated in Figure 7.13, extending from one extreme in which the male and female distributions are identical, and hence become superimposed to give a single normal distribution, to another extreme in which there is virtually no overlap between male and female distributions. It can be seen that bimodality first becomes evident by visual inspection when the male and female means are separated by a distance of 3SD. Hence, sexual dimorphism in individual dimensions will be apparent to the naked eye only in cases where it is relatively pronounced. It is, in any case, difficult to define an objective criterion

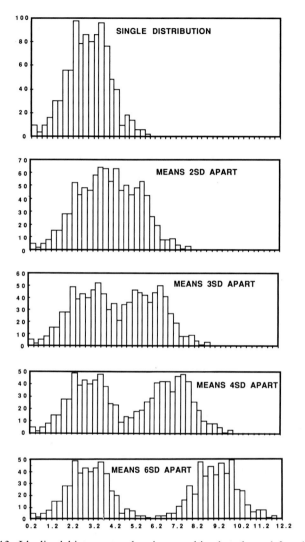

Figure 7.13. Idealized histograms showing combined male and female distributions for a given dimension with identical sample sizes (*n* = 1000 per sex) and identical standard deviations (SD) but with various distances between the mean values for males and females. The range extends from a case in which the mean values for males and females are identical (monomorphism: top histogram) to one in which the male and female distributions are essentially separate, with the means separated by 6SD (extreme sexual dimorphism: bottom histogram). Bimodality (the identifiable presence of two separate peaks) first becomes clearly visible when the means for males and females are separated by 3SD. The overall distribution becomes progressively flattened as the separation between the means increases. (The horizontal axis represents arbitrary units of measurement; the vertical axis indicates frequency.)

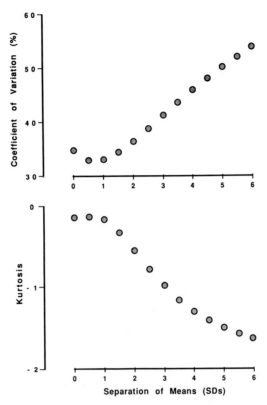

Figure 7.14. Plots of overall values for the coefficient of variation (above) and for kurtosis (below) for the model distributions illustrated in Figure 7.13, taking cases with mean values for males and females separated by distances ranging from 0SD to 6SD. Note that there is an initial dip in the curve for the coefficient of variation and that the overall value first exceeds the level found with a single normal distribution when the means are separated by a distance greater than 2SD. The kurtosis value, by contrast, falls below the level for a single normal distribution as soon as the means are separated by a distance greater than 1SD.

for visual identification of bimodality, as the shape of a histogram for a given data set is sensitive to the size and demarcation of the intervals used.

Taking idealized data of the kind illustrated in Figure 7.13, it is possible to examine the change in value of the coefficient of variation as the separation between the male and female means increases progressively. A plot of values for the coefficient of variation against the distance between male and female means (in SD units) is shown in Figure 7.14 (upper plot). In fact, it can be seen that there is an initial dip in the curve for the coefficient of variation and that the overall value first exceeds the level found with a single normal distribution when the means are separated by a distance greater than 2SD. Hence, this criterion is somewhat more sensitive than simple visual inspection but will detect sexual dimorphism only when the difference between male and female values is quite marked.

Another criterion that could be used to identify the superimposition of two similar distributions with different mean values is the progressive reduction in the central peak of the overall histogram that will occur as the means are separated increasingly (Figure 7.13). This reduction is indicated by a negative value of the *kurtosis* of the histogram, and this measure is also illustrated in Figure 7.14 (lower plot). As can be seen, the kurtosis value remains almost unchanged, initially, but progressively more negative values can be observed when the separation between the means exceeds 1SD. Hence, the kurtosis value can potentially provide a more sensitive indicator than either the coefficient of variation or simple visual inspection for the detection of bimodality in a sexually dimorphic sample. As yet, however, kurtosis has been barely mentioned even in the most detailed quantitative approaches to the detection of sexual dimorphism in the primate fossil record.

The illustrations in Figures 7.13 and 7.14 are based on idealized data sets using large sample sizes (1000 measurements for each sex). In practice, relevant sample sizes for the primate fossil record are usually much smaller (i.e. tens rather than hundreds) and the distinctions shown are much less reliably detectable. In other words, it is likely that only quite extreme cases of sexual dimorphism will be detected by examination of the distributions of values for individual dimensions in a given fossil primate species.

Even if a given fossil sample can be divided into two distinct subsets on the basis of some criterion, it is often difficult to establish whether they represent two sexes or two different species. The problems involved in distinguishing sexual dimorphism from the differences between two closely related species can be illustrated with a test conducted using a sample of dental dimensions from adult gorillas and chimpanzees, first examining the species separately and then examining the data when mixed together. In this test, sample sizes were deliberately kept small to replicate the conditions found typically with the relevant primate fossil samples.

Gorillas show marked dimorphism in adult body weight (male:female body weight ratio = 1.85) and dimorphism in the size of the canine teeth is also pronounced. In a histogram of lower canine breadth, there is no overlap between males and females (Figure 7.15). Sexual dimorphism is, however, far less pronounced in the cheek teeth (premolars and molars).

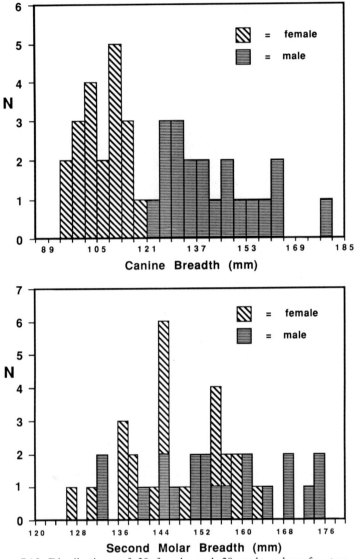

Figure 7.15. Distributions of 20 female and 20 male values for two dental dimensions in gorillas (*Gorilla gorilla*): (above) lower canine breadth (mm); (below) breadth of the lower second molar (mm). There is no overlap between male and female values for canine breadth, but there is extensive overlap in values for second molar breadth.

In a histogram of values for the breadth of the second lower molar, male and female values show extensive mixing, although the mean value for males is greater than that for females (Figure 7.15). In contrast to gorillas, chimpanzees show only mild dimorphism in adult body weight

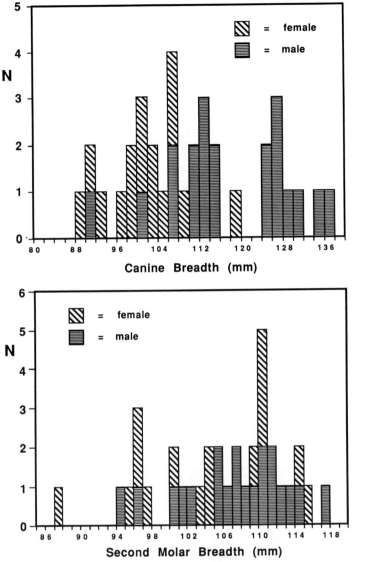

Figure 7.16. Distributions of 15 female and 20 male values for two dental dimensions in chimpanzees (*Pan troglodytes*): (above) lower canine breadth (mm); (below) breadth of the lower second molar (mm). There is moderate overlap between male and female values for canine breadth and almost complete overlap in values for second molar breadth.

(male:female body weight ratio = 1.28) and dimorphism in the size of the canine teeth is only moderate. In a histogram of lower canine breadth, there is a fair degree of overlap between males and females, although the mean value for males is significantly higher (Figure 7.16). There is no clearly detectable bimodality in the histogram. As with the gorilla, sexual dimorphism is even less obvious in the cheek teeth. Male and female values show almost complete overlap for the breadth of the second lower molar and there is a slight difference only between the mean values for males and females (Figure 7.16). These data indicate that dimensions of the cheek teeth are unlikely to provide reliable indicators for bodily size dimorphism, even in cases where it is marked, and that sex difference in the size of canine teeth is only clearly evident in cases of bodily size dimorphism. Cases of moderate sexual dimorphism, as with the chimpanzee, will be exceedingly difficult to demonstrate in the fossil record.

If bimodality is found in a given data set that may contain more than one species, as in Figures 7.17 and 7.18, no conclusion regarding sexual dimorphism is permissible. To illustrate this point, the data sets for gorillas and chimpanzees that were illustrated in Figures 7.15 and 7.16 were mixed together. In the first place, there is extensive overlap between the two species in lower canine breadth, so it would not be possible to distinguish the species on this basis. On the other hand, the histogram (Figure 7.17) could be interpreted as being bimodal, so such a mixture of two species

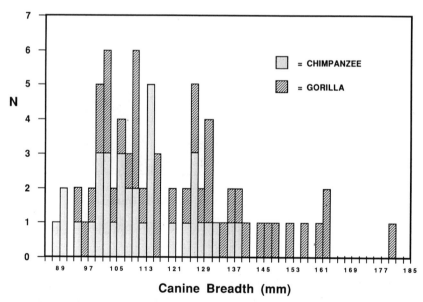

Figure 7.17. Overall distribution for breadth of the lower canine (mm) for the combined sample of chimpanzees (15 females; 20 males) and gorillas (20 females; 20 males). There is extensive overlap between the two species.

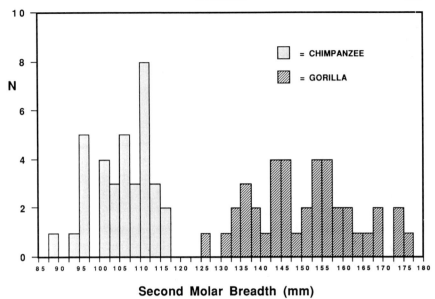

Second Molar Breadth (mm)

Figure 7.18. Overall distribution for breadth of the lower second molar (mm) for the combined sample of chimpanzees (15 females; 20 males) and gorillas (20 females; 20 males). There is no overlap between the two species.

in a fossil sample could easily be misinterpreted as indicating sexual dimorphism in a single species. By contrast, there is complete separation between chimpanzees and gorillas for the breadth of the lower second molar (Figure 7.18), so this could be used as a reliable criterion for distinguishing the species. It should be noted, however, that the combined sample is bimodal and that a pattern of this kind could also be misinterpreted as indicating sexual dimorphism. Overall, therefore, it should be recognized that we must be confident about the identification of individual species before proceeding to any discussion of possible sexual differences.

All the examples provided in Figures 7.15 to 7.18 involve measurements of single characteristics. It is, of course, possible to conduct multivariate analyses in the investigation of sexual dimorphism. One such study has shown that it is possible to achieve a complete separation between chimpanzees and gorillas and between the sexes within each species using a combination of dental measurements. Individual specimens of *Proconsul* were also separated into subgroups on this basis. The problem is that, without some independent basis for species identification of the Miocene ape *Proconsul*, it is still difficult to determine whether separation between subgroups using multivariate analysis is distinguishing sexes or species. Taking the information provided in Figures 7.15 to 7.18, one useful strategy might be first to identify fossil species on the basis of molar dimensions and then to look for sexual dimorphism using canine dimensions. In cases where a bimodal distribution is evident from molar dimensions in a given sample, it would seem to be more likely that a distinction between species, rather than sexes, is involved.

A classic case of the problems involved in separating differences between species from differences between the sexes is provided by fossil hominids of the genus *Australopithecus*. Representatives of this genus are now regarded almost universally as falling into two quite distinct groups: gracile australopithecines (*Australopithecus afarensis, Australopithecus africanus*) and robust australopithecines (*Australopithecus boisei, Australopithecus robustus*). At a stage where the evidence was provided mainly by the two South African species *Australopithecus africanus* and *Australopithecus robustus*, however, some authors interpreted the gracile individuals as being females and the robust individuals as being males of a single species with marked sexual dimorphism. This interpretation was discredited eventually, not least because robust autralopithecines typically come from separate fossil sites and are, on average, considerably younger than gracile australopithecines in geological terms. Currently, most authors accept the existence of the four species named above. Indeed, several authorities feel that the differences between the gracile and robust australopithecines are so pronounced that they should be placed into two separate genera (*Australopithecus* and *Paranthropus*).

The question of sexual dimorphism in australopithecines has been revived recently in another guise, most notably in connection with material allocated to the species *Australopithecus afarensis*. It was at one stage widely accepted that the gracile australopithecines were relatively small, with body weights in the range of 25–35 kg, and that no marked sexual dimorphism was present. Recently, however, there has been a series of publications in which body weight has been estimated for a variety of specimens attributed to *Australopithecus afarensis*. These estimations indicate a very wide range of body weights, which has been explained by the presence of extreme sexual dimorphism. In some cases, average male weight has been calculated as being more than twice as great as female weight, but in a recent, more conservative, analysis, one author has provided estimations for the male to female body weight ratio in *Australopithecus afarensis* of 1.78, using formulae derived from modern apes, and of 1.52, using formulae derived from modern humans. The average value of 1.65 from these two estimations represents quite marked sexual dimorphism, distinctly higher than in modern humans and inter-mediate between chimpanzees and the other great apes (gorillas and orang-utans). On a parallel with modern Old World simians, this apparent evidence of marked sexual dimorphism has been taken as being an indication that the social system of *Australopithecus afarensis* was not monogamous but polygynous, thus implying that polygyny was char-acteristic at least during the early stages of human evolution. Similar claims have been made for *Australopithecus africanus*.

The central problem with the inference of sexual dimorphism in *Australopithecus afarensis* is that it has not been established reliably that the fossil material concerned is all derived from a single species. In fact, several investigations have led to the conclusion that at least two species with fundamentally different morphological characteristics are present in the fossil material allocated to this species.

Another peculiar aspect of the inferred body weight data relating to claims of marked sexual dimorphism is that the mean body weight inferred for the gracile australopithecines (averages for 'males' and 'females' combined) is higher than that inferred for the robust australopithecines. Effectively, this makes nonsense of the terms 'gracile' and 'robust', but this may be because skeletal fragments of a larger-bodied species have been mixed with those of genuine gracile australopithecines, thus making the average body weight greater than for an uncontaminated sample of robust australopithecines.

Sexual dimorphism in brain size in fossil hominids

The question of the body size of gracile australopithecines is linked closely with that of relative brain size. Initially, it was noted that brain size in these early hominids was no greater than that of modern great apes, and, accordingly, it was stated that the expansion of the hominid brain first began some 2 million years ago, with the advent of *Homo habilis*. Then it was pointed out, however, that the gracile australopithecines (with an estimated body weight of 25–35 kg) were considerably smaller in body size than modern great apes and that their brains therefore did show some degree of expansion *relative to body size*. However, if the gracile austra-lopithecines are regarded as being a highly sexually dimorphic species with an average body weight of at least 35 kg and possibly 55 kg or more, brain size relative to body size would seem to lie within the range of the

great apes. Thus, for an entirely different reason, it is once again being
claimed by some authors that the expansion of the hominid brain first
began with *Homo habilis*.

In fact, an analysis of data on brain size from known skulls of
Australopithecus africanus indicates that all the individuals concerned
would probably be classified as 'females' by those who believe that gracile
australopithecines exhibit extreme sexual dimorphism in body weight. All
the relatively intact skulls of *Australopithecus africanus* are obviously
derived from relatively small-bodied individuals, corresponding to body
weight of about 25 kg in all cases, and the range of variation in brain size
is actually somewhat less than that for females of any of the modern great
ape species. An analysis of relative brain size that is restricted to a
comparison between this sample of *Australopithecus africanus* and females
of the modern great ape species shows that relative brain size is
significantly greater in the gracile australopithecine. Therefore, one can
conclude confidently that expansion of the relative size of the hominid
brain had, in fact, already begun at the australopithecine level. Further,
one must either accept that all the skulls of gracile australopithecines
discovered to date have been derived by chance from females or admit
that samples of gracile australopithecines have been mixed up with more
fragmentary remains from a larger-bodied hominid that has yet to be
identified properly. Whatever the answer may be, clearly, one must treat
claims of the existence of marked sexual dimorphism in gracile australopithecines with extreme caution.

Towards a balanced perspective on sexual dimorphism in primates

As stated at the beginning of this chapter, both the sexual competition
hypothesis and the anti-predator hypothesis depend on the assumption
that selection leading to sexual dimorphism acts essentially on males, for
example leading under certain conditions to an increase in male body size
relative to female body size. An alternative approach is to seek selection
pressures acting on both sexes. One suggestion in this direction is that
sexual dimorphism is associated with divergence in feeding habits between
the sexes, resulting in a reduction in competition between them for
resources ('niche expansion hypothesis'). There is, however, only limited
evidence of dietary divergence between males and females in sexually
dimorphic primate species. Further, any such divergence could be a
consequence rather than a cause of sexual dimorphism, as energy
requirements are related directly to body size. In any case, one might

expect the greatest degree of competition for food to exist between the male and female of a monogamous pair with a joint territory, rather than between adults living in polygynous groups. Accordingly, the niche expansion hypothesis would predict that sexual dimorphism should, if anything, be particularly marked in monogamous species, which is the opposite to what is actually observed.

As a general principle, one should at least consider the possibility that sexual dimorphism could arise through selection acting on either sex, and it is important to seek tests to identify the direction of evolution. As noted in the introduction, there is a fundamental problem because we are confronted with an end product of evolutionary change in which one or more differences have arisen between males and females without any direct indication of the manner in which such differences have arisen. In the case of sexual dimorphism in adult body weight, for instance, an increase in male body weight or a decrease in female body weight could produce the same result (Figure 7.2).

A basis for testing the direction of evolution of sexual dimorphism is, in fact, provided by the phenomenon of 'phylogenetic inertia', already mentioned above in another context. This phenomenon is particularly likely to occur with structures that complete most of their growth at an early stage, such as the brain. In all mammals, the brain develops relatively rapidly and has virtually reached its adult size long before the general growth of the body has been completed. As a result, it is likely that, initially, selection to increase or to decrease adult body size may not have much effect on brain size. Hence, for example, a marked and relatively rapid reduction in body size (dwarfing) in the evolution of a species could result in over-scaling of the brain, as the brain might continue initially to develop to match the original body size. A convincing example of such over-scaling of the brain is provided by the diminutive talapoin monkey (*Miopithecus talapoin*), the smallest of the Old World monkeys (adult male: 1.38 kg; adult female: 1.12 kg). Various lines of evidence suggest that the talapoin evolved from a larger-bodied, terrestrial guenon species, and its closest relative in fact seems to be the patas monkey (*Erythrocebus patas*), the largest-bodied member of the guenon group (adult male: 12.6 kg; adult female: 6.3 kg). Dwarfing during the evolution of the talapoin monkey may account for the fact that, among Old World simians, this species has the next highest relative brain size after humans.

Over-scaling of the brain following a reduction in body size, along with its counterpart of under-scaling following an increase in body size, can also provide a test of the direction of change in body size leading to sexual dimorphism. It can be predicted that, if indeed sexual dimorphism in body size is attained typically by an increase in male body size, males of dimorphic species would be expected to exhibit under-scaled brains in comparison to males of monomorphic species. Conversely, if sexual dimorphism in body size is commonly a result of a reduction in female body size, females of dimorphic species would be expected to exhibit over-scaled brains in comparison to females of monomorphic species. Of course, in order to conduct such a test it is necessary to take account of the allometric scaling of brain size to body weight and to take the residual values as being indicators of relative brain size. When this is done, it emerges that the only well-marked tendency is for females of strongly sexually dimorphic species to have over-scaled brains (Figure 7.19, top). Therefore, it can be concluded that sexual dimorphism in adult body size has been achieved mainly through decrease in female body size in the evolution of simian primates, rather than exclusively through increase in male body size. There is also a weak indication that males of strongly sexually dimorphic species may have somewhat smaller brains than expected, and such under-scaling of the brain would indicate a minor role played in the evolution of sexual dimorphism by increase in male body size.

It is possible to cross check this finding by examining another morphological system which, like the brain, develops early in life and, hence, may show phylogenetic inertia following a change in adult body size. The teeth also attain their final size early in life, so similar predictions can be made with respect to the direction of evolution of sexual dimorphism: reduction in female body size should be associated with over-scaling of female teeth, while increase in male body size should be associated with under-scaling of male teeth. It has often been noted that females of sexually dimorphic primate species have relatively large cheek teeth, and this itself might suggest over-scaling in females. When the relative size of the first lower molar (a representative cheek tooth) is examined according to the degree of sexual dimorphism, the pattern that emerges is very similar to that seen with relative brain size (Figure 7.19, bottom). There is a clear tendency for females of sexually dimorphic species (especially for those in which dimorphism is pronounced) to have over-scaled molars. Once again, there is a weak indication that males of strongly sexually dimorphic species may have somewhat smaller molar teeth than expected, and such under-scaling would indicate a subsidiary role played in the evolution of sexual dimorphism by increase in male body size. It emerges overall that females of strongly sexually dimorphic species are characterized by significant

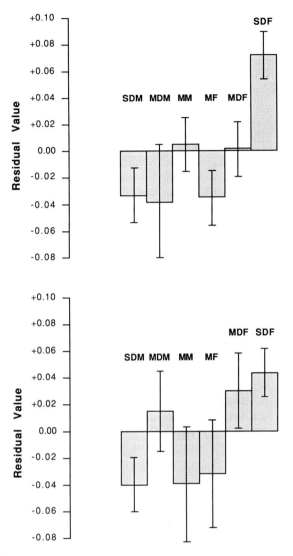

Figure 7.19. (Top) residual values for brain size and (bottom) for size of the first lower molar for males and females with different degrees of sexual dimorphism. SDM, strongly dimorphic males; MDM, mildly dimorphic males; MM, monomorphic males; MF, monomorphic females; MDF, mildly dimorphic females; SDF, strongly dimorphic females. (For strongly dimorphic species, male:female body weight ratio > 1.30; for mildly dimorphic species, male:female body weight ratio > 1.15; for monomorphic species, male:female body weight ratio < 1.15.) In both cases, females of species showing marked sexual dimorphism (SDF) clearly show higher-than-expected values, while males of strongly dimorphic species (SDM) show a weaker trend towards lower-than-expected values.

over-scaling both of the brain and of the molar teeth, whereas under-scaling of these features in males is less clearly indicated. In sum, tests based on the concept of phylogenetic inertia in early maturing organ systems indicate clearly that reduction in female size has played a major, although not exclusive, part in the evolution of sexual dimorphism in adult body size among simian primates.

As indicated by the fact that males of sexually dimorphic species reach sexual maturity later than conspecific females (Figure 7.8), dimorphism has consequences for reproduction. One possibility that arises from this is that female body size may become reduced because of selection favouring more rapid reproduction. Age of sexual maturity is one of the key factors determining reproductive potential among mammals. It is therefore interesting to examine features that might reveal that selection favours increased reproductive turnover of females.

Selection which favours increased reproductive output of females may, among other things, influence maternal investment in individual offspring. A simple measure of maternal investment is the ratio of neonatal weight to gestation period (N/G), which indicates the daily investment of the mother in fetal growth averaged over the course of gestation. The expectation that maternal investment might be higher in species subject to selection for increased reproductive output can be tested by scaling N/G to maternal body weight and by then examining the relationship between the residual values for this measure and the residual values for sexual dimorphism. This particular analysis has been confined to simian species, because it is known that lemurs and lorises have a markedly different pattern of maternal investment and because, in any case, they lack sexual dimorphism. Furthermore, marmosets and tamarins have been excluded because they produce twins, which would complicate the analysis. The data shown (Figure 7.20) have hence been restricted to 27 simian species with single offspring. There is a weak but statistically significant positive correlation ($r = 0.435$; $p < 0.025$) between the two sets of residual values for maternal investment and degree of sexual dimorphism. This finding is compatible with the hypothesis that females in sexually dimorphic species may have been subject to selection for increased reproductive output – contingent on their ability to sustain increased reproductive effort.

The issue of selection for reproductive turnover is linked to the concept of '*r-selection*'. According to this concept, selection may favour increased reproductive potential in species exposed to relatively unpredictable and/or seasonal environments. Thus, it might be expected that the

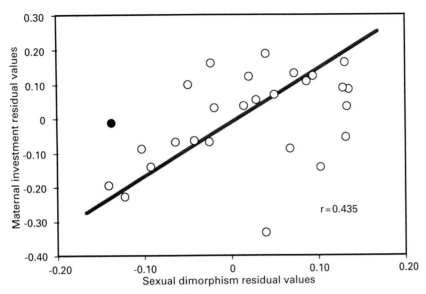

Figure 7.20. Logarithmic values of residuals for maternal investment (ratio of neonate weight to gestation period) plotted against residuals for sexual dimorphism in adult body weight for a sample of 27 simian primate species with single offspring. The black point represents humans and the extreme outlier at the bottom of the graph represents gorillas. There is a weak, but significant, correlation between the two sets of residual values. The best-fit line (hatched) is the major axis.

evolution of smaller female body size in a dimorphic species, associated with more rapid breeding, could be linked to r-selection. This can, in principle, be tested by examining the residual values of r_{max}, the maximum intrinsic rate of population increase, in relation to the degree of sexual dimorphism. The problem is to decide upon an appropriate basis for examining the scaling of r_{max} to body size. If the male body size in fact represents the ancestral body size for both sexes in a species that is now sexually dimorphic, it could be that r_{max} should be scaled relative to male body weight. On the other hand, it is only to be expected that small-bodied females would breed faster than large-bodied females regardless of their evolutionary background, so it might be appropriate to scale r_{max} relative to female body weight. A compromise approach has been adopted in Figure 7.21, in which residual values for r_{max} have been calculated relative to the average body weight for males and females of each species. There is a weak, but none-the-less significant, positive correlation with the residual values for sexual dimorphism in body weight

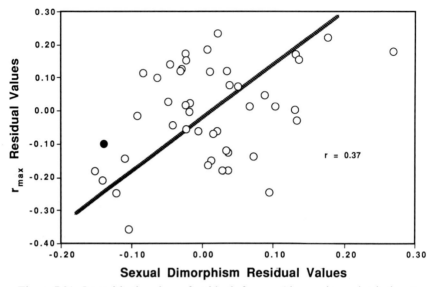

Figure 7.21. Logarithmic values of residuals for r_{max} (the maximum intrinsic rate of population increase) plotted against residuals for sexual dimorphism in adult body weight for a sample of 46 simian primate species. The black point represents humans. There is a weak, but significant, correlation between the two sets of residual values. The best-fit line (hatched) is the major axis.

($r = 0.373; 0.05 > p > 0.01$). If the analysis is repeated with residual values calculated relative to female weight alone, the positive correlation is weaker and not quite significant ($r = 0.251; 0.10 > p > 0.05$), but the expected trend in the data is still present. Hence, despite the fact that the primary response to selection for increased reproductive output of females would presumably be a simple reduction in body size, there is a detectable secondary effect, since females of sexually dimorphic species also have a somewhat higher reproductive capacity than would be expected for their body size. Taken together with the significant evidence for greater maternal investment during pregnancy for females of sexually dimorphic species, this confirms that such females have been subject to selection pressure favouring increased reproductive output.

Discussion and conclusions

Although discussions of the evolution of sexual dimorphism in primates have generally tended to focus on males, it is clear that selection acting on females should also be considered. Explaining the evolution of sexual dimorphism is no easy task, because we are confronted by a difference between the sexes with no independent indication of the direction taken by evolution. As a starting point, it is best to assume that increase in male size and decrease in female size are equally likely, and all potential explanations should be checked carefully for circularity. In addition to the widely recognized effects of competition between males for breeding access to females and defence against predators by large-bodied males, it is also necessary to consider the potential reproductive advantages of smaller, more rapidly breeding females under particular environmental conditions. For instance, relatively open habitats are likely to be associated not only with greater exposure to predation but also with selection for increased reproductive turnover because of the lower predictability of climatic conditions in areas characterized by relatively low rainfall. Thus, the widely accepted connection between terrestrial habits and sexual dimorphism could be explained, at least in part, by selection for more rapid breeding leading to a decrease in female body size. Indications of over-scaling of brain and tooth size in sexually dimorphic females, combined with evidence for increased reproductive investment during pregnancy and an overall increase in reproductive potential, show convincingly that selection for decreased female body size has been a factor in the evolution of sexual dimorphism in primates.

The fact that sexual dimorphism both in body size and in the size of the canines is essentially confined to simians has still to be explained. It has been shown that the degree of sexual dimorphism in both body size and canine size tends to increase with increasing overall size, so it is possible that there is a threshold below which sexual dimorphism is unlikely to develop. As living prosimian primates are generally markedly smaller than living simian primates, such a threshold effect could potentially account for the general lack of marked sexual dimorphism among living prosimians. But we are still left with the problem that sexual dimorphism also appears to have been lacking among large-bodied subfossil lemurs. Further research is required to clarify many of the empirical observations that have been made with respect to sexual dimorphism among primates.

In principle, the course of evolution of sexual dimorphism in primates could be clarified by reference to the fossil record. Interpretation of fossils is, however, complicated by the lack of a fully reliable method for distinguishing differences between species from differences between sexes. Quite generally, fossil species should be identified unambiguously before any interpretation of sexual dimorphism is attempted, and for the time being there is little unquestionable fossil evidence for the evolution of sexual dimorphism.

Acknowledgements

Data on body size (in most cases based on measurements taken on wild animals) were derived from an extensive data base, compiled by the first author in collaboration with Dr Ann MacLarnon with the support of a project grant from the Medical Research Council (UK). Most of the dental data were derived from Daris Swindler (1976), supplemented in a few cases with additional measurements to fill gaps. The approach presented in the paper owes a great deal to discussions with the following people: Dr Leslie Aiello, Professor Andrew Barbour, Dr Paul Harvey, Dr Michael Hills, Dr Ann MacLarnon and Dr Caroline Ross. Thanks are also due to Lukrezia Bieler-Beerli for providing the illustrations of living primates for some of the figures and to Dr G. Anzenberger and Dr Evan Balaban for providing comments on the draft manuscript.

Further reading

Blumenberg, B. (1985) Biometrical studies upon hominid teeth: the coefficient of variation, sexual dimorphism and questions of phylogenetic relationship. *Biosystems*, **18**, 149–84.

Harvey, P. H., Kavanagh, M. and Clutton-Brock, T. H. (1978) Sexual dimorphism in primate teeth. *Journal of Zoology*, **186**, 475–86.

Leigh, S. R. (1992) Patterns of variation in the ontogeny of primate body size dimorphism. *Journal of Human Evolution*, **23**, 27–50.

Leutenegger, W. and Kelley, J. T. (1977) Relationship of sexual dimorphism in canine size and body size to social, behavioural and ecological correlates in anthropoid primates. *Primates*, **18**, 117–36.

Lucas, P. W., Corlett, R. T. and Luke, D. A. (1986) Sexual dimorphism of tooth size in anthropoids. In *Sexual Dimorphism in Living and Fossil Primates*, pp. 23–39. Eds. M. Pickford and B. Chiarelli. Il Sedicesimo, Florence.

Oxnard, C. E. (1987) *Fossils, Teeth and Sex: New Perspectives on Human Evolution*. University of Washington Press, Seattle.

Pickford, M. (1986) On the origins of body size dimorphism in primates. *Human Evolution*, **1**, 111–48.

Pilbeam, D. R. and Zwell, M. (1972) The single species hypothesis, sexual dimorphism and variability in early hominids. *Yearbook of Physical Anthropology*, **16**, 69–79.

Shea, B. T. (1986) Ontogenetic approaches to sexual dimorphism in anthropoids. *Human Evolution*, **1**, 97–110.

Swindler, D. R. (1976) *Dentition of Living Primates*. Academic Press, New York.

Willner, L. A. and Martin, R. D. (1985) Some basic principles of mammalian sexual dimorphism. *Symposium of the Society for the Study of Human Biology*, **24**, 1–42.

Sexual dimorphisms in organ systems

8

Translating gonadal sex into phenotypic sex

JEAN D. WILSON

Embryos of both sexes develop in an identical fashion for the first portion of gestation, and only thereafter do anatomical and physiological developments diverge to result in the differentiation of the male and female phenotypes (see also Chapter 9 by Marilyn Renfree). The fundamental mechanism of sexual differentiation in the eutherian mammal was elucidated by Alfred Jost between 1947 and 1952. He established that a castrated rabbit embryo develops as a female. Male development is induced only by specific hormonal signals arising from the fetal testis. According to the Jost formulation – now the central dogma of sexual development – sexual differentiation is a sequential, ordered and relatively straightforward process. Chromosomal (or genetic) sex, established at the time of conception, directs the development of either ovaries or testes. If testes develop, their hormonal secretions elicit the development of the male secondary sex characteristics, collectively known as the male phenotype. If an ovary develops or if no gonad is present, anatomical development is female in character.

Disturbances that impair any step in this process during embryonic life cause defects of sexual differentiation that can be classified as disorders of genetic or chromosomal sex, disorders of gonadal sex, or disorders of phenotypic sex. The purpose of this review is to focus on the translation of gonadal sex into phenotypic sex in the eutherian mammal, and the primary focus will be on the events in the human.

The background

The concept that the gonadal hormones play a role in sexual differentiation was first proposed in 1903 by Bouin and Ancel. Over the subsequent 30 years, extensive evidence was assembled to indicate that gonadal

203

hormones are critical both for the control of gonadal differentiation and for the development of the sexual phenotypes in amphibians and birds. Furthermore, on indirect grounds, a hormonal theory was believed to be the best explanation for the development of the free-martin in cattle in which virilization occurs in the female calf born as a twin to a normal male calf. Thus, by the 1930s it was assumed widely that sexual differentiation in all species is controlled largely by hormonal factors.

When purified steroid hormones became available, it was possible to define directly the role of hormones in sexual differentiation in the mammal, and at this point the problem became clouded for several reasons. First, although gonadal steroids play an essential role in the maturation of both ovaries and testes, it has never been possible to cause gonadal sex reversal in any eutherian mammal by the administration of hormones. In the mammal, gonadal differentiation is fundamentally under genetic control. Second, the principal mammalian species chosen for the study of differentiation of the sexual phenotypes was the marsupial opossum, *Didelphis virginiana*. This choice was made because the marsupial embryo is sexually immature at the time of birth and because the pouch young are accessible for experimental manipulation after they attach to the nipple. Administration of estrogen caused some feminization of the urogenital tract of the male opossum, and treatment of female pouch young with androgens caused virilization of the urogenital tract. However, no hormone or combination of hormones caused reversal of the basic phenotype or influenced the basic secondary sex characteristics such as development of the pouch in the female or the scrotum in the male. At best, gonadal steroids in the marsupial pouch young, as in pubertal mammals, appeared to act as modifiers rather than as inducers of development. As a consequence, these studies in the opossum cast doubt on the theory that hormones play a primary role in mammalian sexual differentiation.

The Jost experiments with the rabbit embryo were clearcut and established that the development of phenotypic sex in this mammal is exclusively under the control of gonadal hormones. It was immediately and widely recognized that the findings had explanatory potential for understanding sexual differentiation at the basic level and for understanding disorders of human sexual development at the clinical level. It has been the task of subsequent investigators to identify the hormones secreted by the fetal testes, to elucidate the control mechanisms that regulate the rates of secretion of these hormones during the critical phase in embryogenesis, and to characterize at the molecular and genetic levels the mechanisms by which the testicular hormones induce the conversion of the sexually indifferent embryo into the male phenotype. As a consequence, the original formulation of Jost has been confirmed and expanded.

Normal phenotypic development

Phenotypic development involves several processes. The internal urogenital tract arises from the Wolffian and Müllerian ducts, both of which are present in early embryos of both sexes. The Wolffian ducts are the excretory ducts of the mesonephric kidney system and are connected anatomically to the indifferent gonad. The Müllerian duct forms secondarily from the Wolffian duct and is not contiguous with the gonad. Both types of ducts terminate in the urogenital sinus. In the male the Wolffian ducts give rise to the epididymides, vasa deferentia and seminal vesicles, and the Müllerian ducts disappear. In the female, the Müllerian ducts give rise to the fallopian tubes, uterus and upper vagina, and the Wolffian ducts either disappear or persist in vestigial form as Gartner's ducts. Thus, the urogenital tracts in males and females arise from different anlagen. In contrast, the male and female external genitalia develop from common anlagen: the genital tubercle, genital folds and genital swellings. In the female the system elongates but changes little: the genital tubercle becomes the clitoris; the genital folds become the labia minora; and the genital swellings become the labia majora. In the male, fusion and elongation of the genital folds cause formation of the urethra and shaft of the penis and, ultimately, bring the urethral orifice to the genital tubercle (glans penis). The fused genital swellings become the scrotum, and the prostate forms in the wall of the urogenital sinus.

In the absence of the testes, as in the normal female or in male embryos castrated prior to the onset of phenotypic differentiation, all aspects of phenotypic sex, including both the internal urogenital tract and external genitalia, develop along female lines. Thus, formation of the female phenotype does not require hormones from the fetal ovary whereas masculinization of the fetus is the positive result of the imposition of male development by testicular hormones in the sexually indifferent embryo.

Three hormones control the development of the male phenotype. Two (anti-Müllerian hormone (AMH) and testosterone) are secretory products of the fetal testis. AMH is a glycoprotein formed by the Sertoli cells of fetal and newborn testes that causes regression of the Müllerian ducts, probably acting in concert with testosterone, and hence prevents development of the uterus and fallopian tubes in the male.

Testosterone is the principal androgen secreted by the fetal as well as the adult testes. The onset of testosterone secretion occurs prior to the onset of virilization of the male embryo (at about the eighth week of human embryonic development). The factors that regulate the initial secretion of testosterone have not been defined, but the initiation of testosterone biosynthesis (and hence the process of male phenotypic development) appears to be autonomous; control of testosterone formation by pituitary gonadotropins is demonstrable late in embryogenesis.

On the basis of studies of androgen metabolism in embryos of several species, including humans, it was deduced that testosterone promotes virilization of the urogenital tract in two ways (Figure 8.1). Testosterone acts directly to stimulate the Wolffian ducts and to induce development of the epididymides, vasa deferentia and seminal vesicles. Differentiation of the Wolffian ducts into seminal vesicles and epididymides is completed in the human male embryo at 12–13 weeks of development, before the capacity to form dihydrotestosterone is acquired by these tissues. In contrast, in the anlagen of the prostate and the external genitalia, testosterone is converted to dihydrotestosterone, the third hormone of fetal virilization, by 5α-reductase. Dihydrotestosterone acts in the urogenital tubercle, swelling and folds to cause the midline fusion, elongation and

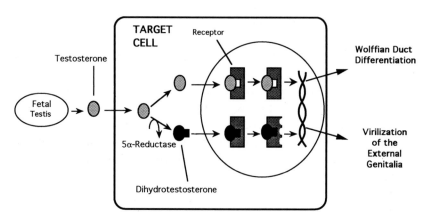

Figure 8.1. A schematic diagram of androgen physiology in the male fetus. Testosterone, secreted by the testis, binds to the androgen receptor in a target cell, either directly or after conversion to dihydrotestosterone. Wolffian duct virilization is mediated by testosterone, and virilization of the external genitalia is mediated by dihydrotestosterone.

enlargement that eventuate in the penis and scrotum and in the urogenital sinus to cause formation of the prostate.

A single high-affinity receptor protein is responsible for mediating the actions of both testosterone and dihydrotestosterone, but it is not clear exactly how two hormones act via a single receptor to produce different actions. As a result of its tighter interaction with the receptor, the formation of dihydrotestosterone may function primarily to amplify a weak hormonal signal. Alternatively, dihydrotestosterone-receptor complexes may be required absolutely for transcriptional control of some genes.

The androgen-receptor mechanism is fundamentally the same in male and female embryos, and exposure of female embryos to androgens at the appropriate time in embryonic development causes virilization of both the Wolffian ducts and the external genitalia in characteristic male fashion. The differences in male and female phenotypic development, therefore, are due solely to the hormone produced by the fetal testes at the critical period of embryonic development and are not due to differences in the receptors for the hormones.

Disorders of androgen or anti-Müllerian hormone action

The separate roles for anti-Müllerian hormone, testosterone and dihydrotestosterone that were deduced on the basis of studies of embryogenesis in normal embryos have been defined further by studies of subjects with inherited defects that impair androgen action or Müllerian duct regression.

5α-Reductase deficiency

An autosomal recessive form of abnormal sexual development was described by Nowakowski and Lenz in 46, XY men in whom the testes, androgen levels, and Wolffian ducts are male in character but the Wolffian ducts terminate in a blind-ending vagina and the external genitalia are female in character. Since this is the phenotype predicted to result from dihydrotestosterone deficiency, it was anticipated and confirmed that this disorder is due to steroid 5α-reductase deficiency. We have characterized the enzyme deficiency in more than 20 affected families from various parts of the world, and a similar number have been described by others. The 5α-reductase enzyme is bound tightly to membranes of the nuclear envelope and the endoplasmic reticulum, and the enzyme has never been solubilized or purified in an active form. The reason for this difficulty became apparent when the cDNA for the rat enzyme was cloned by Andersson and Russell by the indirect technique of expression cloning and it was found that the amino-acid sequence is very hydrophobic.

In studies of human 5α-reductase in cultured genital skin fibroblasts, it was recognized that the enzymes have two pH optima, one at pH 5.5 and a broader optimum in the alkaline range; in most androgen-target tissues such as prostate, the principal enzyme activity is due to the pH 5.5 enzyme. In patients with 5α-reductase deficiency, the pH 5.5 activity is deficient, whereas the more alkaline pH activity may be normal. When the cDNAs for the human enzymes were cloned, the alkaline enzyme was cloned first and is now termed human steroid 5α-reductase 1; the cDNA for the pH 5.5 enzyme, human steroid 5α-reductase 2, was cloned subsequently. As expected, the predicted amino-acid sequence of enzyme 1 is normal in subjects with 5α-reductase deficiency. A deletion of the 5α-reductase-2 gene was documented in one kindred with 5α-reductase deficiency, but most families with 5α-reductase deficiency have point mutations of the 5α-reductase-2 gene, the majority of which result in single amino-acid substitutions in the coding sequence for the enzyme.

Several aspects of the 5α-reductase system are still poorly understood. Although both steroid 5α-reductases are hydrophobic, the amino-acid sequences are only weakly homologous, and exactly how these activities arose during evolution is unclear. Moreover, whereas the pH 5.5 activity plays an essential role in androgen action in the urogenital tract, the function of 5α-reductase 1 is not clear but may involve androgen action in hair follicles and other tissues. The behavioral consequences of 5α-reductase-2 deficiency, namely the fact that some affected individuals change gender role behavior from female to male, and the reason that most patients with 5α-reductase deficiency virilize to some extent at puberty, are not understood. A major imperative in androgen physiology is to define the role of steroid 5α-reductase 1 in androgen action.

Androgen-receptor defects

Mutations of the androgen receptor impair all aspects of androgen action, including gonadotropin regulation, phenotypic differentiation, and maturation at puberty; the degree of impairment depends on the severity of receptor dysfunction. Mutations of the androgen receptor have been identified in individuals whose phenotype varies over a broad spectrum from women with the syndrome of complete testicular feminization to men with ambiguous genitalia to phenotypically normal men with infertility and/or minor degrees of undervirilization. In part because these mutations are X-linked (and hence are expressed in hemizygous 46, XY individuals) defects of this protein are relatively common and are believed to account for three-quarters or more of genetic men who do not virilize normally (male pseudohermaphroditism).

In our laboratory more than 130 families with the clinical, endocrinological and genetic features of androgen-receptor mutations have been studied. On the basis of measurements of androgen binding in cultured skin fibroblasts, the underlying defects have been characterized as binding negative, qualitatively abnormal binding, decreased binding, and a final 10% as androgen binding positive; the latter subjects have clinical and endocrine findings of androgen resistance but no identifiable defect in ligand binding.

We assumed that the functional characterization of receptor binding in these disorders would provide some insight into the underlying mutations, but the true complexity of the system became apparent only when the cDNA for the receptor was cloned by taking advantage of the known X-chromosome linkage of the encoding gene and the predicted homology of the androgen receptor with the progesterone receptor. The structure of the androgen receptor proved to be typical of receptors of the thyroid/steroid/retinoid class of transcription regulatory factors with four identifiable domains – a hormone-binding domain, a hinge region, a DNA-binding region, and an N-terminal region. Three distinct homo-polymeric repeat sequences are present in the N-terminal region – 23 glycines, 9 prolines, and a variable glutamine repeat region with a modal number of 20 or 21.

More than 50 mutations in the human androgen receptor have been reported in detail, and several tentative generalizations about these mutations appear to be warranted.

(1) Major gene rearrangements and/or deletions and defects in RNA splicing are unusual and, at best, account for only about 5% of the total: commonly, single nucleotide substitutions are found in one of the eight coding exons

(2) As predicted from the functional studies, the majority of the mutations are familial in nature, only a few instances of the same mutation in unrelated families having been described. The fact that the mutation in most of the families is distinctive is disappointing, because, at present, the techniques of molecular biology are not useful for diagnosis (the coding sequence is too long and too complex to lend

itself readily to automated sequencing). The diagnosis of new patients for the foreseeable future will continue to require the combination of genetic, endocrine and phenotypic studies of patients, and characterization of receptor function in cultured skin fibroblasts

(3) Most mutations that impair the binding of androgen to the receptor cluster in two specific regions of the androgen-binding domain of the molecule. This finding indicates the critical role of these sequences in receptor function

(4) Most instances of androgen-binding positive mutations are due to mutations in the DNA-binding domain of the molecule. The net consequence is the formation of a receptor molecule that binds hormones normally but cannot act because of failure to attach to hormone regulator elements in DNA

(5) The N-terminal domain appears to be an infrequent site of mutation – only one premature termination codon that causes androgen resistance has been described. Whether this infrequency of mutation is due to biases in the manner in which androgen resistance is ascertained or whether it is due to actual rarity of such mutations is not clear

(6) The N-terminal domain contains the most polymorphic region of the molecule, namely the glutamine polymeric repeat domain in which the number of repeats in normal people varies from 15 to 35; indeed, more than 90% of women are heterozygous for this repeat and the inheritance of this polymorphism (as established using the polymerase chain reaction technique) can be used to trace the inheritance of mutant androgen-receptor genes in many women at risk and for prenatal diagnosis. Polymorphisms that cause lengthening of the region to more than 40 repeats are associated with development of spinobulbar atrophy (the Kennedy syndrome); the nature of this association is unknown

Persistent Müllerian duct syndrome

Persistent Müllerian duct syndrome is an uncommon abnormality of male phenotypic development in which 46, XY individuals with male urogenital tracts have, in addition, a uterus, fallopian tubes and an upper vagina. The disorder is heterogeneous. In one type, one or both testes are descended, and the uterus and ipsilateral fallopian tubes are in the inguinal canal or scrotum. The other type is associated with high bilateral cryptorchidism and no hernias. In both types the vasa deferentia are embedded in the wall of the uterus, a feature that complicates surgical procedures designed to preserve potential fertility. Most subjects have uninformative family histories, but in some families the condition appears to be inherited as an autosomal recessive mutation. Because the external genitalia are well developed and the patients masculinize normally at puberty, it is assumed that the fetal testes produced a normal amount of androgen during the critical stage of embryonic sexual differentiation and that Müllerian regression does not occur, either because of failure of the fetal testis to produce AMH or because of failure of the tissues to respond to this hormone.

Utilizing immunoassays, patients with persistent Müllerian duct syndrome can also be divided into two groups. In the AMH-positive group, serum and testicular levels of immunoreactive AMH are normal, and it was postulated that the AMH was either ineffective or the AMH receptor was defective. One AMH-positive subject had a nucleotide change in the coding sequence that caused a single amino-acid substitution in the protein, which was believed to render the protein functionless. In the AMH-negative group, serum AMH is absent or decreased, and it was postulated that the mutation was due to failure to produce AMH. One AMH-negative subject was found to have a single nucleotide substitution that changed a normal codon to a stop codon, which arrested translation and caused the formation of a truncated protein.

Summary

The overall scheme by which hormones virilize the male during embryogenesis and during postnatal life has been substantiated by studies at the molecular level of intersex men whose disorder was due to mutations of one of the three critical proteins involved: the steroid 5α-reductase that converts testosterone to dihydrotestosterone; the androgen receptor that mediates the intracellular action of both androgens; and the anti-Müllerian hormone that prevents development of a uterus in men. As a result of these studies, virilization of the male embryo of the eutherian mammal is probably as well understood as any aspect of embryogenesis, but the mechanisms by which the various hormone signals trigger cellular differentiation and function of the urogenital tract still remain to be delineated.

Further reading

Donahoe, P. K., Cate, R. L., MacLaughlin, D. T., Epstein, J., Fuller, A. F., Takahashi, M., Coughlin, J. P., Ninfa, E. G. and Taylor, L. A. (1987)

Müllerian inhibiting substance: gene structure and mechanism of action of a fetal regressor. *Recent Progress in Hormone Research*, **43**, 431–62.

French, F. S., Lubahn, D. B., Brown, T. R., Simental, V. A., Quigley, C. A., Yarbrough, W. G., Tan, J., Sar, M., Joseph, D. R., Evans, A. J., Hughes, I. A., Migeon, C. J. and Wilson, E. M. (1990) Molecular basis of androgen insensitivity. *Recent Progress in Hormone Research*, **46**, 1–42.

George, F. W. and Wilson, J. D. (1994) Sex determination and differentiation. In *The Physiology of Reproduction*. Eds. E. Knobil and J. Neill. Raven Press, New York, pp. 3–28.

Griffin, J. E. and Wilson, J. D. (1992) Disorders of sexual differentiation. In *Campbell's Urology*, 6th edn, pp. 150–4. Eds. P. C. Walsh, A. B. Retik, T. A. Stamey and E. D. Vaughan. W. B. Saunders, Philadelphia.

Griffin, J. E., McPhaul, M. J., Russell, D. W. and Wilson, J. D. (1994) The androgen resistance syndromes: 5α-reductase deficiency, testicular feminization, and related syndromes. In *The Metabolic Basis of Inherited Disease*, 7th ed. Eds. C. R. Scriver, A. L. Beaudet, W. S. Sly and D. Valle. McGraw-Hill, New York, in press.

Grumbach, M. M. and Conte, F. A. (1992) Disorders of sex differentiation. In *Williams Textbook of Endocrinology*, 8th edn, pp. 853–952. Eds. J. D. Wilson and D. W. Foster. W. B. Saunders, Philadelphia.

Josso, N., Boussin, L., Knebelman, B., Nihoul-Fekete, C. and Picard, J. Y. (1991) Anti-Müllerian hormone and intersex states. *Trends in Endocrinology and Metabolism*, **2**, 227–33.

Jost, A. (1972) A new look at the mechanisms controlling sex differentiation in mammals. *John Hopkins Medical Journal*, **130**, 38–53.

McPhaul, M. J., Marcelli, M., Tilley, W. D., Griffin, J. E. and Wilson, J. D. (1993) Androgen resistance caused by mutations in the androgen receptor gene. *Journal of Clinical Endocrinology and Metabolism*, **76**, 17–23.

Thigpen, A. E., Davis, D. L., Milatovich, A., Mendonca, B. B., Imperato-McGinely, J., Griffin, J. E., Francke, U., Wilson, J. D. and Russell, D. W. (1992) Molecular genetics of steroid 5α-reductase 2 deficiency. *Journal of Clinical Investigation*, **90**, 799–809.

Wilson, J. D. (1989) Sexual differentiation of the gonads and of the reproductive tract. *Biology of the Neonate*, **55**, 322–30.

9

Sexual dimorphisms in the gonads and reproductive tracts of marsupial mammals

MARILYN B. RENFREE

The overt sexual dimorphisms of mammals overlie a fundamental difference beginning at fertilization, when the male and female pronuclei fuse to form a new individual which bears either an XX or an XY chromosomal constitution. The fascination of studying mammalian sexual differentiation in marsupials is that they differ in the nature, the timing and the extent of the development of sexual dimorphisms of the external and internal genitalia. In marsupials, the young are born with relatively undifferentiated gonads, and testicular descent is not complete until some weeks after birth. Thus, studies of fetal and neonatal stages of gonadal and genital differentiation in marsupials can contribute to our understanding of how sexual dimorphisms arise.

As the name implies, most female marsupials possess a pouch (or marsupium) on the ventral abdominal wall that encloses the teats and mammary glands, although in several genera, such as the South American *Monodelphis*, the pouch is absent. The pouch and teats are absent in all adult males.

Internally, all female marsupials possess two separate uteri, which open by separate cervices into the vaginal culs-de-sac. Two vaginae run lateral to the ureters to open posteriorly into the urogenital sinus; these lateral vaginae are used exclusively for sperm ascent. Birth occurs through a third structure, the median vagina or birth canal, which also connects the vaginal cul-de-sacs with the urogenital sinus. This median vagina becomes patent at the time of parturition and closes up again afterwards in most species, although it remains open after the first parturition in the macropodids (kangaroos and wallabies). Male marsupials are unlike eutherians because their scrota are anterior to their penises, and their vasa deferentia open into the urethra without first having to curve around the ureters.

Unlike the analogous structures in most eutherians, the urinary and

213

genital passages of all marsupials open into a common urogenital sinus, thence to a single external anogenital orifice in both sexes. The penis of the male protrudes through this when erect. Although the rectum shares this common external orifice, the rectum and the urinogenital openings have separate sphincters.

In all the species of marsupials that have been investigated, a mere handful of the 249 extant species, it has been difficult to sex the offspring at birth by external phenotypic characteristics. Gonadal differentiation continues after birth, something that is unheard of in eutherians, and is followed by differentiation of the Wolffian (mesonephric) and Müllerian ducts, and testicular migration and descent.

American pioneers in the study of mammalian sexual differentiation, Carl Hartman, R. K. Burns and C. R. Moore, realized that marsupials provide a unique opportunity for studying sexual differentiation in what amounts to an exteriorized fetus. The marsupial neonate is not under the influence of any placental hormones whilst it is in the pouch, where gonadal differentiation is completed (although the placenta is capable of steroid synthesis when gonadal differentiation begins), and this relative independence suggests that it might be possible to modify the gonads by administration of sex steroids, as is the case in fish and amphibians, where steroids can induce a complete, functional gonadal sex-reversal. Steroids do not cause gonadal sex-reversal in eutherian mammals if administered to the mother, but partial female-to-male gonadal sex-reversal does occur in cattle freemartins when a female fetus shares a blood supply with its male co-twin.

Although early investigators in the United States of America realized the experimental advantages of marsupial pouch young, which are readily accessible for administration of hormones or surgical manipulation during several critical stages of sexual differentiation, the only indigenous marsupial in North America was the omnivorous Virginia opossum, *Didelphis virginiana*. Despite having multiple young (3–13), it is a seasonal breeder, and even by marsupial standards it has a very short gestation (13 days), and the newborn young weigh a mere 125 mg, too small for experimental convenience. After the Second World War, work on *Didelphis* virtually ceased, and our understanding of its basic endocrinology now lags far behind our understanding of that of some of the Australian marsupials.

The best understood of Australian marsupials is the tammar wallaby, *Macropus eugenii*, a medium-sized (5 kg) herbivorous macropodid now largely confined to islands off the south and south-west coasts of Australia. It is also a seasonal breeder, as befits its southerly latitude, normally giving birth to a single offspring, which weighs about 400 mg, approximately a month after the summer solstice, with pouch exit late in the following spring, about 9–10 months later. Much of what follows is based on the study of this species.

Anatomy and evolution of the mammalian urogenital system

In most classes of vertebrates, sperm produced in the testes are conducted to the outside by a duct system. This system consists of tubules that originate as the pronephric and mesonephric tubules of the embryonic kidney. In most anuran amphibians (frogs and toads) the Wolffian duct serves as both the urinary duct and the seminal duct. In reptiles, birds and mammals (all amniotes), the Wolffian duct has lost its connection with the mesonephros and has become the ductus deferens. The tubules of the mesonephros become incorporated into the epididymis. The mesonephros degenerates in the fetus or neonate and a metanephric kidney assumes all excretory functions.

It is the relative position of the urinary and genital ducts that most clearly separates the major groups of living mammals. The ureters of monotremes open into the urogenital sinus opposite the urethral openings of the bladder. In marsupial and eutherian mammals, the ureters migrate from a dorsal position adjacent to the Wolffian duct to a direct connection to the bladder, but the ureters pass medially between the genital duct in marsupials, and laterally in eutherians (Figure 9.1). This very minor difference in migration of the ducts had profound consequences for the development of the reproductive tract. These consequences are most dramatic for females, but the anatomical differences persist also in males.

In eutherian mammals, fusion of the posterior end of the paired oviducts forms a single vagina, but the vaginae remain paired in monotremes and marsupials. This is because the ureters of marsupials pass between and not around the developing oviduct, and so fusion into a single vagina is not possible. The retention of the paired lateral vaginae and the development of a third, median birth canal, results in the bizarre, tripartite arrangement of marsupials (Figures 9.1 and 9.2). In male marsupials there is another anatomical difference: the penis is posterior (caudal) to the scrotum.

Hormones and sexual differentiation

In mammals, all somatic sexual dimorphisms have been assumed to be a consequence of gonadal hormone action. The conventional view of

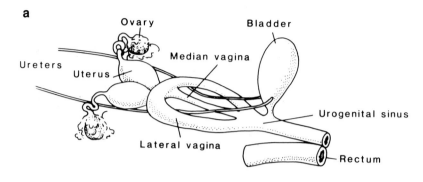

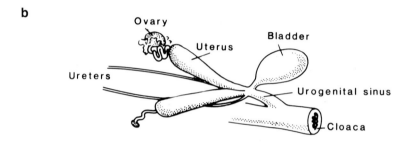

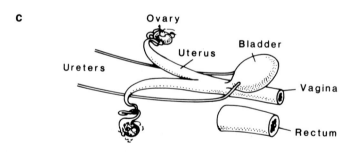

Figure 9.1. The anatomy of the urogenital systems of the three groups of living mammals (a) marsupial (b) monotreme and (c) eutherian. Note the entry of the ureters relative to the position of the bladder and of the genital ducts. In marsupials, fusion of the lateral vaginae to form a common vagina would occlude the ureters. (From Renfree, M. B. (1993) Ontogeny, genetic control, and phylogeny of female reproduction in monotreme and therian mammals. In *Mammal Phylogeny*, pp. 4–20. Eds. F. S. Szalay, M. J. Novacek and M. C. McKenna. Springer Verlag, New York.)

Figure 9.2. Anatomy of (a) the female and (b) the male reproductive system of the tammar wallaby, *Macropus eugenii*. (From Renfree, M. (1993), see Figure 9.1.)

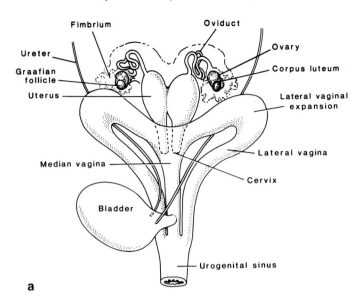

a

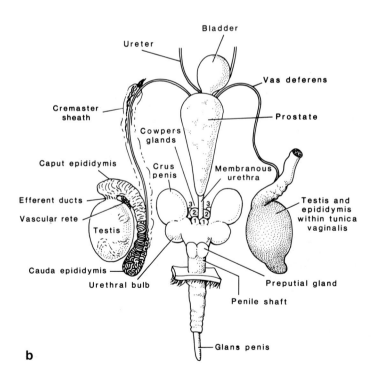

b

mammalian sexual differentiation has been that a gene or genes on the Y chromosome causes the indifferent gonad to develop into a testis, which then secretes two classes of hormone, the androgen testosterone, of Leydig cell origin, and the polypeptide Müllerian inhibitory substance (MIS), of Sertoli cell origin. These are responsible respectively for masculinizing the Wolffian duct derivatives and the external genitalia to form the male reproductive tract, and for bringing about atrophy of the Müllerian duct derivatives. In the absence of a Y chromosome, the indifferent gonad develops into an ovary which is endocrinologically quiescent to begin with, so that the Wolffian duct derivatives atrophy, the male external genitalia fail to develop, and the Müllerian ducts persist to form the female reproductive tract. Alfred Jost's extensive studies, in which testis grafts adjacent to ovaries produced a localized stimulation of the Wolffian duct and a localized regression of the Müllerian duct on the side of the graft, provided compelling evidence of the paramount importance of hormones of the testis in inducing maleness (see Chapter 8 by Jean Wilson). The marsupial studies did not completely fit with this paradigm, and the early workers, so persuaded by the idea of omnipotence of hormones, failed to realize the significance of their results. Recent renewed interest in marsupial sexual differentiation has led to some surprising findings which have begun to explain these apparent anomalies.

Hormone-dependent and hormone-independent sexual dimorphisms in marsupials

Testicular and ovarian structure

Marsupials, like eutherians, normally require the presence of a Y chromosome for testicular formation. A male-specific homologue of the putative testis-determining gene SRY (sex-determining region of the Y chromosome) has been identified recently in marsupials (see Chapter 18 by Jennifer Graves), but as yet the products of this gene have not been isolated. The testis differentiates during late fetal and early neonatal life. The genital ridge, which becomes the gonad, forms on the medial side of the mesonephros, the functional kidney of marsupial neonates. Germ cells migrate to the genital ridge before and during differentiation of the gonad into a testis or ovary. At day 21 of the $26\frac{1}{2}$-day gestation period in the tammar wallaby the genital ridge consists of an undifferentiated gonadal primordium. It is not until 23–24 days of gestation that the blastemal core is organized into early gonadal cords.

In male gonads there is a hint of an increased cytoplasmic/nuclear ratio

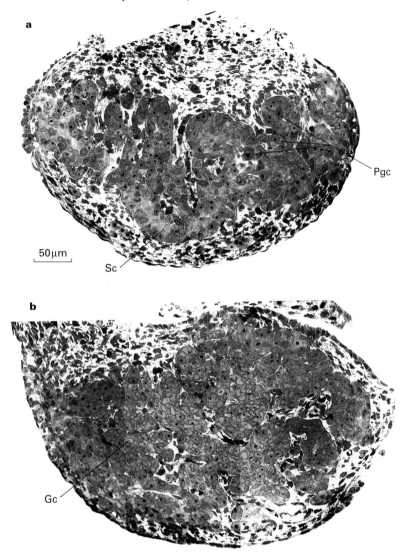

Figure 9.3. Ultrastructure of the gonads of developing marsupials. Representative 1 micron plastic sections of the gonads of tammar pouch young on day 2 postpartum. In (a) the male gonad, the cords form long hairpin loops, characteristic of the testis, and many of the Sertoli cell nuclei within the cords have a basal or peripheral nucleus. By contrast, in (b) the female gonad, the cords are more convoluted and not so well defined, and the cell nuclei are arranged randomly. The capsule under the surface epithelium which will form the tunica albuginea in the male is generally thicker than the equivalent region in the female. Abbreviations: Pgc, primordial germ cells; Sc, seminiferous cords; Gc, gonadal cords. (Scale bar = 50 μm.)

in a few selected areas. By 25 days of gestation in males, the nucleus of the pre-Sertoli cells becomes more peripheral, whereas in females it remains central. At birth, the male gonad has features characteristic of a testis, with a multi-layered tunica albuginea, and a parenchyma consisting of branching seminiferous cords in which primitive Sertoli cells are often positioned perpendicular to the limiting basal lamina. By day 2 post-partum the testes have clear seminiferous cords, with putative dark Leydig cells frequently appearing in the interstitium (Figure 9.3). In the female, ovarian development lags behind that of the testis. Gonadal cords are also present, branching with cells oriented randomly throughout each cord. The cytoplasmic/nuclear ratio is markedly less than in males. There is a gradual orientation of the cells to form the ovarian cortex by about 8 days postpartum.

This sequence of events differs from that of the mouse and rat not only in the timing, but also in the fact that in eutherians the scrotum, labia and mammary glands are sensitive to the presence or absence of testicular hormones and differentiate accordingly. In marsupials there are at least four sexually dimorphic characteristics that differentiate *before* this process of gonadal differentiation is complete, namely the scrotum, the mammary anlagen, the gubernaculum and the processus vaginalis. These structures, together with the later development of the pouch, appear to be under direct genetic control, rather than secondary hormonal control as in eutherians (see Figure 9.8).

Gonadal hormone production

Scrotal bulges and mammary anlagen appear in the tammar fetus four to five days before birth, at the time when the gonadal ridge is clearly undifferentiated and consists only of two to three cell layers (see Table 9.1 and above). The scrotum arises as two laterally placed bulges which do not fuse centrally until 9 days postpartum (Figure 9.4). In the tammar at the light microscope level, the gonad does not develop clear testis cords until 2 days postpartum (Figure 9.3). This is in contrast to the didelphids, *Didelphis virginiana* (the North American opossum) and *Monodelphis domestica* (the South American grey short-tailed opossum) in which seminiferous cords are seen in males on the day of birth. On the day of birth, the mean volume of the gonads of the tammar is not significantly different in males compared to females, but it does differ in the grey short-tailed opossum, although there appears to be more variation in neonatal body weights of opossums than tammars. In both

Table 9.1 *Stages in the differentiation of sex in the tammar wallaby,*
Macropus eugenii

Stage (days relative to birth)	Female	Male
−4 (fetus)	Indifferent gonadal ridge Mammary anlage is visible	Indifferent gonadal ridge Scrotal anlage is visible
−1 (fetus)		Sertoli cells with greater nuclear cytoplasmic ratio than in ovaries at electron microscope level
0 (newborn)	Ovary is undifferentiated at light microscope level	Testis is undifferentiated at light microscope level
2		Testis cords differentiated
7	Ovarian cortex is distinct	Müllerian duct regression has commenced
8	Pouch is visible	Scrotal sacs fuse medially
9–10		Onset of transabdominal testicular migration
25	Advanced regression of Wolffian ducts	Testes are inguinal Prostate buds start to form Müllerian ducts are vestigial
50–60		Testicular descent is complete

Source: From Renfree *et al.* (1992).

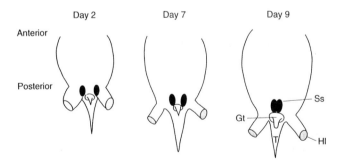

Figure 9.4. The relative movement of the two separate scrotal bulges in a neonatal tammar to the final, fused scrotal sac by 9 days postpartum. Testicular descent to the scrotum is not complete until around 60 days postpartum. Note that the scrotum is cranial to the phallus. Abbreviations: Gt, genital tubercle; Hl, hind limb; Ss, scrotal swellings; T, tail.

American and Australian marsupials, however, it appears that there is no gonadal hormone production prenatally.

The gonads of pouch young of the North American opossum aged less than 10 days can synthesize progesterone, and the testes produce testosterone several days before the virilization of the Wolffian ducts, urogenital sinus and genital tubercle. Furthermore, the opossum scrotum has very low 5α-reductase activity, which is consistent with the idea that its differentiation is not androgen dependent. Similarly, testes of tammars aged 5–10 days contain significant amounts (0.9 ng/mg protein) of testosterone, but neither testes nor ovaries of tammars aged 0–4 days have appreciable testosterone (0.2 ng/mg protein) (Figure 9.5). Testosterone remains high for 40 days after birth when the male genitalia are differentiating, then falls to values similar to that seen in ovaries. As in the opossum, the tammar scrotum has little 5α-reductase activity, in contrast to relatively high activity in the phallus, prostate and urogenital sinus of both species. Thus, in both species the differentiation of the scrotum precedes significant testosterone synthesis.

Effects of exogenous steroids

Administration of androgens has no influence on the ovary of female neonatal marsupials, but oestrogens can disrupt the normal development of the testis. Burns, using massive doses of hormones of questionable purity, showed that development of neonatal gonads in most cases was unaffected by treatment with testosterone. However, male opossums which were treated repeatedly with 0.2–0.3 μg oestradiol dipropionate every two to three days from birth for 20 to 30 days, underwent varying degrees of gonadal sex reversal, depending on the dose and duration of treatment. In some cases the reversal was so complete that the resultant ovotestis was almost indistinguishable from a normal ovary. Treatment of male young, when initiated several days after birth, was without effect.

This remarkable result has been studied further using the grey opossum, *Monodelphis domestica*. Treatment of genetic male young on days 1 and 3 postpartum with oestradiol led to complete suppression of testicular development, as determined at the time of autopsy 22 weeks later, but there was no transformation into an ovary. In these oestrogenized males, the internal and external genitalia were completely female, although the animals retained their scrota. Likewise, treatment of *Didelphis* pouch young with testosterone led to retention of the Wolffian duct in

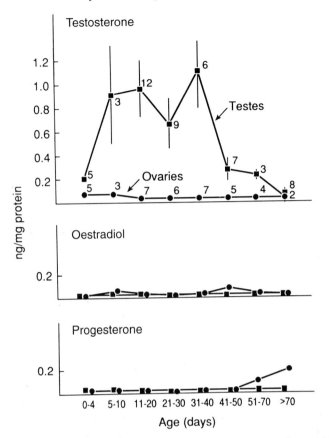

Figure 9.5. Gonadal steroid hormone content as a function of age. The steroid hormones indicated were measured by radioimmunoassays. The numbers on the figure indicate the number of individual pools of ovaries (filled circles) or testes (filled squares). This analysis omits two ovarian oestrogen samples that were thought to be high due to contamination. The bars represent standard errors. As measured by a Welch approximation to the *t*-test, testes have significantly higher testosterone content at ages between 11 and 70 days of pouch life. (From Renfree *et al.*, 1992.)

females, and development of an epididymis. Paradoxically, treatment with oestradiol induced hypertrophy of the Wolffian duct in both sexes. Treatment of neonatal male *Monodelphis* with testosterone, or treatment of females with oestradiol, was apparently without effect on gonadal or genital differentiation. However, neither mammary anlagen nor scrotal bulges are influenced by exogenous hormones, and, as they clearly differentiate before the gonads produce endogenous hormones, they appear to be hormone-independent structures.

Less dramatic effects of exogenous oestrogen on testicular development occur in the tammar. Oestrogen inhibited testicular development and there was no real transformation into an ovary. Feminization of the internal genitalia did not occur. In female tammar pouch young treated with testosterone, prostatic buds were induced in the urogenital sinus and the Wolffian duct was hypertrophied, as Burns had found in the opossum. Similarly, adult female tammars, which had received testis grafts at 10 days postpartum, developed normal prostates. These animals had the normal female development of the pouch and mammary glands, and of Müllerian duct derivatives (the vaginal complex, uteri and the oviducts). Ovarian follicles and oocytes were reduced in number or were absent in the ovaries of these animals. In addition, however, Wolffian duct derivatives were retained, and a prostate and penis developed as in normal males. Castration of male tammar wallaby pouch young at 10 days postpartum resulted in poorly developed internal genitalia, although the scrotum and the gubernaculum developed normally. Thus, normal masculinization of the marsupial Wolffian duct is dependent on androgens, and Müllerian regression is probably dependent on MIS production by the developing testis. MIS and androgens are also presumably essential for testicular migration and descent.

In the tammar, the gubernaculum in male neonates is long and thin, and together with the deep processus vaginalis it reaches the inguinal region between the pubis and the base of the scrotal anlagen (Figures 9.6 and 9.7). In females, the gubernaculum is short and thick, entering a shallow processus vaginalis which reaches only as far as the abdominal traversus muscle (Figure 9.7). Thus, it is sexually dimorphic before gonadal hormone production begins. However, the gubernaculum appears to be under both genetic and hormonal control. Although treatment of females with testosterone for 25 days did not induce ovarian descent, and the gubernaculum remained merged with the abdominal wall without a distinguishable processus vaginalis as in normal females, oestradiol treatment of males affected the penetration of the gubernaculum into the centre of the scrotum, and instead it terminated outside the scrotum and inhibited testicular descent. Oestrogen treatment of male fetuses also inhibits testicular descent in eutherians. It is not yet clear how testicular descent is controlled in marsupials, but it is clear that androgens alone do not mediate transabdominal migration and gubernacular penetration of the scrotum.

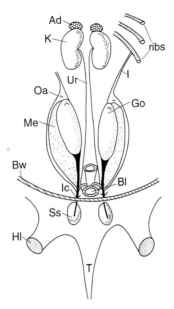

Figure 9.6. Anatomy of the pouch young of a male tammar 10 days postpartum. The testes are not yet descended into the scrotal sacs; the mesonephros, although still large, is no longer the primary excretory organ; and the metanephros has moved to its final (cranial) location. The testes are already rounded up. Abbreviations: Ad, adrenal; Bw, body wall; Bl, bladder; Go, gonad; Hl, hind limb; Ic, internal inguinal canal; K, kidney; L, anterior ligament; Me, mesonephros; Oa, ostium abdominale; Ss, scrotal sac; T, tail; Ur, ureter.

The genital tubercle of marsupials is also sensitive to androgen as predicted from the Jost model. Testosterone stimulates penile development in the female, and oestradiol inhibits it, presumably because of the testicular inhibition. It is remarkable that neither androgen nor oestrogen have any effect on pouch or scrotal development even in massive doses. In other words, marsupials obey the hormone-dependence rule for all Wolffian and Müllerian duct derivatives, and the prostate and phallus, but the gubernaculum, processes vaginalis, scrotum, pouch and mammary anlagen are clearly hormone-independent structures (see Figure 9.8).

Scrotum, pouch and mammary anlagen in intersexual marsupials

Further evidence in support of the hormonal independence of the scrotum, mammary anlagen, pouch and gubernaculum comes from Geoff Sharman's studies of spontaneously occurring intersexes in several Australian species of marsupial. One tammar was Klinefelter-like with a 17, XXY karyotype

Fig. 9.7. Scanning electron micrograph of the urogenital organs of a tammar (a) female and (b) male pouch young, about 10 days postpartum. Note the differing shapes of the ovary (fusiform) and testis (rounded). The gubernaculum in the male induces a 'W' shape in the Wolffian duct as it passes through the internal inguinal canal. Abbreviations: Bl, bladder; Co, colon; Gu, gubernaculum; Ic, internal inguinal canal; Md, Müllerian duct; Me, mesonephros; Ov, ovary; T, testis; Wd, Wolffian duct.

Table 9.2 *Phenotypic characteristics of intersex marsupials*

XO	XXY or XY
Ovaries or streak gonads	Intra-abdominal testes
Müllerian ducts present	Müllerian ducts absent
Wolffian ducts absent	Wolffian ducts present
No penis	Penis developed
	Pouch and mammary glands
Scrotum present	No scrotum

Source: Adapted from Shaw *et al.* (1990).

(normal diploid number is 16) and had intra-abdominal testes, complete masculinization of the male reproductive tract internally and a normal penis, but externally there was a pouch, mammary glands and no scrotum. A similar picture was found in two XY individuals. On the other hand, a 15, XO individual (akin to Turner's syndrome) had hypoplastic ovaries, normal development of the female reproductive tract internally, but an empty scrotum (Table 9.2). Thus, the scrotum can develop in the absence of a testis, whereas the pouch and mammary glands can develop in the presence of one.

Now, it appears that the differentiation of the scrotum and pouch depends on an X dosage mechanism (Figure 9.8). Scrotal development occurs when only a single X chromosome is present (XY, XO), whilst two X chromosomes are necessary for pouch formation (XX, XXY). Although marsupial scrotal development is determined genetically, whereas eutherian scrotal development is under hormonal control, it seems unlikely that the marsupial scrotum has evolved completely independently from the eutherian scrotum. Since adult marsupials have a single urogenital opening which is identical in appearance for both males and females, there are no distinctive female external genitalia, and hence no female genital homologue of the scrotum. However, the eutherian scrotum arises during embryogenesis from the labio-scrotal swelling. Thus, the labia majora of the vulva are homologous with the scrotum, and in the presence of androgen a scrotum will form in a genetic female. In marsupials, the scrotum arises from the abdominal cavity cranial to the phallus (Figure 9.4), and androgen or oestrogen treatment has no influence on its differentiation. It has been suggested that the pouch is the female homologue of the scrotum, but it seems to be more likely that the pouch and scrotum arise from different primordia in the same morphogenetic field.

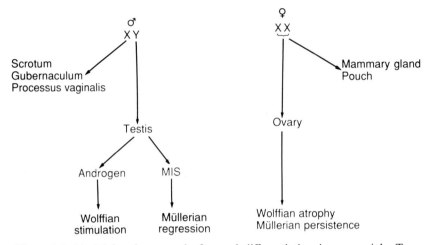

Figure 9.8. Model for the control of sexual differentiation in marsupials. Two X-chromosomes are necessary to code for a mammary gland, pouch and ovary, whereas a single X-chromosome codes for the scrotum, gubernaculum and processus vaginalis. (Redrawn from Shaw *et al.*, 1990.)

Mammary glands in male marsupials

Many male eutherian mammals have mammary glands, and retain the ability to lactate if provided with the appropriate hormonal stimulus. Although a large number of Australian marsupials have been surveyed, there is no evidence of mammary or teat tissue in the male pouch young, juveniles or adults. By contrast, there are mammary anlagen in male didelphid pouch young. Mammary gland rudiments are found in fetuses and newborns of both sexes of four species of American opossums. These rudiments remain for at least 60 days postpartum. The reason for this dichotomy is unknown, but there appears to be a fundamental distinction between American marsupials, which have mammary anlagen in both sexes, and Australian marsupials, in which no evidence of mammary anlagen can be found in males. Mammary anlagen are not formed in male tammars at any stage of gestation or during pouch life, even after prolonged oestrogen treatment of male pouch young. There is evidence that the same is probably true for males of the Australian dasyurid, *Antechinus stuartii*, the brown marsupial mouse. By contrast, in the North American opossum and the South American grey opossum, mammary anlagen are clearly discernible in males and females on the day of birth. However, males of both species have less than one-third of the full female complement of mammary anlagen, always located in the most cranial positions. Scrotal anlagen were present in all male, but no female, neonates sampled on the day of birth of these two species, and they were always caudal to the normal site of the mammary anlagen in the female. At no stage during postnatal life do males of either opossum have a full female complement of mammary primordia. Males of both species have primordia in positions equivalent to the more anterior of those in females. The possession of mammary anlagen by these didelphid males suggests that, in this aspect of sexual differentiation, the Australian and American marsupials have evolved separately.

Conclusions

Although the sequence of events in marsupial sexual differentiation is similar to that in eutherian mammals, there appear to be a number of important differences. The results of hormonal administration to pouch young, and a study of spontaneous intersexes, make it clear that androgen is apparently incapable of stimulating scrotal development or of inhibiting pouch or mammary development in females. Likewise oestrogen is apparently incapable of inhibiting scrotal development or inducing pouch or mammary development in males. In eutherians, on the other hand, scrotal development and inhibition of mammary development are all thought to be androgen-dependent events. In marsupials, development of the teat, pouch, mammary gland and scrotum are not under hormonal control, and instead appear to be under direct genetic control. Undoubtedly, Burns induced transformation of the indifferent gonads of male Virginia opossums with oestrogen, but comparable results have not been obtained in the tammar and grey opossum. The abnormal gonads of oestrogenized male marsupials deserve further study, especially with respect to the fate of the germ cells.

Finally, there is the age-old question about the embryological origin of the marsupial's pouch. Pouched marsupials are thought to have evolved from pouchless forms, so it would appear that the pouch is of more recent origin than the scrotum. This hypothesis makes it improbable that the pouch and scrotum have evolved as strictly homologous structures, unlike the labia majora and scrota of eutherians. It seems safer to conclude that the pouch and scrotum may share a partial homology, arising from different parts of the same morphogenetic field. While both mammary anlagen and scrotum may be present simultaneously in the pouch young of some species (such as *Didelphis*), in adult animals, the pouch and scrotum appear to be mutually exclusive. However, in rare cases of marsupial bilateral gynandromorphs the animals have a hemiscrotum on one side of the body, and a hemi-pouch and mammary gland on the other!

Marsupials and eutherians have evolved separately for the last 100 million years following the breakup of Gondwana, and although the basic mechanisms involved in sex differentiation have remained the same in both groups, marsupials seem to have given more emphasis to genetic rather than hormonal mechanisms for controlling some of their sexual dimorphisms. Whether this is a mere accident, or is of adaptive significance, remains to be determined.

Further reading

Cooper, D. W. (1993) The evolution of sex determination, sex chromosome dimorphism, and X-inactivation in therian mammals: a comparison of metatherians (marsupials) and eutherians ('placentals'). In *Sex Chromosomes and Sex Determining Genes*, pp. 183–98. Eds. K. C. Reed and J. A. M. Graves. Harwood Academic Publishers, Chur, Switzerland.

Jost, A. (1970) Hormonal factors in the sex differentiation of the mammalian foetus. *Philosophical Transactions of the Royal Society of London*, **B259**, 119–30.

O. W.-S., Short, R. V., Renfree, M. B. and Shaw, G. (1988) Primary genetic control of sexual differentiation in a mammal. *Nature*, **331**, 716–17.

Renfree, M. B. (1992) The role of genes and hormones in marsupial sexual differentiation. *Journal of Zoology*, **226**, 165–73.

Renfree, M. B. and Short, R. V. (1988) Sex determination in marsupials: evidence for a marsupial-eutherian dichotomy. *Philosophical Transactions of the Royal Society of London*, **B322**, 41–54.

Renfree, M. B., Wilson, J. D., Short, R. V., Shaw, G. and George, F. W. (1992) Steroid hormone content of the gonads of the tammar wallaby during sexual differentiation. *Biology of Reproduction*, **47**, 644–7.

Sharman, G. B., Hughes, R. L. and Cooper, D. W. (1990) The chromosomal basis of sex differentiation in marsupials. *Australian Journal of Zoology*, **37**, 451–66.

Shaw, G., Renfree, M. B. and Short, R. V. (1990) Primary genetic control of sexual differentiation in marsupials. *Australian Journal of Zoology*, **37**, 443–50.

Shaw, G. S., Renfree, M. B., Short, R. V. and O, W.-S. (1988) Experimental manipulation of sexual differentiation in wallaby pouch young with exogenous steroids. *Development*, **104**, 689–701.

Tyndale-Biscoe, C. H. and Renfree, M. B. (1987) *Reproductive Physiology of Marsupials*. Cambridge University Press.

10

Regulation of sexual dimorphism in rat liver

JAN-ÅKE GUSTAFSSON

Although it may not be generally recognized amongst scientists interested
in sex differentiation, many of the metabolic events in rodent liver are
sexually dimorphic and their regulation occurs via a sophisticated interplay between hypothalamic, pituitary and gonadal hormones. In fact, this
system for control of sexually differentiated liver functions constitutes a
novel endocrine system in its own right, namely the hypothalamo-
pituitary-liver axis.

Sex differences in hepatic steroid metabolism

Our interest in the sexually dimorphic liver metabolism in rats originated
in the late 1960s, when we performed gas-chromatographic–mass-
spectrometric analysis of steroid profiles in faeces and urine from germ-
free and conventional rats. During this work, which defined a plethora
of previously unknown steroid metabolites, we were impressed with
the very significant sex differences in the steroid excretory profiles,
particularly with regard to corticosterone metabolites. Thus, the pre-
dominating metabolites excreted by female rats, 3α, 11β, 15β, 21-tetra-
hydroxy-5α-pregnan-20-one and 3α, 15β, 21-trihydroxy-5α-pregnane-11,20-
dione, excreted as monosulphates and disulphates, were practically absent
from excreta of male rats. These metabolites are formed by 15β-hydroxy-
lation of corticosterone derivatives and, later, we showed that sulphury-
lation of the 21-hydroxyl group precedes the 15β-hydroxylation. We
became interested in regulation of these sex differences in steroid metabolism with the idea that such a system could be used for understanding
the mechanism for the action of sex steroids.

Later, it was possible to show that sexual dimorphism in steroid
excretory profiles was mainly due to sex differences in liver metabolism.

231

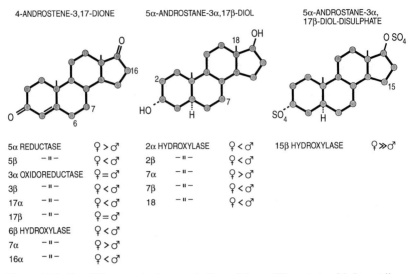

5α REDUCTASE	♀ > ♂	2α HYDROXYLASE	♀ < ♂	15β HYDROXYLASE	♀ ≫ ♂
5β — " —	♀ < ♂	2β — " —	♀ < ♂		
3α OXIDOREDUCTASE	♀ = ♂	7α — " —	♀ > ♂		
3β — " —	♀ < ♂	7β — " —	♀ < ♂		
17α — " —	♀ < ♂	18 — " —	♀ < ♂		
17β — " —	♀ = ♂				
6β HYDROXYLASE	♀ < ♂				
7α — " —	♀ > ♂				
16α — " —	♀ < ♂				

Figure 10.1. Sex differences in the metabolism of three different steroids by rat liver.

In particular, the microsomal fraction from liver homogenates was very active in metabolizing *in vitro* a variety of steroid substrates. Usually, male rats were more efficient in hydroxylating neutral steroids such as 4-androstene-3,17-dione and 5α-androstane-3α,17β-diol. However, when 5α-androstane-3α,17β-diol-disulphate was added to the incubation mixture, we observed a very specific 15β-hydroxylation in female rat liver microsomes (Figure 10.1). We could show that the enzyme was of the cytochrome P450 type and that it was active in the liver of both sexes during a brief period in 3–4-week-old rats but that at the time of puberty it disappeared in male rats and increased in activity in female rats.

During these years at the beginning of the 1970s, we also made the interesting observation that the sexually dimorphic liver metabolism was imprintable by neonatal androgen, in much the same way as sexual behaviour can be masculinized by single doses of androgens administered to rats during the first few days of life. During the first days of life, administration of one dose of testosterone propionate to neonatally gonadectomized male rats resulted in complete abolishment of 15β-hydroxylase activity in adult rats, whereas neonatally gonadectomized male control rats that were given only the vehicle displayed female levels of 15β-hydroxylase activity in the adult period. This is a case of 'negative' imprinting; other P450-dependent hydroxylases which are male-specific in adult rats (e.g. 4-androstene-3,17-dione 16α-hydroxylase) are 'positively' imprinted, i.e. they are turned on in neonatally gonadectomized male or female rats who have been given one single dose of testosterone propionate during the first few days of life.

Pituitary control of hepatic steroid metabolism

We also made another interesting observation: when the pituitary gland was removed from female rats their livers turned 'masculine' with regard to steroid metabolizing enzyme activities. The corresponding operation in male rats had little effect on liver metabolism. Based on these observations and on experiments involving transplanted pituitaries below the kidney capsule as well as pituitary stalk transections and various lesions in the brain we suggested that the female pituitary secretes a factor, the 'feminizing factor', which feminizes a basically masculine type of liver metabolism. According to this concept, the liver itself, without influence from the pituitary (as in hypophysectomized rats), shows a constitutive male-type metabolism. A similar situation exists in male rats where the secretion of the feminizing factor from the pituitary is inhibited by a feminizing factor inhibiting factor, turned on by neonatal androgenic imprinting (Figure 10.2). We spent a considerable time attempting to identify the feminizing factor and could finally show, by fractionating rat pituitary glands and infusing each of the fractions separately into hypophysectomized male or female rats via osmotic minipumps, that the feminizing factor activity cochromatographed with growth hormone. Using reference preparations of pituitary hormones, we could show that this effect of growth hormone was associated entirely with the somatogenic properties of growth hormone, whereas lactogenic principles such as prolactin had no regulatory function on liver metabolism.

Sex steroids administered to rats affect liver metabolism, so that male sex-hormones masculinize a female liver whereas female sex-hormones feminize a male rat liver. These effects, however, are seen in intact animals only and are eliminated completely in hypophysectomized rats which have a masculinized liver metabolism independent of sex hormones. The simplest interpretation of these data is that sex steroids affect the hypothalamo-pituitary axis, modulating secretion of growth hormone in such a way that liver metabolism is changed (Figure 10.3).

At this point in time it became evident that growth hormone conveys signals from sexually differentiated brain centers to the liver so that hepatic metabolism also becomes sexually dimorphic. The explanation is that the growth-hormone secretory pattern in itself is sexually differentiated.

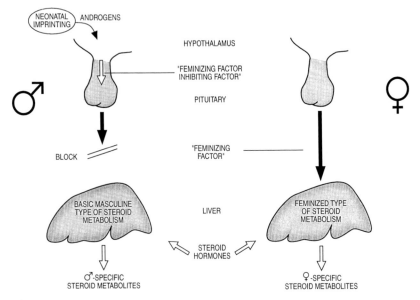

Figure 10.2. Original concept of a pituitary feminizing factor (FF) and a hypothalamic feminizing factor inhibiting factor that is in control of sex-dependent liver metabolism.

In male rats, growth-hormone levels in blood are characterized by high peaks occurring with a frequency of about three to four hours and low troughs during which the growth-hormone level decreases to zero. In female rats, on the other hand, the secretory pattern is more irregular and rather oscillates around a mean value showing much lower peaks and more shallow troughs than in male rats. The female growth-hormone secretory pattern can be mimicked by infusion of growth hormone in osmotic minipumps. When growth hormone is administered in this way to hypophysectomized rats their liver metabolism becomes feminine. On the other hand, intermittent injections of growth hormone to hypophysectomized rats turn on the male-type enzyme activities. Apparently, the organism has preferred to use one hormonal principle which the target tissue may recognize in two different modes rather than to exert regulation via two distinct balancing factors. According to this notion, female sex-hormones modulate the growth-hormone secretory pattern in male rats so that hepatic metabolism becomes feminine and vice versa (Figure 10.4). Furthermore, we were able to show that neonatal imprinting by androgen as described above did not affect the liver *per se* but changed the growth-hormone secretory pattern so that

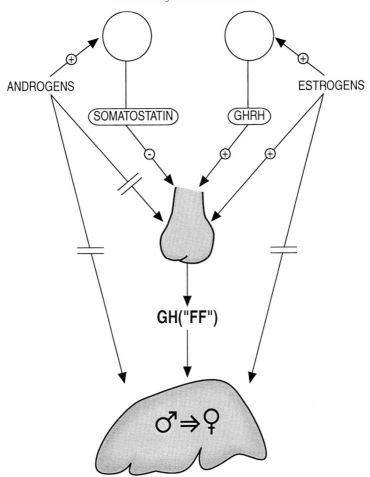

Figure 10.3. Sex steroids affect liver metabolism via an effect on the central nervous system resulting in changed secretion of growth hormone from the pituitary. (Broken lines indicate no effect. GH, growth hormone; GHRH, growth hormone releasing hormone; FF, feminizing factor.)

it became more male-like and, accordingly, the liver became masculinized in adult animals.

The role of growth hormone in hepatic sexual differentiation

Since it has been demonstrated that the secretory pattern of growth hormone is of importance for maintaining sexual dimorphism in the rat's

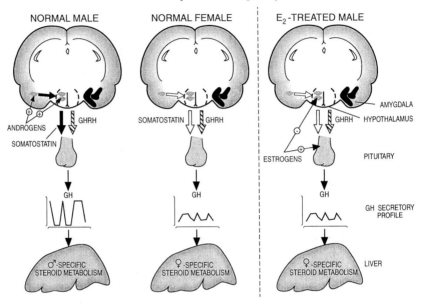

Figure 10.4. Sex steroids affect liver metabolism via a modulation of the sex-dependent growth-hormone secretory pattern (see text for further explanation). (GH, growth hormone; GHRH, growth hormone releasing hormone.)

liver, we tried next to understand how this hormone worked, and paid particular attention to the opposing effects of different secretory patterns of this hormone. In order to obtain molecular tools to allow a detailed dissection of the mechanisms involved, we cloned several of the major sexually dimorphic isozymes of cytochrome P450 in rat liver. These include the male-specific P450 2C11 (16α-hydroxylase) and 2C13 (6β-hydroxylase) as well as the female-specific 2C12 (15β-hydroxylase) and the female-predominant 2C7. We obtained cDNAs and genomic clones, and developed assays for measurement of mRNA levels of these enzymes following various endocrine manipulations. Using a solution hybridization assay, we showed that P450 2C11 and 2C13 mRNA started to appear in male rat livers at the time of puberty, whereas no corresponding messages were seen in female rats. Conversely, P450 2C12 mRNA could not be detected in male rat livers and started to appear in female rat livers at the time of puberty. Interestingly, dwarf rats, which have a deficient secretion of growth hormone and show a severe growth retardation, were sexually dimorphic in their livers, demonstrating female-specific levels of 2C12 mRNA and male-specific levels of 2C11 mRNA, respectively. Careful measurements of their growth-hormone secretory patterns utilizing a very sensitive radioimmunoassay technique showed that, although their growth-hormone levels in general are about two orders of magnitude lower than in normal rats, they do show a sexually dimorphic growth-hormone secretory pattern of principally the same type as in normal rats. Apparently, very low levels of growth hormone are sufficient to regulate P450 levels in the liver. This is consistent with the previous findings that a pituitary transplanted under the kidney capsule (which is known to secrete very small amounts of growth hormone) in intact male rats or in hypophysectomized rats was capable of completely feminizing liver metabolism in recipient animals.

Further studies using run-on assays indicated that the level of sexual dimorphism was at the rate of initiation of transcription. Thus the 2C11 and 2C13 genes were transcribed only in male rat livers, whereas the 2C12 gene was transcribed only in female rats. The 2C7 gene was transcribed more efficiently in female rat livers than in those of male rat livers. Furthermore, utilizing hypophysectomized animals and administering growth hormone either via osmotic minipumps (female-type pattern) or by intermittent injections (male-type pattern), it was shown that female-type growth-hormone administration resulted in an increased rate of transcription of 2C12 as well as of 2C7 genes with no response of 2C11 or 2C13 genes. Intermittent injections of growth hormone, on the other hand, turned on transcription of 2C11 and 2C13 genes but did not affect the transcription of 2C12 or 2C7 genes. These results demonstrate clearly that growth hormone indeed has completely different effects on expression of one and the same gene depending upon how the target tissue is exposed to the hormone. It is interesting to speculate that secretory patterns of other peptide hormones may also be of great significance for their biological action in their target tissues.

Sexual differentiation of hepatocytes *in vitro*

In order to study the mechanism of the growth hormone's action under somewhat more controlled conditions than in whole animals we have made extensive use of primary hepatocytes in tissue culture. These cells are notoriously difficult to work with, since they lose their more differentiated functions very easily, e.g. expression of cytochrome P450 and hormonal responsiveness. Cultivating hepatocytes on a special type of collagen-containing matrix (matrigel), however, results in cells that are responsive to growth hormone, cells which we have used to study how the hormone acts. It was possible to obtain both time- and dose-dependent increases in cytochrome P450 2C12 mRNA as well as IGF-I mRNA, quantified by solution hybridization assays, following addition of growth hormone to the medium. Both messages reached steady state levels about 12 hours after adding growth hormone to the medium. Interestingly, addition of insulin to the medium enhanced the 2C12 mRNA stimulating effect of growth hormone. Using a growth-hormone-receptor cDNA probe it was shown that the effect of insulin was not at the level of growth-hormone-receptor mRNA, since this parameter was unaffected by addition of insulin. When, however, specific binding of growth hormone was measured, it was demonstrated clearly that insulin resulted in greatly enhanced growth-hormone-receptor protein levels. Consequently, insulin would seem to enhance growth-hormone effects on the liver through a posttranscriptional effect on the expression of growth-hormone receptor. Furthermore, the effect of growth hormone could be shown to be caused by the somatogenic properties of this hormone: prolactin had no effect on expression of 2C12. Other factors that were inefficient in inducing 2C12 mRNA in primary hepatocytes in culture were IGF-I and IGF-II, indicating that growth hormone regulates female-specific P450 through a mechanistic pathway separate from the classical one via IGF-I. Interestingly, cycloheximide blocked the effect of growth hormone on 2C12 mRNA, but not on IGF-I mRNA. This indicates that induction of 2C12 mRNA requires ongoing protein biosynthesis, a requirement not shown for the effect of growth hormone on IGF-I mRNA. This result emphasizes further the different mechanisms of growth-hormone action in the regulation of 2C12 and IGF-I, respectively.

As is obvious from the data presented above, it is possible to mimic the feminizing effect of growth hormone on liver enzymes by its addition to primary hepatocytes in culture. Attempts to mimic the male type of growth-hormone secretory pattern by alternate addition/withdrawal of growth hormone to the tissue culture medium have met with little success. The male-specific P450 2C11 and 2C13 messages in primary hepatocytes reach similar levels to those in livers of hypophysectomized rats after a few days in culture. Interestingly, continuous presence of growth hormone resulted in down regulation of 2C11 and 2C13 mRNA, mimicking the effects of growth hormone in animals *in vivo* where infusion of growth hormone via osmotic minipumps feminizes liver metabolism both in male rats and in hypophysectomized rats of both sexes.

When run-on assays were performed on hepatocytes in culture, growth hormone was shown to affect the rate of transcription of the regulated genes in the same way as in the liver *in vivo*. This indicates that the P450 genes under study can be used as model genes to study which transcription factors are involved in growth-hormone regulation of the genes. As discussed above, we have cloned genomic DNA for all the P450 genes under investigation and, for some time, we have been attempting to identify growth-hormone-responsive elements in the genes. This is done by placing various parts of the 5' flanking regions or introns of the genes in front of reporter genes, transfecting these constructs into primary hepatocytes and investigating whether any of the constructs would respond to growth hormone by increased reporter gene activity. So far this approach has not lead to identification of any distinct growth-hormone-responsive elements.

We are also taking another approach to understanding the mechanisms of growth-hormone action on the hepatocyte. Growth hormone that was added to hepatocytes was found to lead to an intracellular accumulation of diacylglycerol peaking about one hour after addition of the hormone. This indicated a possible involvement of protein kinase C (PKC) in the mechanism of action of growth hormone, a notion which was tested further by adding PKC-inhibiting concentrations of phorbol esters. Phorbol acetate was found to block growth-hormone effects on both 2C12 mRNA and IGF-I mRNA. Also staurosporine, another inhibitor of PKC which inhibits other kinases as well, was found to block the effects of growth hormone on 2C12 mRNA and IGF-I mRNA. Even though all these results would support an involvement of PKC, lower concentrations of phorbol myristyl acetate, which induces c-fos mRNA, were inefficient in enhancing the accumulation of 2C12 mRNA or IGF-I mRNA. Accordingly, PKC does not seem to be involved directly in growth-hormone effects on the genes but rather has a permissive role. In other experiments, protein kinase A (PKA) did not seem to be involved in growth-hormone induction of 2C12 mRNA whereas both 8-bromo-cAMP and forskolin induced IGF-I mRNA, possibly indicating some role of PKA in growth hormone regulation of IGF-I mRNA.

In summary, rat liver is sexually dimorphic with regard to several biochemical parameters including certain isozymes of cytochrome P450, the prolactin receptor, as well as many other proteins. The physiological role of a sexually differentiated liver is unclear. Similar sex differences have been described in mouse liver and sexually dimorphic liver characteristics have also been found in hamsters and even in humans. It has been speculated that certain effects of sex steroids on target tissues might be mediated, or at least enhanced, by growth hormone. Thus, the effects of anabolic steroids on muscle growth could possibly occur in conjunction with sex-steroid-controlled changes in the growth-hormone secretory pattern. Growth hormone is necessary for normal expression of the estrogen receptor in rat liver and, thus, for the estrogen responsiveness of rat liver. Hypophysectomized rats do not show the estrogen-induced increase in low density lipoprotein (LDL) receptor concentration seen in normal rats but, following substitution of growth hormone, hypophysectomized rats again respond to estrogen by an increase in the LDL receptor mRNA and protein. Interestingly, growth hormone was found to induce LDL receptors also in human beings.

Acknowledgements

This work was supported by a grant from the Swedish Medical Research Council (No. 03X–06807).

Further reading

Edén, S. (1979) Age- and sex-related differences in episodic growth hormone secretion in the rat. *Endocrinology*, **105**, 555–60.

Gustafsson, J.-Å. and Stenberg, Å. (1974) Irreversible androgenic programming at birth of microsomal and soluble rat enzymes active on 4-androstene-3,17-dione and 5α-androstane-3α,17β-diol. *Journal of Biological Chemistry*, **249**, 711–18.

Gustafsson, J.-Å., Gustafsson, S. A., Ingelman-Sundberg, M., Pousette, Å., Stenberg, Å. and Wrange, Ö. (1974) Sexual differentiation of hepatic steroid metabolism in the rat. *Journal of Steroid Biochemistry*, **5**, 855–9.

Gustafsson, J.-Å., Mode, A. and Norstedt, G. (1983) Central control of prolactin and estrogen receptors in rat liver – expression of a novel endocrine system, the hypothalamo-pituitary-liver axis. *Annual Review of Pharmacology and Toxicology*, **23**, 259–78.

Gustafsson, J.-Å., Mode., A., Norstedt, G. and Skett, P. (1983) Sex steroid changes in hepatic enzymes. *Annual Review of Physiology*, **45**, 51–60.

Guzelian, P. S., Li, D., Schuetz, E. G., Thomas, P., Levin, W., Mode, A. and Gustafsson, J.-Å. (1988) Sex change in cytochrome P-450 phenotype by growth hormone treatment of adult rat hepatocytes maintained in a culture system on matrigel. *Proceedings of the National Academy of Sciences, USA*, **85**, 9783–7.

Jansson, J.-O., Ekberg, S., Isaksson, O., Mode, A. and Gustafsson, J.-Å. (1985) Imprinting of growth hormone secretion, body growth and hepatic steroid metabolism by neonatal testosterone. *Endocrinology*, **106**, 306–16.

Legraverend, C., Mode, A., Wells, T., Robinson, I. and Gustafsson, J.-Å. (1992) Hepatic steroid hydroxylating enzymes are controlled by the sexually dimorphic pattern of growth hormone secretion in normal and dwarf rats. *FASEB*, **6**, 711–18.

Legraverend, C., Mode, A., Westin, S., Ström, A., Eguchi, H., Zaphiropoulos, P. G. and Gustafsson, J.-Å. (1992) Transcriptional regulation of rat P-450 2C gene subfamily members by the sexually dimorphic pattern of growth hormone secretion. *Molecular Endocrinology*, **6**, 259–66.

Morgan, E. T., MacGeoch, C. and Gustafsson, J.-Å. (1985) Hormonal and developmental regulation of expression of the hepatic microsomal steroid 16α-hydroxylase cytochrome P-450 apoprotein in the rat. *Journal of Biological Chemistry*, **260**, 11895–8.

Tollet, P., Enberg, B. and Mode, A. (1990) Growth hormone (GH) regulation of cytochrome P-450 IIC12, insulin-like growth factor-I (IGF-I), and GH receptor messenger RNA expression in primary rat hepatocytes: a hormonal interplay with insulin, IGF-I, and thyroid hormone. *Molecular Endocrinology*, **4**, 1934–42.

Tollet, P., Legraverend, C., Gustafsson, J.-Å. and Mode, A. (1991) A role for protein kinases in the regulation of cytochrome P450 C12 and insulin-like growth factor I messenger RNA expression in primary adult rat hepatocytes. *Molecular Endocrinology*, **5**, 1351–8.

Zaphiropoulos, P. G., Mode, A., Norstedt, G. and Gustafsson, J.-Å. (1989) Regulation of sexual differentiation in drug and steroid metabolism. *Trends in Pharmacological Sciences*, **10**, 149–53.

11

Sex differences in sounds and their causes

EVAN BALABAN

It is early summer in the temperate zone, with concerts at dawn and at dusk. In the morning, a babble of sparrows, finches, starlings and warblers greets the early riser. The rich tapestry of sound woven around a late evening stroller is no less palpable: the eerie whine of katydids, the gentle chirping of crickets, the visceral croaking of frogs. These animal recitals bear a striking resemblance to many concert hall performances: both are predominantly the products of male organisms. While symphonic societies have seen the error of their sexist ways, and the sex ratios of performing groups have made significant progress towards equality in the last two decades, the population of animal sound *artistes* is unlikely to undergo a radical demographic shift in the near future. This is because the assignment of a performing role depends on evolutionary forces that work over an entirely different time-scale and by means that are different from those mediating employment decisions in musical organizations.

Bioacoustics (the study of animal sounds) is the field charged with elucidating nature's concert bookings. It is one of the most popular areas of modern behavioural biology for two good reasons. The first has to do with technology. It is difficult to take non-subjective behavioural measurements that are readily convertible into the numbers that scientists like dealing with. It is very easy to make measurements of sounds – much easier than quantifying movements or subtle variations in other behaviour. This is because sound waves can be converted readily into electrical signals by a microphone, and can be broken down into fundamental components (frequency, amplitude and phase) via specialized instruments such as frequency analysers, sound spectrographs and digital computers. These components can be assigned precise numerical values, and can even be manipulated and reassembled for use in behavioural experiments.

The second and more compelling reason is the pivotal role that sounds

243

play in the courtship and social behaviour of vast numbers of vertebrates and invertebrates. Sounds serve a fundamental role in mediating reproductive activity both in species where males and females lead rather solitary lives and gather together only for the purposes of mating (like the fruitfly *Drosophila melanogaster*), as well as in species with more complex forms of social organization (as in elephants, red deer, non-human primates and humans). This chapter introduces the diversity of sex differences in sound communication, and current enquiry regarding the causes of these differences. In the first section, we will explore two general patterns of sex difference in sound communication, with examples. The second section will consider whether these two general patterns are the result of very different biological processes, or are instead the product of a single underlying cause.

Patterns of bioacoustic sex differences

Animal communication sounds can be broadly thought of as falling into three different categories: sexual advertisement, warning and social communication. Sex differences in all three types of signals have been reported in different species, but most of the studies are focused on sexual advertisement signals. This is because they are technically the easiest to record – they are typically very loud. Advertisement sounds also have the additional advantage of evoking measurable behavioural responses in listeners.

Biological correlates of sex dimorphism in sound communication are of two types. The examples that tend to get the most attention from scientists are those in which sex differences in sound are mirrored by spectacular structural differences in male and female communicating apparatus (*overt dimorphisms*), as seen in the loud clicks or whines of male cicadas. Yet a similar number of dramatic sex dimorphisms in sound production occur without overt differences in male and female morphology (*covert dimorphisms*), as in fruitfly courtship song. Both types of dimorphism can be found in animals from every major class of vertebrates (except hagfishes, lampreys and sharks) and in most major groups of insects.

The fruitfly and the cicada: equivalent sound dimorphisms achieved by different means

While we usually think of flies as creatures living in a chemical world, sounds have been shown to play a vital role in fly courtship. They are

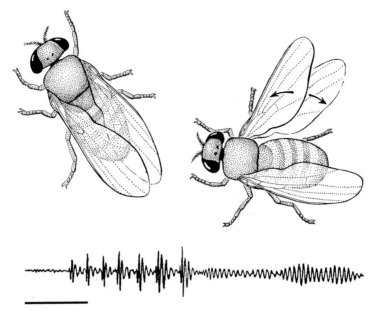

Figure 11.1. (Top) A male fruitfly (*Drosophila melanogaster*; right) courts a female with song. Note the male's wing (extended at a right angle to his body) which is producing the acoustic signal. (Based on Bennet-Clark, H. and Ewing, A. (1970) The love-song of the fruitfly. *Scientific American*, **223**, 84.) (Bottom) A representation of the variation in sound amplitude with time, called an oscillogram. (Time bar = 0.1 s.) (Modified from Wheeler, D., Kulkarni, S., Gailey, D. and Hall, J. (1988) Spectral analysis of *Drosophila* courtship songs. *Behavior Genetics*, **18**, 697.)

produced when the male holds a wing out to the side of his body and vibrates it with a characteristic temporal pattern (Figure 11.1). The fact that flies do not have ear-like structures anywhere on their bodies (unlike other insects such as crickets) actually led early investigators to conclude that these 'wing-waving' displays were visual and not acoustic. The feathery antennae of flies, called aristae, turn out to be the sound receiving organs. The male tries to orient its body so that the sound energy from wingbeating will be loudest at the female's head. Recording the sounds that male flies make by beating their tissue-like wings was at first a challenging technical problem. One set of researchers constructed a heavily insulated underground recording bunker to record the faint buzzing noises with conventional microphones. However, if sound is recorded not as variation in the air pressure impinging on a conventional microphone, but rather as variations in the velocity of the air molecules produced by the beating wings (as recorded by a special transducer called a displacement or ribbon microphone) then the sounds can be picked up quite well in a normal room. The standard situation for studying fly courtship consists of a small plexiglass chamber containing male and female flies paced on top of a displacement microphone.

Usually, female flies are not silent throughout this process. There are sets of sounds they give to reject the male's advances. In addition, both sexes make similar sounds in a variety of other contexts. Females do not produce male courtship sounds in spite of the fact that male and female wing musculature is very similar. While no detailed studies have yet been conducted to ascertain whether there are subtle male–female differences in the relative mass of wing muscles used intensively in male courtship song, there is no apparent reason why females should not be able to produce male song.

The calling system of the Australian bladder cicada (*Cytostoma saundersii*) stands in stark contrast to that of the flies. Males have a special sound generator called a tymbal, a cuticular structure formed from the modified first abdominal segment (Figure 11.2). Females do not have this structure; while the development of cicada tymbals has not been studied in detail, older descriptive work suggests that males and females start out with a similar abdominal plate which is elaborated during male development to become the tymbal and which atrophies in females. The tymbal works like the deformable piece of metal found on the 'cricket clicking' toys favoured by many children. A large muscle attached to structures that are linked mechanically to the tymbal deforms it in the same way that pushing in on the free end of a 'cricket clicking' toy deforms the metal which produces the click. The muscle that causes tymbal deformation can contract and release at rates exceeding 100 times per second. When the muscle relaxes, a second click is produced as the tymbal snaps back into shape. While calling, male bladder cicadas distend large abdominal airsacs which act as resonators and amplify the clicks. There is nothing surprising about this sex difference in communication: female bladder cicadas just do not retain and elaborate the male cuticular structures necessary for sound production.

Patterns of overt acoustic sex dimorphisms

Overt dimorphisms like those found in the bladder cicada may vary among species in the degree to which male sound-producing structures are elaborated, but from the point of view of sex differences these systems are boringly repetitive. One sex has a monopoly on making a particular

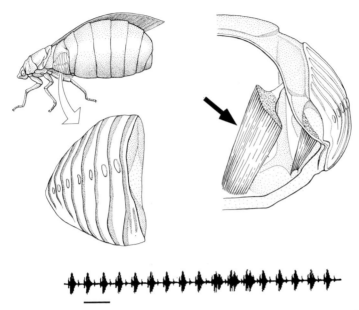

Figure 11.2. (Top left) A male bladder cicada (*Cytostoma saundersii*) producing courtship song. Note the tymbal on the first abdominal segment behind the legs, and the inflated abdomen. (From Bailey, W. J. (1991) *Acoustic Behavior of Insects.* Chapman and Hall.) (Centre: left and right) A magnified view of the tymbal together with a schematic drawing of a cross-section showing the tymbal muscle (arrow). (From Simmons, P. and Young, D. (1978) The tymbal mechanism and song patterns of the bladder cicada. *Journal of Experimental Biology*, **76**, 27–45.) (Bottom) The courtship song of the Australian bladder cicada. (Time bar = 0.04 s.) (From ibid.)

kind of sound, and is in sole possession of the apparatus to do so. While the degree of elaboration of the sound-producing organs may tell us something about the evolutionary forces shaping male communication behaviour, it does not necessarily explain their sex-exclusivity.

In some species the two sexes do not make different kinds of sounds, but rather they differ in how effectively they can broadcast the same sounds. In the South American howler monkey (*Alouatta seniculus*) both males and females produce similar vocalizations, including loud territorial barks and roars. However, it is possible to hear the loud calling of males at much greater distances (up to 1 km in dense forest) than the calling of females. Male calls are amplified by a set of male-specific bony resonators in the throat (Figure 11.3), formed in part by a spectacular modification of the male hyoid bone into a huge air-filled chamber like the body of a lute. Elephants are able to transmit their sounds over even

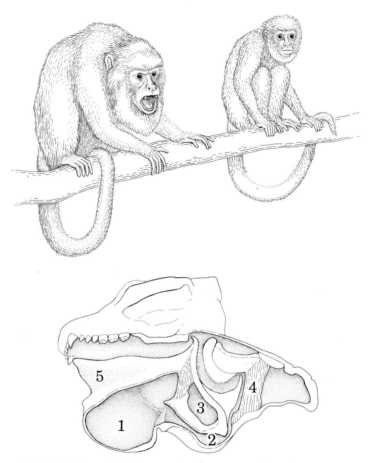

Figure 11.3. (Top) A male (left) and female (right) Howler monkey (*Alouatta seniculus*). Note the prominent bulge in the male's throat that is completely absent in the female. (Bottom) The internal anatomy of the male vocal tract. 1, hyoid bulla; 2, thyrohyoid canal; 3, lateral air sac; 4, vocal cord with high vocal lip; 5, tongue. (Based on Shön-Ybarra, M. (1986) Loud calls of adult male Red Howling Monkeys. *Folia primatologica*, **47**, 204–16.)

more impressive distances, but this ability does not seem to be sexually dimorphic (see Chapter 14 by Joyce Poole).

Overt vocal dimorphisms are not always so spectacular. During the breeding season, male red deer (*Cervus elaphus*) defend harems of females from encroachment by other males (see Chapter 15 by Tim Clutton-Brock). Part of this defense depends on the ability of males to engage in 'roaring contests' in an attempt to intimidate rivals; roars are also thought to stimulate female breeding condition. While females are able to produce a roar-like vocalization during calving, it is not exactly the same as the male roar. Males and females both use a set of vocal folds in their larynges to produce these vocalizations, but the folds are much larger in breeding males, presumably due to some developmental effect of hormones on the mass of muscle and tissue in the vocal folds. A similar situation exists in the vocal organ of male passerine (song) birds, called the syrinx. During the breeding season – the primary period for male singing – the muscular mass of the male's syrinx increases dramatically under the influence of testosterone secreted by the gonad. Indeed, the syringeal muscles are full of hormone receptors, leading to extensive sex dimorphism in syringeal muscle mass. In the non-breeding season, male and female syringeal muscle mass is much less different.

Patterns of covert acoustic sex dimorphisms

Covert dimorphisms like those found in the fruitfly parallel overt systems in all details. Males can produce sounds never produced by females, or both sexes can produce the same basic kinds of sounds that differ in subtle ways. Although both male and female gibbons (*Hylobates lar*) in Malaysia can produce long and elaborate territorial song displays that propagate over the same distances and can last for nearly half an hour, the vocal performances of the two sexes are subtly different. Male displays are less repetitive and consist of more complicated vocal elements than those of females (Figure 11.4).

It is not enough just to make sounds: you must also have the apparatus to hear them. In animals like frogs and crickets, where males and females live largely separate lives, one would expect all kinds of specializations to enhance the ability of the sex which normally listens (females) to locate and to evaluate the sounds made by the sex that usually does the calling (males). Yet hearing is one domain where overt dimorphisms between the sexes are strikingly rare. This is noteworthy because sex differences in perceptual structures used for mate localization are legion in other sensory modalities. One has to think only of the huge feathery antennae of male moths, which are capable of detecting the presence of single molecules of female pheromones at astonishingly low dilutions. Why is it not possible for ears to do the same? It seems just as advantageous for a female cricket to be able to home in on a calling male from a greater distance as it is for a male moth to home in on a perfumed female. It is possible to produce hypothetical explanations for why sound should be different from other

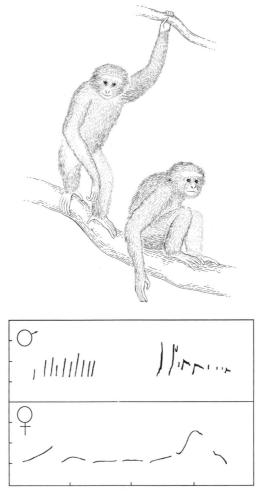

Figure 11.4. (Top) Male (left) and female (right) gibbon (*Hylobates lar*). Although the sexes can differ in colour, they are otherwise monomorphic. (Bottom) Representations of male and female long call segments. These representations show the frequencies at which sound energy is present on the *y*-axis and time on the *x*-axis, and are called spectrograms. (Time divisions in 5 s steps on the *x*-axis. The three frequency ticks on the *y*-axis in the calls of each sex represent 500, 1000 and 1500 Hz.) (Modified from Cowlishaw, G. (1992) Song function in gibbons. *Behaviour*, **121**, 131–53.)

sensory modalities, but none of them are very convincing. Clearly, this subject needs more attention.

While sound perception is as integral a part of acoustic sex dimorphism as is production, it is much harder to study. Production can be measured by just recording and quantifying sounds. Perception can be studied only indirectly: by forcing animals or people to act on their perceptions. To complicate matters, it is frequently not possible to use the same technique to measure the perceptions of males and females. While it is usually straightforward to find a way of studying call perception using members of the sex that normally does the listening, it can be quite challenging to find analogous measures for the call-producing sex.

The female animals whose call perceptions have been looked at in the greatest detail are the overtly dimorphic European field crickets *Gryllus bimaculatus* and *Gryllus campestris*. Here, as in most cricket species (and unlike grasshoppers), females lack the scraper-and-file mechanism on the tegmina of their wings that males use to produce song, even though, in aggressive situations, females can be observed to raise their wings and move them back and forth as males do when they sing. In experiments admired for their technical virtuosity, one group of researchers has elegantly defined the song parameters that females tune into by putting them on a spherical treadmill and by examining which of two sound sources they try to approach (Figure 11.5). Unfortunately, males do not try to track the song, and so it is not currently possible to study sex dimorphisms in this way. This is generally true for most overtly dimorphic signalling systems: males will not respond to presentation of male calls.

For a few dimorphic signalling systems such as bullfrog (*Rana catesbiana*) mate calling and the territorial songs of passerine birds discussed by Manfred Gahr and John Wingfield (see Chapters 12 and 13), it is possible to find assays for both male and female call perception. Female bullfrogs will approach recordings of male calls, and males will call in response to recorded bullfrog croaks, but not in response to other sounds. In birds, it is the males who aggressively approach a loudspeaker broadcasting the vocalizations of another male of their own species. If adult females are held in isolation during the breeding season and are administered a dose of the steroid hormone 17-β oestradiol, when hearing a taped presentation of male song, they will give a posture that, normally, is given to solicit a male to copulate. These copulation solicitation displays are given only to species songs, and the ability of a song to elicit such postures is used as a measure of how acceptable a song is perceived to be.

Interpreting data obtained with these techniques is not straightforward. We do not know whether the results of experiments testing sex dimorphism reflect true differences in perception, or differences in the workings of the behavioural tests themselves. William Searcy and Elliot Brenowitz have

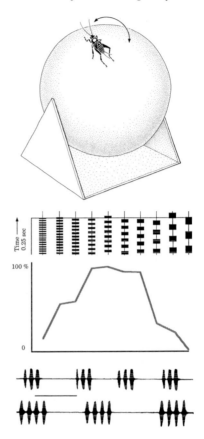

Figure 11.5. (Top) A female field cricket and the spherical treadmill used to test its song preferences. Two speakers on either side of the large foam ball play songs, and a computer adjusts the movement of the ball to counterbalance exactly any directional movements made by the cricket. Thus, the cricket stays in one place, and the computer uses the alterations it has to make to keep the cricket in one place as a measure of the cricket's speaker choices. (Based on Weber, W. and J. Thorson (1989) Phonotactic behavior of walking crickets. In *Cricket Behavior and Neurobiology*, p. 313. Eds. F. Huber, T. Moore and W. Loher. Cornell University Press.) (Centre) The proportion of time during a behavioural test (*y*-axis) that female field crickets spend approaching particular song patterns (illustrated above the curve). This is a measure of how attractive the songs are: it is assumed that females spend a greater proportion of the available time approaching songs that are more attractive to them. Note that females seem to be attuned preferentially to songs with certain temporal characteristics located in the middle of the figure. (Modified from ibid., p. 320.) (Bottom) Calling songs of the European field crickets *Gryllus campestris* (above) and *Gryllus bimaculatus* (below). (Time bar = 0.2 s.) (Modified from Bennet-Clarke, H. (1989) Songs and the physics of sound production. In *Cricket Behavior and Neurobiology*, p. 230. Eds. F. Huber, T. Moore and W. Loher. Cornell University Press.)

shown that red-winged blackbird (*Agelaius phoenicus*) females do not give copulation solicitation displays to slightly abnormal species songs; males will vigorously attack a loudspeaker broadcasting the same songs. The authors of this study proposed that males and females differ in their song recognition, with females being more discriminating listeners than males. One would expect that males are especially anxious to guard their turf against any potentially threatening intruder. They are primed to investigate anything that remotely resembles another male, even a slightly weird song played through a loudspeaker. The test has the opposite bias for females. Because it takes special housing and hormonal conditions for females to deliver copulation displays to songs in the first place, you might expect that females will display their sexual receptivity only when they know that there is a normal, healthy male nearby. Both males and females may realize that a stimulus song is strange, but the constraints of the testing situation force them to act differently on the basis of this information. Perceptual differences between the sexes are perhaps the most interesting aspect of covert bioacoustic dimorphisms, but they must be studied very cautiously.

The human voice: a mixture of overt and covert dimorphisms

Documenting human sex differences in speech is not difficult. In every language in the world, men and women express themselves differently. While some people believe that such relatively subtle differences in the way that men and women speak may ultimately be due to differences in the biology of male and female brains, there is no convincing evidence on this point. The function of all these covert dimorphisms remains shrouded in obscurity.

But covert dimorphisms in linguistic features like speech patterns, sentence structure and vocabulary all pale beside one very overt dimorphism that is a regular feature of human vocal communication. This is the difference in perceived pitch of an octave or more between most male and female voices, and the corresponding externally observable difference in the structure of the larynx.

One laryngeal structure called the thyroid cartilage is shaped differently in men and women (Figure 11.6). Male and female thyroid cartilages remain similar until puberty, growing at the same rate as other parts of the body. Sometime between the ages of 10 and 14 years, the growth rate of the male thyroid cartilage appears to increase dramatically, whereas the female cartilage continues to grow at the same rate as the body. The

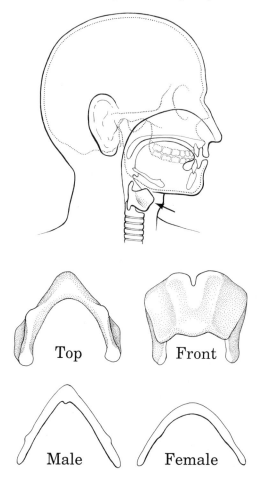

Figure 11.6. (Top) The human vocal tract in a schematic cross-section. The location of the thyroid cartilage is indicated by the arrow. (Centre) Two views of the thyroid cartilage. (Bottom) A schematic illustration of the sex difference in the angle of the thyroid cartilage. (Top, centre and bottom: modified from Dickson, D. and W. Maue-Dickson (1982) *Anatomical and Physiological Bases of Speech*, pp. 60, 140–2. Little, Brown & Co.)

rapid growth rate in the male cartilage presumably leads to a difference in the angle at which the two quadrilateral plates or laminae fuse ventrally to form a protrusion. In males, the two plates come together to form a rather sharp angular projection of about 90 degrees; in females, this projection is smoother and more rounded, and the plates tend to join each other at an angle of about 120 degrees. It is the angular projection of the thyroid cartilage that shows through the skin of the throat as the 'Adam's apple'.

The sex difference in the morphology of the thyroid cartilage is not itself the source of the pitch difference between male and female voices. Rather, the rapid growth that produced it leaves males with proportionally larger larynges than females. Inside the larynx there is a pair of muscular flaps called the vocal folds which are responsible for producing all the voiced sounds that humans make when they speak. They work by vibrating back and forth to open and close a passageway that the airstream follows when coming from the lungs. The puffs of air that get through the vibrating vocal folds are the source of voicing in speech, and the vocal folds' rate of opening and closing (called the 'fundamental frequency') is one part of what determines the relative quality of the male and female voice. The larger larynx of adult males is parallelled by a longer and more massive set of vocal folds which opens and closes the air passage at a slower rate, hence lowering the frequency of the vibrational source.

Vocal folds are sexually monomorphic before puberty, but contain receptors which respond to male-specific hormonal signals commencing at adolescence. Male vocal-fold tissues then undergo an abrupt increase in growth rate in much the same way as the thyroid cartilage does. At this time the pitch of the voice of the individual starts to change, and goes through an unstable period until the larynx and vocal folds slow their rate of growth. Females with ovarian tumours or who undergo anabolic steroid therapy go through a similar pattern of vocal changes.

The hormonally induced sex difference in the pattern of laryngeal development has had disastrous consequences on the reproductive success of a subgroup of European males. Frequently, boys with beautiful singing voices were castrated with the permission of their greedy guardians in the belief that this would prevent their voices from changing hence spoiling a lucrative talent, a practice which continued well into the nineteenth century. Charles Burney, the gossipy chronicler of eighteenth century musical life, described the practice in this way:

I enquired throughout Italy at what place boys were chiefly qualified for singing by castration but could get no certain intelligence. I was told at Milan that it was Venice; at Venice that it was at Bologna; but at Bologna the fact was denied, and I was referred to Florence; from Florence to Rome, and from Rome I was sent to Naples ... before the operation is performed they [the boys] are brought to the Conservatorio to be tried as to the probability of voice ... It is said, however, to be death by the laws to all those who perform the operation, and

excommunication to everyone concerned in it, unless it be done, as is often pretended, upon account of some disorders which may be supposed to require it, and with the consent of the boy... it is my opinion that the cruel operation is but too frequently performed without trial, or at least without sufficient proofs of an improvable voice; otherwise such numbers could never be found in every great town throughout Italy, without any voice at all, or at least without one sufficient to compensate such a loss. Indeed all the *musici* [male sopranos or contraltos] in the churches at present are made up of the refuse of the opera houses, and it is very rare to meet with a tolerable voice upon the establishment in any church throughout Italy. The virtuosi who sing there occasionally, upon great festivals only, are usually strangers, and paid by the time.
(Scholes, P. (Ed.) (1959) Dr Burney's Musical Tours in Europe. *Vol. I, pp. 247–8.*
Oxford University Press)

The last known castrato was Professor Alessandro Moreschi (1858–1922), who served as the musical director at the Vatican for Pope Pius X, and actually made a few gramophone recordings.

Other European musical traditions have risen and fallen on the pattern of male laryngeal development. Records from J. S. Bach's all-male choir at the Leipzig Thomasschule in 1744 have been used to estimate the age at which voices broke in the eighteenth century. By scrutinizing the age distribution of altos and their conversion to other voice types, it appears that this age was 17 years, as compared to 13.3 years in London in 1959. This acceleration of puberty, if true, has dire implications for the maintenance of modern all-male choirs. The justification for such groups has always been that the vocal quality of equally talented prepubescent boys and girls is different to the trained musical ear. Yet the structure of the larynx of prepubescent humans does not seem to be sexually dimorphic, nor is the frequency of vocal-fold vibration of the two sexes different before puberty. Are such claims simply chauvinistic, or are there other differences in male and female voices?

The frequency of vibration of the vocal folds is only half the story of how speech sounds are produced. The other half lies with the air passages in the head through which the sounds emanating from the laryngeal source must pass. These passages act as a kind of filter whose characteristics can be changed by varying the length and shape of the throat, mouth and sinuses (using the tongue, lips, cheeks, jaws and throat musculature).

The existence of sex differences in supralaryngeal filtering is most clearly shown by experiments on whispered speech. Here the vocal folds do not interrupt the flow of air at all, and are thus irrelevant; yet listeners can classify correctly the sex of the speaker 75–85% of the time from recordings of whispered speech. Listeners also do very well at classifying the sex of preadolescent children's speaking voices. The proponents of all-male choirs may be acting partially out of chauvinism, but there seems to be a grain of truth to their claim of voice differences.

The suggested mechanism for these supralaryngeal differences is learning. It is claimed that the difference in the measured filtering properties of adult male and female English speakers is greater than would be predicted by anatomy alone. For males and females with equally sized supra-laryngeal vocal tracts, the filters of males are always set to lower frequencies. Males manipulate their vocal tracts to make their voices sound lower, and females perform analogous manoeuvers to raise the apparent pitch of their voices. Children, being keen observers and imitators of all adult fads and foibles, start this practice long before they reach adolescence. With training, young male singing voices could probably sound just like those of young females and vice versa.

Causes of bioacoustic sex differences

What makes the sexes sound so different in the first place?

If an acoustic sex difference is to evolve, there has to be some reproductive benefit derived by males and females who make different sounds (or by calling males and silent females) relative to more monomorphic males and females. If the sexes lead ecologically different lives, then habitat differences (habitats with different kinds and densities of vegetation favouring the propagation of different types of sounds) or differences in the ears of predators that specialize on males or females could be causative factors that select for sex differences in sounds. But this kind of selection does not usually explain acoustic sex differences. Male advertisement calls are social signals whose efficacy depends on the responses they evoke in listeners. Calls which are audible over longer distances are easier to localize, or those whose sounds are recognizably different from the norm (but not too different) may boost the reproductive success of their bearers because of their greater potential to attract females and to scare away other males. Sexual selection is thus a major force which drives the elaboration of these sounds (discussed in Chapter 3 by John Reynolds and Paul Harvey).

A spectacular case illustrating how natural and sexual selection can shape the properties of a bioacoustic system is shown by the South American túngara frog (*Physalaemus pustulosus*). Males call at night to attract females to breeding pools. The male's call has two different

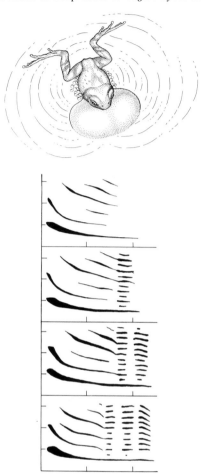

Figure 11.7. (Top) A male túngara frog (*Physalaemus pustulosus*) mate calling. Note the inflated vocal sac that acts as a resonator. (Based on Ryan, M. (1985) *The Túngara Frog.* University of Chicago Press.) (Bottom) Spectrograms of a whine (uppermost) and successive additions of chucks in a descending series. (Ticks on the y-axis represent frequency in 1 kHz steps from 0–8 kHz; x-axis ticks are 0.2 s each.) (Modified from Ryan, M. and Keddy-Hector, A. (1992) Directional patterns of female mate choice and the role of sensory biases. *American Naturalist*, **139**, S4–S35.)

components, a low-pitched downswept 'whine' followed by a variable number of higher pitched 'chuck' sounds, all in less than one second (Figure 11.7). Females preferentially approach males that give calls with a greater number of chucks, and, therefore, these males should enjoy a mating advantage. The problem is that calls containing chucks make it easier for the frog-eating bat (*Trachops cirrhosus*) to localize its favourite food and to make short work of it. Rarely is the choice between sex and death so clear cut. The structure of the male's call literally walks a tightrope between selection due to predation to make only a whine, and sexual selection from female choice for calls with chucks.

Are different mechanisms responsible for overt and covert sex differences in sound communication?

While natural and sexual selection may give us the ultimate reasons why males and females sound differently, it is of equal value to understand the mechanisms through which these evolutionary forces exert their effects. It is only when we can elucidate both the evolutionary reasons for divergence and the proximal means by which the divergence is carried out that we can more fully understand and explain sex differences in communication.

Currently, there is a consensus that different developmental mechanisms are responsible for overt and covert sex differences in sounds. Overt dimorphisms are thought to result primarily from the action of selection on external morphology. Covert dimorphisms are thought to be the products of selection acting directly on brain structures that make males and females sound different (sex differences in the brain are covered in detail in Chapter 12 by Manfred Gahr). In what follows, I argue that this conventional view is wrong, and that overt and covert differences stem from the same source: selection acting at the level of external morphology. I also suggest experimental tests to assess the validity of this position.

How does selection produce overt and covert acoustic sex differences?

If male and female behaviour is a direct consequence of having different morphology, sound communication is usefully viewed as being another secondary sex characteristic like the rooster's comb and wattles, the male widowbird's long tail, or a male red deer's antler rack. The behavioural difference shown by the two sexes is simply a consequence of their being provided with different bodily 'machinery', and their being raised in a social world where individuals with different morphology are treated differently or tend to imitate different behavioural models. Changes in the brains of males and females do not appear to be necessary. I mentioned above that female European field crickets will give the same wing movements that males use to produce sound in aggressive contexts, even though females entirely lack sound producing structures on their wings (these animals are nocturnal and it is unlikely that such movements are visual aggressive signals). Apparently, females have the neural potential to make sounds: it would be intriguing to glue male wing parts artificially onto a female and to see what she does with them. We know from experience that in spite of the difference in the pitch and timbre of human male and female voices, both sexes use them in incredibly similar ways. Having and using a sex-limited effector organ like a wing scraper-and-file mechanism or even a larynx with more massive vibratory tissues may be all that is necessary to develop male bioacoustic behaviour. (Note that I am not saying that neural changes do not occur. In bladder cicadas, where females lack the specialized tymbal and its associated muscular hypertrophy, we would expect to see differences in the nervous system of the two sexes, but we do not think of these as being the 'primary' cause of the differential ability of males and females for sound production. Here we would expect that a female with the organ *and* its associated muscle development would probably have a nervous system much like that of a male, for reasons that are described in the text which follows.)

It is usually easy to see how morphological differences between the sexes, like the difference in human laryngeal structure that gives male and female voices their distinct pitch and timbre, could have become elaborated due to selection. Commonly, males and females of a species differ in body size for complex reasons (see Chapter 3 by John Reynolds and Paul Harvey and Chapter 7 by Robert Martin and colleagues for further discussion of the role of sexual selection in this type of difference). When there is a body size difference between males and females, the larger sex would have larger laryngeal vocal folds which vibrate less rapidly. Thus, members of the physically larger sex would tend to produce lower fundamental vocal frequencies. In addition, the volume of the cavities in the head would provide lower resonating filter frequencies. What could have been initially a small difference between male and female voices based on body size could be worked on by selection to produce the more exaggerated differences we see today by the modification of laryngeal growth curves. We would still have to explain why the sexes differ in body size (see Chapter 7 by Robert Martin and colleagues and Chapter 4 by J. William Stubblefield and Jon Seger), and why selection exaggerated the laryngeal difference between the sexes, but the starting point (variation due to a sexual size dimorphism) and the proximal mechanisms acted upon by selection (differential growth) are relatively simple to interpret for these types of overt dimorphisms.

But what about covert sex differences? According to the conventional view, a piece of the nervous system is the equivalent of the red deer's antler rack: one sex has it, the other does not, and, since it is the nervous system that ultimately produces acoustic behaviour, this must explain any sex differences in sound not explained by external morphology.

It is not as easy to see how a difference in male and female vocal behaviour that is solely a result of brain anatomy or physiology could become elaborated by selection. There are no known general neural differences between the sexes (akin to the correlates of body size discussed above) that could provide selection with reliable material to elaborate upon. One must assume that a brain difference existed a priori in a more subtle form, or that it is especially likely to arise by chance. An examination of how brain differences between the sexes developed does not shed much light on how selection could have produced them. While it is easy to understand why so many of them are dependent on hormonal signals (which are already being used to drive the development of the reproductive tract), such mechanisms require the expression of particular enzymes and hormone receptors in particular brain regions to be effective. It is unclear how or why males and female should start out with some subtle sex-linked difference in enzyme or hormone-receptor-gene expression in particular regions of their nervous systems that selection could exploit.

One resolution of this dilemma is to propose that it is not actually a difference in the nervous system which starts the ball rolling. Subscribers to the 'brain-as-antlers' (BAA) model have rarely looked to see whether there are subtle sex differences in the form of sound-producing organs or their associated musculature. Such subtle differences in the periphery can lead to the development of substantial differences in the nervous system of males and females because of a feature of brain development that sets it apart from other sexually dimorphic structures.

Neural development is highly plastic, and nervous systems maintain their plasticity into adulthood in organisms from sea slugs to humans. Neural plasticity provides opportunities for small differences in anatomy, behaviour and the experiences of individual organisms to influence the structure and function of their adult nervous systems in major ways. This is radically different from the development of other sexually dimorphic structures. It is possible that the amount of head-butting engaged in by a young red deer male could exert some kind of influence over how its antlers would grow in the future. But any such influence would be slight compared to the major changes in the structure and function of the developing central nervous system produced by differences at the sensory and motor periphery (an excellent popular account of this aspect of brain development is provided by Dale Purves).

Rather than selection acting directly on the brain to produce sex differences in acoustic communication, I propose that it acts instead on subtle differences in the form and/or use of peripheral structures whose effects are amplified by the process of neural development. It is these 'amplified' brain differences working in concert with the subtle peripheral differences which give rise to the fully fledged sexual dimorphism. Let us call this the 'neurally amplified morphological evolution' (NAME) model. It proposes that acoustic sex dimorphisms, overt or covert, both result from the same process: selection acting on peripheral differences. This model changes our view of the primary locus of selection (from the central nervous system to morphological structures at the periphery), and changes our interpretation of the proximal 'cause' of a difference in acoustic behaviour (from the structure of a piece of brain tissue to a combination of peripheral morphology, neural development, and functional use).

Such a scheme has two advantages over the more traditional BAA model: it can explain how selection finds any differences to act on in the first place, and it can explain many of the developmental details of sexually dimorphic communication systems.

The NAME model and frog bioacoustic sex differences

A sizeable body of work on frog mate-calling provides strong support for a view of sex dimorphisms in communication driven by subtle differences in peripheral structures. Imagine the ear of female frogs as being a thin wall that sounds can pass through. Some sounds penetrate the wall more easily because of the structure of the wall: as any apartment dweller who has a neighbour with a loud stereo knows, it is easier to hear a bass player than a vocalist. A female frog has to find a mate at night, and sounds which she can hear more easily will make it more likely that she can orient herself to their source. One simplistic idea is that the male whose call blasts the most sound energy through the wall that is the female's ear is the one that she will find the most attractive.

Work done on túngara frogs suggests that the ear of the female does seem to function like our hypothetical wall. The sensitivity of the female túngara frog's ear to sounds of different frequencies was compared with the properties of male calls and the behavioural preferences of females for male calls within a population. Female call preferences were ascertained by giving females a choice between calls played from two loudspeakers to see which call they would approach most reliably. As discussed previously, females need to have a whine in the call to respond to it (see Figure 11.7), but they prefer calls with chucks in them to calls with whines alone.

Frog ears have two inner ear structures that are sensitive to airborne vibrations: the amphibian and basilar papillae (we have just one: the cochlea). These two organs contain receptor cells that respond to different frequencies of sound: the basilar papilla is more sensitive to higher frequencies, and the amphibian papilla to lower frequencies. The amphibian papilla appears to be primarily sensitive to the whine portion of the call, and the basilar papilla mainly to the chuck. The pitch at which males give chucks is correlated with their body size, and females prefer to mate with larger males.

The distribution of energy in the chucks of a population of males was compared to the most sensitive frequency of the basilar papillae of females from the same population. Female ears as a group were most sensitive to frequencies slightly below those that are found commonly in male calls. Females prefer natural calls with lower-pitched chucks, but there are many ways that calls can be manipulated artificially to induce females to prefer them. For instance, if a call is made which contains an unadorned whine that is louder than a call with chucks, the females can be induced to prefer the simple whine. This seems to work with any manipulation that causes one call to deliver more total sound energy through the female ear than the other. Note that the call still must have a normal whine to be recognized as a species-typical call.

In similar studies on cricket frogs (*Acris crepitans*), a correlation was found between a female's body size, the pitch of the calls she prefers, and the most sensitive frequency of her basilar papilla. Since, in most frog species, females are larger than males, differences in body size alone may explain why female basilar papillae are most sensitive to sounds that are slightly lower than those made by most males. Male calls in general correspond to the most sensitive frequencies of their basilar papillae, which are higher than in females.

Male–female size dimorphisms in frogs are thought to result from differences in how fecundity changes with body size in the two sexes. In general, larger females can produce more eggs than smaller females, but size has less influence over male sperm production (even small males can make enough sperm to fertilize huge numbers of eggs). A sex difference in size exists independently of mate calling.

Within the constraints provided by the general tuning of the ear due to body size, many different 'preferred' call structures are possible; however, the reason why the number of call structures is limited may be associated with constraints on the physiology of male sound production which make certain types of calls much more likely to evolve than others. Comparative data have been advanced to support the idea that the tuning of the female ear is phylogenetically older than the call structure of the males of any particular species. The tuning of the basilar papilla of a remotely related sister species to the túngara frog is no different from that of the túngara frog, but the structure of the male call is different, consisting of just a whine without any chucks.

That peripheral differences in ears and laryngeal structure may drive sex differences in frog communication rather than some independent neural dimorphism receives support from work by Darcy Kelley and colleagues on male mate-calling in the African clawed frog (*Xenopus laevis*). Here, both male and female brains seem to be able to produce the neural signals required for male call production, but features unique to the male larynx prevent females from producing male calls. Male laryngeal muscles are coupled electrically and are predominantly of a different muscle fibre type from those of females, which results in a very different pattern of muscular responses when activated by nerves. Such normally invisible functional differences in the cellular properties of muscle underline the potential subtlety of covert peripheral features that neural development can amplify. It is well known that muscle fibre type and firing properties exert a considerable influence over the development of nerve circuits connected directly and indirectly with muscles in insects, fish, amphibians, birds and mammals.

Evaluating NAME and BAA in higher vertebrates

It is instructive to take an example of a well-characterized bioacoustic sex difference in a higher vertebrate with a more complicated communication system and to subject it to a critical analysis in which the NAME and BAA models can be contrasted. We will take a recent example from the bird vocalization literature which claims to demonstrate a simple neural difference that leads to a subtle difference in the vocal behaviour of males and females.

All species of birds have vocalizations that are given equally by both sexes. One type is called the 'contact' or 'distance' call, which is given by individuals of a pair bond when they are out of visual contact with one

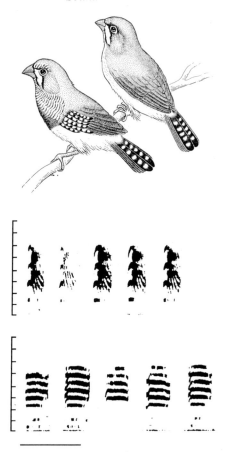

Figure 11.8. (Top) A male (left) and female zebra finch (*Taeniopygia guttata*). (Bottom) Spectrograms of male (upper) and female (lower) distance calls. Note the two main structural differences: the female calls are longer, and the male calls have a rapidly descending set of harmonics during the second half of the call, which is lacking in females. (Time bar = 0.5 s; frequency is marked in 1 kHz steps from 0–8 kHz.) (Modified from R. Zann (1984) Structural variation in the zebra finch distance call. I. Effects of crossfostering to Bengalese finches. *Zeitschrift für Tierpsychologie*, **66**, 328–45.)

another. These calls have been studied intensively in the zebra finch (*Taeniopygia guttata*), a small, monogamous seed-eating inhabitant of Australia and Indonesia that is a favourite cage bird because it adapts so readily to captive conditions. Zebra finches have sexually dimorphic distance calls (Figure 11.8). Male zebra finch distance calls in the wild contain a tonal element with harmonics (energy bands at integral multiples of the lowest frequency) overlapped by a rapid frequency downsweep (with more closely spaced harmonics). Female calls lack this rapid frequency downsweep, and are about 25% longer in duration than male calls. Studies on birds in captivity indicate that these acoustic differences are readily perceptible to both adult male and adult female zebra finches.

Songbirds like the zebra finch are renowned for their vocal learning abilities. Could learning be responsible for the sex difference in contact calls? This possibility was tested by using a cross-fostering experiment, which involved placing small groups of eggs from wild zebra finches into nests of the related Bengalese finch (*Lonchura striata*). When the zebra finch eggs hatch, they are reared by the Bengalese finch parents together with their own Bengalese finch offspring. Bengalese finches also display sex differences in distance calls, but their structure is different from the calls of zebra finches, and they are pitched higher. If sex differences in call structure are learned, then the cross-fostered zebra finches should produce calls of the sex-appropriate foster parent. Clearly, male zebra finch distance calls were altered by cross fostering: out of 19 foster-reared males, 3 produced the call of either the foster mother or foster father Bengalese finch, 11 gave calls without the frequency downsweep, and the remaining 5 animals gave other abnormal versions of the zebra finch call. In contrast, all 12 females reared by Bengalese finches gave calls that looked normal, except that, as a group, their calls were given at a higher pitch than normal (we will return to this detail below). All the cross-fostered males had calls which lacked the frequency downsweep but were still shorter than the calls of cross-fostered females. Female calls attained a stable structure by about day 35 posthatching, but male calls did not stabilize until 60–80 days of age. It was suggested that the male's call, but not that of the female, is open to environmental influence.

The theme of learning males and non-learning females was carried further by a different group of researchers, who suggested that there are anatomically separate regions of the nervous system which control distance call production in males and females. This was demonstrated by lesioning selected areas of the nervous system in males and females. It was claimed that lesions disrupted the distance call production of males but not that of females (Figure 11.9). It was also suggested that early treatment of females with a male-like hormonal regime masculinized their distance calls. If areas affecting male call structure were lesioned in these 'masculinized' females, the structure of their distance calls became disrupted (remember that these same lesions were claimed to have no

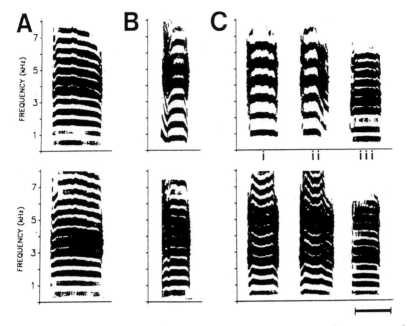

Figure 11.9. An example of the effect of nervous system lesions on the structure of zebra finch distance calls. In this case, the nerves controlling many of the syringeal muscles have been interrupted. The calls in the upper part of the figure were all recorded before nerve section, those in the lower part of the figure were recorded afterwards. A is a female, B is a male, and C is a series of three males: i and ii were classified by the investigators as being male-like, iii as being female-like. Although the investigators claimed that the lesions affect the male-like calls only, note that the duration is not affected in any of the cases, and that the distribution of energy (where the darkest harmonics are located, indicative of the most intense acoustic energy) appears to be altered in all calls. (Time bar = 0.15 s.) (Modified from Simpson, H. B. and Vicario, D. (1990) Brain pathways for learned and unlearned vocalizations differ in zebra finches. *Journal of Neuroscience*, **10**, 1541–56.)

effect on normal females). The conclusion of these studies was that male zebra finch brains have something that female brains lack which is responsible for the sex difference in calling.

I think that there are three aspects of these studies which do not fit with the model of a simple brain difference between males and females, but which could be explained by a peripheral difference amplified by neural development.

The first has to do with similarities in the control of male and female calls. Zebra finches control the way they allocate sound energy to different harmonic bands in their vocalizations, and appear to need some practice to accomplish this. Birds of both sexes are able to perceive differences in energy allocation within an otherwise identical call. Remember also that zebra finch females raised with Bengalese finches gave distance calls at a higher pitch than normally reared zebra finches. Such a change could be attained by shifting the distribution of energy in their calls to give a closer approximation to the higher pitched calls of their Bengalese finch foster parents. Although hard to judge from published figures, lesion treatments which affect the structural feature possessed only by males also seem to affect the pattern of energy allocation in both male and female calls. The fact that lesions may affect shared male and female call features in the same way is very important to our understanding of how male and female call control differs. In fact, these results suggest that males and females control their distance calls in a neurally similar way. This does not sit comfortably with the idea of a globally causal neural dimorphism. It fits in quite well, however, with the idea that the two systems develop in a similar way except for changes induced by subtle differences in structure or usage of peripheral musculature.

The longer duration of the female distance call is the most reliable sexually dimorphic feature of wild and laboratory-reared zebra finches. Lesions significantly shortened the distance calls in four out of nine female birds that received them in these experiments (the same treatments lengthened the distance calls in 11 out of 15 males). The proportion of males whose calls were affected by these operations (73%) is higher than, but not statistically different from, the proportion of affected females (44%). With this degree of difference between males and females, it would be necessary to examine about 40–50 lesioned males and an equal number of lesioned females to see whether the females' calls are affected less readily. But the fact is that almost half the females had their calls shifted toward a male duration by lesioning, and a clear majority of males had their calls shifted toward female values. The original view of a simple neural difference was based on the idea that lesions affect only the calls of one sex, when, in fact, they affect the calls of both sexes. This suggests that some subtle feature of call control is differentiated equally in males and females. Could this reflect subtle differences in muscle morphology or syrinx use by the two sexes? The original hypothesis that female brains are simplified structures whose functions can be mimicked by damaging the brains of males receives little support.

Finally, the researchers who carried out these investigations emphasized the effectiveness of hormone treatments in inducing females to produce the call feature that, normally, is seen exclusively in males, but failed to point out that the treatments do not consistently affect other call features that display sex differences. As a population, supposedly masculinized females had distance call durations that were no different from normal females but were significantly longer than normal males. While this does show that early hormone treatments have a statistical tendency to make females incorporate a uniquely male component into their calls, hormonal differences between males and females do not by themselves explain sex differences in the more reliable feature of call duration. If a simple brain dimorphism whose development is due to hormonal differences in male and female embryos or juveniles is responsible, we would expect both call parameters to be affected equally in this experiment.

The point of this example is to indicate that data advanced in support of the BAA model of sex differences in sound communication is highly equivocal. I would not argue that there are no covert differences in communication which are solely the result of differences in the central nervous system, but rather I would observe that such cases may be rare exceptions to a general rule. Both overt and covert sex differences have a common mechanism – they are due to a large or small difference at the periphery. It is a pity that no one has yet examined the male and female zebra finch syrinx in great detail. Rather than focusing exclusively on the nervous system, researchers interested in covert bioacoustic sex differences should spend more time in looking at peripheral differences and in exploring their developmental implications.

An experimental approach to the causation of bioacoustic sex differences

One promising approach to sorting out how selection and development interact to produce sex differences in sound communication systems could be the use of surgical chimaeras. These are produced by substituting precursors of appropriate tissues in the early embryos of one type of animal for the same tissues in another type of animal embryo. Although embryologists have been making these surgical chimaeras for almost a hundred years, it is only recently that such techniques were applied to the study of behaviour. In collaboration with Nicole Le Douarin and Marie-Aimée Teillet of the CNRS Institute of Cellular and Molecular Embryology, I have applied chimaeric analyses to understanding the relative role that peripheral and brain differences play in determining the form of species differences in unlearned vocal behaviour in birds. The results are quite striking: it is possible to transfer aspects of acoustic behaviour which differ between species with the transplant of defined regions of the presumptive brain. Although the resulting animals behave normally in other respects, it is possible to transplant either the characteristic temporal pattern of crowing, or the species-distinctive posture it is delivered with, from Japanese quail (*Coturnix coturnix japonica*) into domestic chickens (*Gallus gallus domesticus*) (Figure 11.10).

Chimaeras could be used effectively to examine the role that peripheral structures play in determining bioacoustic sex differences. Female brains could be placed into male bodies to see whether they are equally capable of producing normal male acoustic behaviour in the developed animal. One could also transplant male peripheral structures into female bodies, and treat these tissues locally with hormones to assure their appropriate sexual differentiation. The effect of male peripheral structures on the development of sex differences in sound production could then be studied. Similar experiments with female sensory structures could also be carried out.

Specific evolutionary scenarios could be tested by transplanting tissue between species. If the characteristics of the ears of female frogs are highly evolutionarily conserved, then it should be possible to swap them among species with little disruption in acoustically mediated approach behaviour.

All these experiments would partially test the hypothesized role of the nervous system as an amplifier of peripheral differences. But more direct tests are also possible, such as putting brain regions of males and females from a monomorphic species (which normally exhibit no sex dimorphism in sound producing structures or the brain regions controlling them) into the bodies of male and female members of a dimorphic species. NAME may or may not prove to be a general feature of covert sex dimorphisms, but at least it is open to refutation. Quite large evolutionary changes in acoustic behaviour may be entirely due to selection acting at the level of peripheral morphology, if brain development proves plastic enough to adapt itself to peripheral demands.

Chimaeras provide a viable and exciting experimental pathway for examining the interplay between evolutionary forces and developmental programs which produce different levels of behavioural sex dimorphism. Hopefully, such studies will contribute to a fuller understanding of some small part of the bewildering array of sex dimorphisms in acoustic communication, which will continue to play a dominant role in the biological analysis of behavioural diversity.

Acknowledgements

This manuscript benefitted from the comments of Roger Short, David Crews, Ani Patel, Marc Hauser, Beth Brainerd and Susan Ramsey. Special thanks to Laszlo Meszoly for drawing the figures, and to Al Coleman

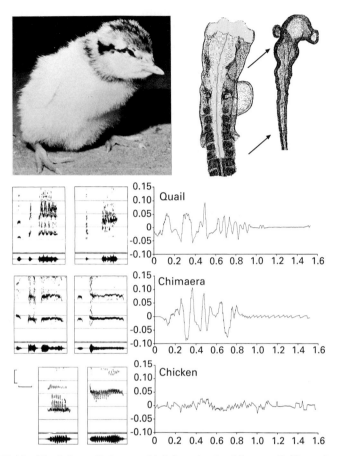

Figure 11.10. (Top) A quail-donor, chick-host brain chimaera (left) produced by transplanting the entire neural primordium (called the neural tube) to the level of the cervical spinal cord during the second day of incubation (out of a 21-day incubation period). The operation is illustrated schematically on the right. Note the quail-like feather pigmentation in the region containing the graft. (From Balaban (1990).) (Bottom) On the left are two examples each of spectrograms of testosterone-induced juvenile crowing vocalizations from normal quail (uppermost), quail-donor, chick-host chimaeras (centre) and normal chickens (lower). Note that the chimaeras have crows whose durations and temporal patterns correspond to normal quail. (Time and frequency bars for spectrograms: 0.2 s and 2 kHz.) On the right are measurements of a species difference in crowing posture: quail bob their heads rapidly up and down, and chickens do not. These postural differences were measured with a head movement tracking system which displays head movements as a change in voltage on the y-axis. The uppermost trace shows normal quail head movements during crowing; the central trace shows the head bobs of a quail-donor, chick-host chimaera, and the lower trace shows the lack of head movement shown by chickens during crowing. The transplantable species differences in sound patterning and head movement are controlled by different parts of the brain. (Balaban, unpublished results.)

and John Lupo who aided in figure production. The research on chimaeras is supported by National Institutes of Mental Health FIRST award MH47149 to Evan Balaban.

Further reading

Bailey, W. J. (1991) *Acoustic Behaviour of Insects: An Evolutionary Perspective.* Chapman and Hall.

Balaban, E. (1990) Avian brain chimeras as a tool for studying species behavioural differences. In *The Avian Model in Developmental Biology: From Organism to Genes*, pp. 105–18. Eds. N. Le Douarin, F. Dieterlen-Lievre and J. Smith. Editions du CNRS.

Kelley, D. (1986) Neuroeffectors for vocalization in *Xenopus laevis*: hormonal regulation of sexual dimorphism. *Journal of Neurobiology*, **17**, 231–48.

Purves, D. (1988) *Body and Brain.* Harvard University Press.

Ryan, M. (1985). *The Túngara Frog: a Study in Sexual Selection and Communication.* Chicago University Press.

Ryan, M. (1990) Sexual selection, sensory systems, and sensory exploitation. *Oxford Surveys in Evolutionary Biology*, **7**, 157–95.

Sassoon, D., Gray, G. and Kelley, D. (1987) Androgen regulation of muscle fiber type in the sexually dimorphic larynx of *Xenopus laevis*. *Journal of Neuroscience*, **7**, 3198–206.

Searcy, W. and Brenowitz, E. (1988) Sexual differences in species recognition of avian song. *Nature*, **332**, 152–4.

Sebeok, T. (Ed.) (1979) *How Animals Communicate.* Indiana University Press.

Tobias, M. and Kelley, D. (1987) Vocalizations by a sexually dimorphic isolated larynx: peripheral constraints on behavioural expression. *Journal of Neuroscience*, **7**, 3191–7.

12

Brain structure:
causes and consequences of brain sex

MANFRED GAHR

Some of the most striking sex differences in behaviour shown by male and female organisms are seen in the realm of reproduction, where behavioural sex roles tend to be defined very clearly. Males display, mount, and deposit sperm, and females select a male, solicit copulations, and donate eggs. Such clearcut behavioural differences between the sexes have led many investigators to hypothesize that males and females must differ fundamentally in the construction and/or function of the machinery that produces all behaviour: the central nervous system (CNS).

Although sex differences in the brain had been a matter of speculation since the early 1900s, it was not until the pioneering work of Raisman and Field in 1971 that a dimorphism was demonstrated conclusively (in the anatomy of the preoptic area of male and female rats). Since that time, countless reports on sexual dimorphisms in the nervous systems of vertebrate and invertebrate species have been published. Such differences turn up at many levels: gross-anatomical, ultrastructural, biochemical and molecular. They include: total brain size and weight; the size of certain brain regions or subregions called brain nuclei; the numbers of neurons, glial cells and axons in certain brain areas; the size of the cell bodies in the same brain region in males and females; dendritic branching patterns of the same neurons in males and females; patterns of synaptic organization; the structure of the neuronal plasma membrane; and the distribution and relative concentration of enzyme activities, neuro-peptides, neuropeptide receptors, steroid hormone receptors, and cyto-skeletal proteins in male and female brains. Table 12.1 lists some of the better characterized examples of sexual dimorphisms in the vertebrate central nervous system but it is by no means complete.

There is now a wide consensus that sex differences in the brain are ubiquitous, but there is less agreement about how male and female brains

Table 12.1 *Structural dimorphisms in the vertebrate central nervous system*

Brain weight		
	Rat, meadow vole	T
Volume of brain areas		
Medial preoptic – anterior hypothalamic area	Rat, ferret, gerbil, hamster, human, montane vole, lizard	E, T
Bed nucleus of the stria terminalis	Human, montane vole, guinea pig	E, T
Hippocampus	Rat, meadow vole	
Visual cortex/neocortex	Rat	
Ventromedial hypothalamus	Rat, lizard	E
Spinal nucleus of the bulbocavernosus	Rat, mouse, monkey, dog, human	T, DHT
Neuron number		
Hippocampus	Rat, meadow vole, house mouse	
Visual cortex	Rat	
Spinal nucleus of the bulbocavernosus	Rat	E, T
Forebrain song nuclei	Canary, zebra finch	E, T
Neuron size		
Ventromedial hypothalamus	Rat	E, P
Medial preoptic area	Rat, gerbil	E, T
Spinal nucleus of the bulbocavernosus	Rat	E, T
Neuron plasma membrane		
Arcuate nucleus	Rat	E
Infundibular hypothalamus	African green monkey	
Dendritic branching		
Medial preoptic area	Hamster	T
Visual/prefrontal cortex	Rat	
Spinal nucleus of the bulbocavernosus	Rat	
Song nucleus RA	Canary	T
Synaptic organization		
Medial preoptic area	Rat	T, E
Arcuate nucleus	Rat	
Song nuclei RA, HVC	Canary	T
Neuropeptide distribution		
Serotonin	Rat (preoptic area)	E
Vasopressin	Rat (lateral septum)	T, E
Vasotocin	Zebra finch (lateral septum, archistriatum)	T
Met-enkephalin	Rat (preoptic area)	
Galanin-like peptide	Fish, *Poescilia latipinna*	
Neurotensin/neuromedin N	Rat (preoptic area)	

Table 12.1 – (*cont.*)

Enzyme activity		
Aromatase	Ring dove (preoptic area)	T, E
	Quail (preoptic area)	
	Zebra finch (caudal forebrain)	
Receptor content and distribution		
Progesterone receptor	Rat (cortex)	
Gaba$_A$ receptor	Rat (preoptic area, mediobasal hypothalamus)	
μ-Opiate receptor	Rat (medial preoptic area)	
Oestrogen receptor	Canary (song nucleus HVC)	
Gene expression		
NGF receptor	Rat (basal forebrain)	
Oestrogen receptor	Rat (mediobasal hypothalamus)	E
GAP-43	Rat (cortex, ventromedial hypothalamus)	E
Proenkephalin	Rat (ventromedial hypothalamus)	E
Proopiomelanocortin	Rat (ventromedial hypothalamus)	E
Preprocholecystokinin	Rat (medial amygdala)	
Growth hormone releasing hormone	Rat (arcuate nucleus, ventromedial hypothalamus)	
Somatostatin receptor	Rat (periventricular nucleus)	

E, sensitivity to oestradiol; T, sensitivity to testosterone; P, sensitivity to progesterone; DHT, 5α-dihydrotestosterone; HVC, nucleus hyperstriatalis ventrale, pars caudale; NGF, nerve growth factor; RA, nucleus robustus archistriatalis.

become different and to what extent these differences in brain features translate into differences in behavioural output. We examine both these issues in the following pages. While we initially discuss both vertebrate and invertebrate examples of sexual dimorphisms in the CNS, the bulk of this chapter is concerned with vertebrate dimorphisms, primarily because these systems have been characterized in greater detail.

The causes of brain sex

Sex differences in the brain have traditionally been thought of as arising in one of two ways. Such differences could be 'genetically programmed' to develop in a cell-autonomous manner, uninfluenced by differences in the environment around the cell (above and beyond requirements for substances that are necessary for cellular survival and growth). Sex differences could also be due to extracellular ('epigenetic') factors such as the influence of hormones on particular cell types which are initially similar in males and females. Many people have tended to regard these explanations as being mutually exclusive alternatives: genetic 'nature' versus epigenetic 'nurture'. However, any such clear dichotomy between nature and nurture quickly becomes lost, because epigenetic mechanisms (such as hormone dependence) that determine the fate of a given group of cells could themselves be genetically determined, and phenotypes which at first glance appear to develop in a cell-autonomous manner may be influenced in subtle but important ways by the physical and social environments.

Historically, work on genetic influences shaping the fate of nerve cells has been carried out primarily in insects because of the possibility of applying genetic techniques that are not realizable in vertebrates. Work on vertebrate development has tended to emphasize epigenetic explanations. Yet in both insects and vertebrates, sex differences in the nervous system seem to rely on a combination of genetic determination of cell fate (hormone secreting cells) and epigenetic signalling (changes in the development of nerve cells regulated directly or indirectly by hormones). Hormones are the central organizing feature in which nature and nurture meet head on. First, from insects and higher vertebrates, some examples of CNS sexual dimorphisms are given to provide the reader with some feeling for the diversity of the differences that have been discovered so far. Next, we explore the unique niche that hormones occupy in the causal pathways leading to sex differences in the brain.

The development of sexual dimorphisms in insect brains

Sexual dimorphisms are well known in insect nervous systems. In the visual system of the blowfly *Calliphora erythrocephala*, males have an area of high visual acuity, and 12 male-specific visual interneurons in their optic lobes with dendritic fields that serve the high acuity region. Males are better than females at locating and intercepting small rapidly moving targets. In the moth *Manduca sextans*, both sexes have antennal olfactory receptors which project to specialized balls of neural cells called glomeruli in the olfactory lobe of the moth brain. But males have specialized olfactory glomeruli that contain neurons which are selectively responsive to female pheromones (see Chapter 16 by Jean-François Ferveur). In the gustatory system of the fruitfly *Drosophila melanogaster*, there is a sexual dimorphism in the connectivity of gustatory axons to the nervous system, as well as a distinct region of the central nervous system (called the mushroom bodies) where females have 10% more Kenyon-cell fibres than males. In honeybees *Apis mellifera*, the mushroom bodies of sterile female workers are considerably larger than those of male drones.

Sex differences in the nervous system of insects fall into two patterns. In the first pattern, identifiable neurons are uniquely present in one sex and absent in the other. In the second pattern, homologous neurons that are present in both sexes differ morphologically. These patterns suggest that at least two different mechanisms are responsible for structural differences in the insect nervous system. Neurogenesis or cell death could be controlled in a sex-specific way to produce the first type of difference, and neuron-autonomous growth control would produce the second. In both cases, the regulating agents appear to be products of sex-determining genes which act epigenetically.

Two examples of the first pattern are found in the nervous system of *Manduca sextans*. Recall that males and females have different neural structures connected to their antennae. If the imaginal disk containing antennal precursor cells is transplanted from a male to a female larva, a male antenna will develop as well as a male-like olfactory glomerulus with pheromonally responsive neurons. It is not known whether some early hormonal difference between male and female larvae causes the imaginal disk cells to be different before the transplant is performed, or whether male and female differences emerge cell autonomously.

Many *Manduca sextans* larval motoneurons persist through metamorphosis into adulthood and innervate the skeletal muscles of the adult. The survival of some of these motoneurons is correlated with the sex-dependent survival of the muscles that they innervate. In the genital segments, some hindgut motoneurons change their innervation patterns during metamorphosis and connect with the oviduct in adult female moths. These motoneurons are lost in adult males because they do not have the oviduct muscles. The mechanisms controlling the fate of the oviduct muscles are not known.

An example of the second pattern is seen in *Drosophila melanogaster*, where the action of a set of regulatory genes controls all aspects of somatic cell differentiation. Normal function of some of these genes (such as the transformer (*tra*) and transformer-2 (*tra-2*) genes) are necessary to guide female sexual differentiation and to prevent male sexual differentiation in females. Chromosomal females (XX) with mutations in *tra-2* develop into phenotypic males. Temperature-shift experiments with particular strains of *tra-2* mutants, whose gene products function normally within one range

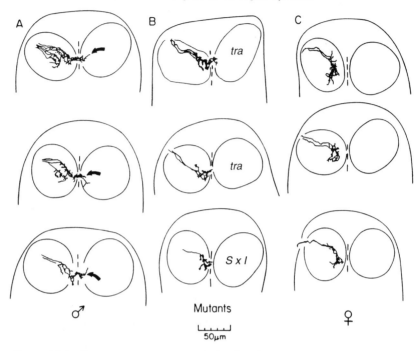

Figure 12.1. The gustatory system of the fruitfly is sexually dimorphic. The branching pattern of gustatory sensory neurons differs between male (A) and female (C) fruitflies. In males, the neuron crosses the body midline (arrow) while it is restricted to one side in the female. Fruitfly mutants (B) with the mutation transformer (*tra*) or sex-lethal (*Sxl*) develop a male phenotype. (Taken from Possidente and Murphy (1989).)

of temperatures and abnormally in another, demonstrate that *tra-2* function is required continuously in all cells during development for the occurrence of female sexual differentiation. Behavioural experiments with these temperature-sensitive *tra-2* mutants show that the regulatory genes controlling sexual differentiation continue to function in the adult CNS (see Chapter 20 by Claude Pieau and colleagues).

As mentioned above, the morphology of gustatory sensory neurons differs between male and female *Drosophila melanogaster* (Figure 12.1). Females with mutations in sex-determining genes develop a male phenotype so that the gustatory neurons form a male-typical branching pattern. However, it is not known whether the expression of these sex-determining genes affects the appropriate parts of the developing brain directly (in a cell-autonomous manner) or indirectly (via some chemical signal from another cell).

The development of sexual dimorphisms in vertebrate brains

For most sexually dimorphic vertebrate brain areas, the relevant brain structures are monomorphic in early development. At some later point during ontogeny, hormonal factors (primarily gonadal steroids) induce the differentiation of these areas. The action of steroid hormones in the CNS requires the production of the steroids in the gonad (or elsewhere), the passive or active transport of steroids to the brain, the tissue-specific presence of metabolizing enzymes in the brain, and cell-specific expression of the proper steroid receptor. Sex-specific gonadal development could lead to a sex-specific hormonal profile and, as a consequence, to sex-specific brain differentiation in areas with a sexually monomorphic distribution of steroid receptors and steroid metabolizing enzymes.

Gonadal steroids affect all levels of sexual dimorphism listed in Table 12.1. To illustrate where, when, and how gonadal steroids induce sex differences in brain phenotype, I will focus on three vertebrate systems that show large steroid-dependent differences in neuroanatomy. These are the song-control system of songbirds (see Figure 12.2), and two different sexually dimorphic areas of the rat nervous system, the motor nucleus of the bulbocavernosus (SNB) in the spinal cord, and the sexually dimorphic nucleus (SDN) of the preoptic area of the hypothalamus.

Where do steroids act?

Steroids affect cells via two main pathways, by binding directly to intracellular receptor proteins or by binding to cell membrane proteins after being converted to so-called 'neurosteroids'. Since we do not yet know whether the second pathway affects cellular differentiation, the subsequent discussion is limited to the steroid–steroid receptor pathway.

Among the large number of different steroids and related substances, only gonadal steroids such as testosterone, 17β-oestradiol and progesterone appear to be important for sex-specific brain development. Cells that express the proper receptor for a certain gonadal steroid are primary targets for steroid action (Figure 12.3a). Within a cell, the primary action site is the genome, since steroid receptors are all ligand-dependent DNA transcription factors.

The hormone secreted by the gonad may not be the same as the hormone that activates the steroid receptors inside neurons. There are enzymes produced only in certain populations of brains cells which mediate the conversion of one steroid hormone into another. The simultaneous presence of an enzyme called aromatase, which mediates the

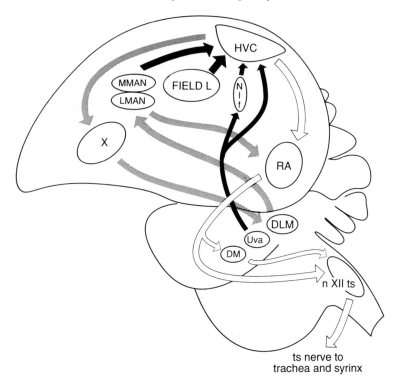

Figure 12.2. The neural song-control system. Sagittal section of an adult songbird brain, showing the major pathways involved in song control. Inputs to the HVC (nucleus hyperstriatalis ventrale, pars caudale) are shown in black. The descending pathway that innervates the sound-producing muscles (syrinx) is shown in white. A recursive loop is shown in grey. Field L is the auditory projection of the caudal neostriatum. DLM, medial portion of the dorsolateral thalamic nucleus; DM, pars dorsalis medialis of the nucleus intercollicularis; LMAN, lateral magnocellular nucleus of the anterior neostriatum; MMAN, medial magnocellular nucleus of the anterior neostriatum; Nlf, nucleus interface; RA, robust nucleus of the archistriatum; n XII ts, pars tracheosyringealis of the hypoglossal nucleus; ts nerve, tracheosyringealis nerve; Uva, nucleus uvaeformis; X, area X.)

conversion of testosterone to oestrogen, is necessary to render the cells responsive to the hormone (see Chapter 20 by Claude Pieau and colleagues). Thus, cells with oestrogen receptors may respond to testosterone secreted by the gonad if these or neighbouring cells contain aromatase activity. One must know both the spatial distribution of hormone receptors and the spatial distribution of metabolizing enzymes to understand hormone-induced changes in the brain.

Only a fraction of all brain cells contain receptors for gonadal steroids. These cells are not distributed randomly in the brain but occur in higher numbers in certain brain structures (such as the hypothalamic and preoptic areas) in species of all vertebrate classes (Figure 12.3b). Steroids seem to induce direct effects in their primary target cells that contain receptors, and they seem indirectly to affect cells that lie in close proximity to receptive cells or that share connections with them. Either the indirect effects of steroid action could be mediated by chemical 'trophic' factors released by the primary target cells or they could be induced electrically.

In the song-control nucleus HVC (nucleus hyperstriatalis ventrale, pars caudale) of the canary, 15–20% of all neurons contain oestrogen receptors. Cells that have no oestrogen receptors are very often clustered (in close spatial proximity) with cells that contain oestrogen receptors. Under oestrogen action almost all cells of the HVC change their morphology, indicating both direct and indirect action of steroids in HVC. Similarly, the SNB motoneurons of the rat exist in clusters. Gap junctions (channels which form direct electrical links between different cells) have been found between these motoneurons, and the number of gap junctions is androgen-dependent. Since only a subpopulation of the SNB motoneurons contain androgen receptors in adulthood, the androgen-receptor-containing motoneurons change the morphology of all SNB motoneurons directly or indirectly.

The song-control nucleus area X of the canary contains no oestrogen receptors but undergoes oestrogen-dependent differentiation. HVC neurons that contain oestrogen receptors project to area X and appear to mediate the differentiation of area X indirectly by electrical contact or release of trophic factors. Apparently, gene products that are the result of steroid action in the primary target cells are transported up nerve cell axons and induce effects across nerve terminals, as is also found in cells projecting from the ventromedial hypothalamus to the midbrain central grey area in the female rat.

The concept of the critical period in the action of steroids

Experimental manipulations of the hormonal milieu show that elevated steroid levels (primarily those of testosterone and its oestrogenic and androgenic metabolites) during a certain period of ontogeny are crucial for the development of permanent structural sexual dimorphisms in the brain. In rats, the critical period for the SDN and the SNB seems to start at embryonic day 17 or 18 with a surge of plasma testosterone in the male

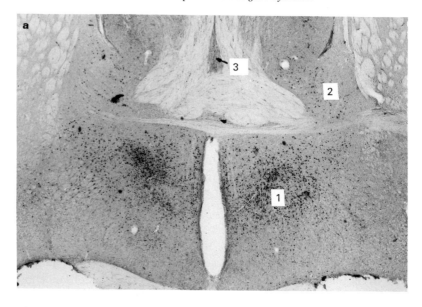

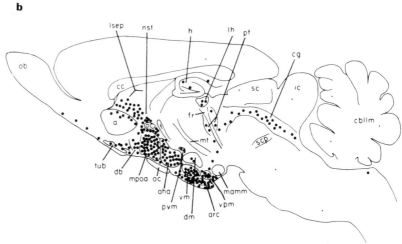

Figure 12.3. (a) Localization of oestrogen target cells in the brain. Oestrogen target cells are identified in the brain by looking for the presence of oestrogen receptors. The cross section of the mouse preoptic area shows cells that are stained with an antibody for oestrogen receptors (darkly stained cells). Oestrogen receptors occur mainly in the medial-preoptic (1) area, in the bed nucleus of the striaterminalis (2), and in the nucleus triangularis septi (3) in this section. (Photograph by M. Gahr.) (b) Sex steroid receptors are expressed in cells in particular areas of the vertebrate brain. Oestrogen receptors occur in high densities in the preoptic and hypothalamic areas. Few oestrogen receptors are found in the cerebellum and in the cortex of adult rats. The distribution of oestrogen-concentrating neurons in the brain of female rats is represented schematically

fetus and ends perinatally. In the female zebra finch, oestrogens induce male-like differentiation of the song areas between day 1 and day 30 posthatching. Later oestrogen treatments result in a reduced degree of masculine development of the song areas, indicating a gradual decline in steroid responsiveness. Oestrogen treatments of adult female zebra finches have no effect on the song system (Figure 12.4).

The events that determine the end of critical periods are unknown. For the female zebra finch, oestrogen receptors in and around the song nuclei HVC and RA (nucleus robustus archistriatalis) that mediate masculinization of these areas disappear at the same time as the cells in the song control areas begin to die. It is thought that the downregulation of oestrogen receptors and the subsequent cell death in the female HVC and RA are due to the lack of some necessary hormone during a critical period. In the SDN of female rats, oestrogen receptors persist despite a sexually dimorphic differentiation of the SDN. This indicates that the lack of oestrogens in the preoptic area, rather than the lack of oestrogen receptors, regulates the feminine differentiation of SDN neurons.

There are different critical periods for the differentiation of particular brain structures in the same individual. For example, the steroid-dependent masculinization of the song system in female zebra finches is temporally independent of the differentiation of brain areas that control female copulatory behaviour and egg laying. Critical periods for steroid-dependent brain differentiation are area specific.

In contrast with these permanent early events, there are steroid-induced changes in the brain at later stages of life which are reversible and maintained only under certain hormonal profiles. In songbirds such as the canary, testosterone and its metabolites alter the soma size, dendritic length, synapse morphology, and phenotypic appearance of the neurons

---

in a sagittal plane. Locations of these neurons are represented by black dots. (a, Nucleus accumbens; ac, anterior commissure; aha, anterior hypothalamic area; arc, arcuate nucleus; cbllm, cerebellum; cc, corpus callosum; cg, central grey; db, diagonal band of Broca; dm, dorsomedial nucleus of the hypothalamus; f, fornix; fr, fasciculus retroflexus; h, hippocampus; ic, inferior colliculus; lh, lateral habenula; lsep, lateral septum; mamm, mammillary bodies; mpoa, medial preoptic area; mt, mammillothalamic tract; nst, bed nucleus of the stria terminalis; ob, olfactory bulb; oc, optic chiasm; pf, nucleus parafascicularis; pvm, paraventricular nucleus (magnocellular); sc, superior colliculus; scp, superior cerebellar peduncle; tub, olfactory tubercle; vm, ventromedial nucleus; vpm, ventral premammillary nucleus.) (Taken from Pfaff, D. and Keiner, M. (1973) Atlas of estradiol-concentrating cells in the CNS of the female rat. *Journal of Comparative Neurology*, **151**, 121–58.)

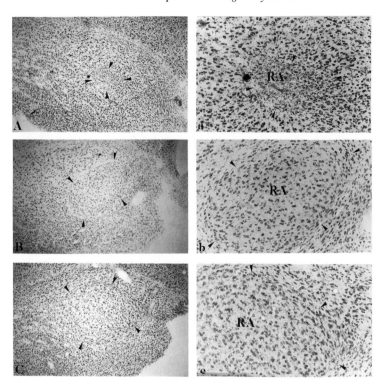

Figure 12.4. Oestrogen-dependent differentiation of song-controlling brain areas in the female zebra finch. Treatment of juvenile female zebra finches induces male-like differentiation of the song-control area RA (arrows). (A, a) RA of a female zebra finch treated with oestradiol in adulthood. (B, b) RA of a female zebra finch, treated with oestrogens during ontogeny. (C, c) RA of the adult male zebra finch. RA of females treated with oestrogen in ontogeny is much larger compared to untreated females or females treated in adulthood only. Oestrogens spare cells in RA from cell death during ontogeny. A, B, C show low-power photomicrographs (200 ×); a, b, c show high-power photomicrographs of the Nissl-stained sections (500 ×).

in the song areas HVC and RA. In the SNB of adult rats, the size of the motoneurons, their dendritic length, and the number of gap junctions and chemical synapses they make are testosterone-dependent. In the preoptic area of Japanese quails and ring doves, aromatase activity is testosterone-dependent. In the ventromedial hypothalamus of adult female rats, the level of progesterone receptors and of oxytocin receptors is oestrogen-dependent. The induction of reversible quantitative differences in adulthood can occur in the same structures that underwent permanent sex-specific differentiation earlier in development. Typically, the adult alterations strengthen the degree of sexual dimorphism achieved by these structures.

Steroids and neuron number: a common vertebrate brain dimorphism

As seen in Table 12.1, sex differences in the number of neurons in specific brain areas have been reported in mammals, birds, lizards, frogs and fishes. Neuron number is governed by: how many cells arise in a given brain region; how many move into it or leave it; and how many of the cells which try to stay in this region survive. In most areas that have a sexual dimorphism in neuron number in adulthood, there is no such dimorphism in early ontogeny. One of the factors affecting neuronal survival is the connectivity that cells in a given region are able to maintain with their synapse partners (both within the central nervous system and with the afferent and efferent periphery).

Sexual dimorphisms in the CNS could be induced by sexual dimorphisms in the afferent and efferent periphery. Muscles may be fewer in number or completely absent in one sex, as in the sonic muscles of the sound-producing teleost plainfin midshipman fish *Porichthys notatus*, in the laryngeal muscles of the frog *Xenopus laevis*, in the syringeal muscles of the zebra finch, or in the bulbocavernosus muscles (BC) of the rat.

The BC muscles together with their motor nucleus, the SNB, control penile erection and are essential for successful insemination. At birth, the BC muscles are present in both males and females, but they degenerate as females develop. The neurons of the SNB of adult rats are only one-third as numerous in females as in males, yet the number of SNB neurons is not sexually dimorphic in newborn rats. Androgen treatment of female rats saves the BC muscles and saves the SNB neurons from cell death. Since only the BC muscles, but not the SNB neurons, contain androgen receptors, the androgen-dependent survival of the appropriate target of SNB neurons leads to the survival of the moto-neurons. In females, what happens to the higher neural circuits that drive copulation in male rats? The degree to which peripheral effects of steroids on BC muscle survival feed back to higher brain areas has not yet been determined. The effects that sex-specific survival of the rat penile musculature impose on the structure of the nervous system could serve as a general model for relating peripheral neuron survival, structure or function to sex-specific changes in the organization of other regions of the higher nervous system (see Chapter 11 by Evan Balaban for further development of this theme).

More centrally located synaptic changes may also modulate cell number. The song-control system of songbirds has a chain of interconnected brain nuclei that innervate the vocal organ called the syrinx (see Figure 12.2). The forebrain song areas, HVC and RA, are much larger and contain many more neurons in adult male zebra finches compared to adult female zebra finches. These sex differences in neuronal number appear to be the result of neonatal effects of oestrogens.

The song nucleus RA develops in a similar fashion in males and females up to day 30 posthatching. At this time, neurons in female RA and HVC start to atrophy and to die, whereas neurons in male HVC and RA continue to grow. Male HVC neurons start to innervate RA; female HVC neurons do not. Oestrogen treatment of juvenile female zebra finches induces their HVC neurons to innervate RA and prevents both HVC and RA neurons from atrophy and death (Figure 12.4). Since most of the neurons in RA are generated before hatching, and because females lose many of these neurons during the juvenile period, differential neuronal death is believed to be responsible for the development of this sex difference. Although the oestrogen-dependent mechanisms that prevent cell death are unknown, the innervation of RA by HVC neurons appears to be a crucial event in the survival of both song nuclei (the HVC of the male zebra finch degenerates if HVC neurons are prevented from innervating RA). If HVC fibres that grow towards RA are cut or if HVC is lesioned, oestrogen cannot prevent RA neurons from cell death or atrophy. During this period in development, oestrogen receptors are found in and around HVC and close to where HVC axons innervate RA. Oestrogens could, therefore, induce the growth of HVC neurons and induce HVC axons to grow into RA.

Cell death is also likely to be responsible for the sexually dimorphic development of the anteroventral nucleus of the preoptic area (AVPv–POA) of the rat and gerbil. The AVPv is larger and has a greater density of cell bodies in females. Neonatal testosterone treatment of female rats induces an increased rate of cell death and degeneration in the AVPv compared to untreated females, which results in a male-like phenotype of this brain area.

The SDN of adult rats is sexually dimorphic in volume and in neuronal number. This dimorphism is first detectable postnatally (5–10 days after birth) while most of the SDN neurons are present before embryonic day 18. At birth, the SDN phenotype is not sexually dimorphic. Steroid-dependent neurogenesis is not a likely cause for the SDN dimorphism, since oestrogen treatment alters neuron number after neurogenesis has ceased. No oestrogenic effects on the migration pattern of SDN neurons have been found; cell respecification or cell death may account for the male–female difference. The SDN is connected with another sexually dimorphic area, the bed nucleus of the stria terminalis, and perinatal testosterone exposure induces a male-like pattern of synaptic terminations from the stria terminalis to the medial-preoptic area. Because both the bed nucleus of the stria terminalis and the SDN contain oestrogen receptors in male and female rats, oestrogen could affect the SDN directly or via afferent innervation.

A final mechanism that generates sex differences in neuron numbers is continuous cell birth. Neurons are incorporated in the song nucleus HVC of the zebra finch throughout life, but the rate of incorporation is highest during the initial differentiation of the song system when birds are still juveniles. Many more new cells are incorporated in the male HVC at this time as compared to the female HVC.

These examples indicate that regions may vary with respect to the mechanisms that underlie sexual differentiation of neuron number. Sex-specific survival of neurons in CNS nuclei and/or sex-specific respecification of neurons appear to be a common way to shape a sexual dimorphism in neuron numbers. These different mechanisms could also act synergistically.

Intracellular events mediating steroid-dependent brain sex

The chain of events induced by sex steroids leading to structural and neurochemical sex differences is not known in detail. Do steroids trigger their effects directly or induce expression of certain proteins that themselves initiate changes in cell growth or phenotype? Do different steroids induce their cellular effects independently or synergistically? Do steroids just initiate cellular changes, or are they required at different key steps in a cascade of events?

Phenotypic neural changes can occur at widely varying times after systemic steroid treatments, from 30 minutes to several weeks. This makes it difficult in many cases to distinguish between primary steroid-dependent molecular and structural events and subsequent events that may not be directly steroid-dependent.

Since steroid receptors are all ligand-dependent gene transcription factors, sex steroids are likely to trigger cascades of events in the brain similar to those induced in peripheral tissues such as the oviduct and the liver (see Chapter 10 by Jan-Åke Gustafsson). Indeed, sex-steroid treatment does increase protein production in sexually dimorphic brain areas such as the ventromedial hypothalamus of the female rat and the forebrain song-control areas of songbirds. Primary genes that are turned on via steroid receptors must have hormone-responsive elements, but these have been found only in a few genes thus far. Steroids could induce the expression of 'early genes' such as *c-fos* that themselves trigger a cascade of biochemical events which initiate changes in cell phenotype. Alternately, sex steroids could regulate the expression of receptors or enzymes directly. Examples are: the testosterone-dependent upregulation of aromatase in the hypothalamus and preoptic areas of the ring dove (*Streptopelia risoria*) and the Japanese quail (*Coturnix coturnix japonica*); the oestrogen-dependent downregulation of oestrogen receptors in the zebra finch forebrain and in the hypothalamic ventromedial and arcuate nuclei in rats; and the oestrogen-dependent upregulation of progesterone receptor mRNA in the ventromedial hypothalamus of the female rat.

In addition to their effects on regulatory gene products, steroids also affect the synthesis of proteins participating in the structural alterations necessary for cellular growth. *Growth-associated-protein-43* is upregulated by oestrogens in the ventromedial hypothalamus and the cortex of the rat, and the cytoskeletal protein tubulin is upregulated by neonatal testosterone in the rat hypothalamus.

Sex differences in steroid-dependent gene expression are ultimately a consequence of two different processes. The first process regulates the distribution of steroid receptors and steroid metabolizing enzymes in the brain as well as the distribution of factors that interact with steroid receptors such as heat-shock proteins. The second process is the sex-specific production of gonadal steroids. The pattern of expression of steroid receptors in the brain, and the pattern of gonadal steroid production have so far proved to be unresponsive to the experimental manipulation of environmental factors during early ontogeny, when permanent differences in steroid-dependent brain differentiation are taking place. Although we are gaining some understanding of the genetic control of gonadal hormone secretion (see Chapter 8 by Jean Wilson, Chapter 9 by Marilyn Renfree and Chapter 18 by Jennifer Graves), we have little specific information on how the patterns of steroid-receptor and steroid-metabolizing enzyme expression are established in the developing brain.

Other, non-steroid factors may also be important for sexual differentiation of the brain. The differentiated gonads of males and females secrete various non-steroid humoral factors at the time when morphological differentiation of the brain starts. Müllerian-inhibiting-substance (MIS) is one such factor which causes regression of Müllerian ducts in males. MIS is a peptide growth factor which binds to a putative member of the transforming-growth-factor-α (TGF-α) family of growth factor receptors, and was found recently in various parts of the developing rat brain where its function is currently not known. The testis-determining gene SRY may also play a role in the differentiation of sex-specific brain areas. SRY was shown recently to bind to the promoter of the gene for aromatase and to regulate the transcription of this enzyme. Since aromatase is expressed in the brain of vertebrates and since SRY is also expressed in the brain (at least in mice), SRY could affect aromatase expression directly (see Chapter 18 by Jennifer Graves and Chapter 19 by Karl Fredga).

Environment and behaviour can influence brain sex

The hippocampus and the cerebral cortex of rats are sexually dimorphic in size and dendritic properties. In an experiment examining the effects of the environment on the development of sex differences in the brain, juvenile rats were placed in environments of different complexity at weaning and they were left there for one month. Rats that were housed in standard laboratory cages (a relatively boring environment) showed little diversity of behaviour in either sex, and no sex differences were found in the behaviour of these rats. If rats were housed in more 'complex' environments which had extra objects for sensory stimulation and were allowed to play, about 2% of their behaviours were sexually dimorphic, and males showed significantly larger total dendritic length of neurons within their visual cortex compared to females raised in the same environment. Furthermore, males in this 'complex' environment had increased dendritic length compared to males raised in the simple environment. Thus, the environment has a profound influence on the emergence of sex differences in the visual cortex. It is unclear why males and females respond differently to the more complex environment, and what factors mediate the structural difference that this environment produces in their brains.

In group-living animals, social ranking is important for gaining mating opportunities. In extremely competitive social systems such as those found in voles, hamsters and certain monkeys, dominating females suppress oestrus in subordinate females. Copulatory behaviour can also influence hormone secretion. Ovulation occurs subsequent to an oestrogen-dependent surge of luteinizing hormone (LH) in spontaneously ovulatory species such as hamsters, or after an intromission-dependent LH-surge in reflex ovulators such as voles. In male rabbits and voles, sexual arousal and/or intromission causes a marked increase in serum testosterone and dihydrotestosterone. If mating is suppressed then no ovulation, copulation or subsequent related behaviours occur in females or males. Since increased levels of steroids are involved in structural alterations in sexually dimorphic brain areas of adult males and females, differences in the mating behaviour of individuals could affect the sex-specific cytoarchitecture and neurochemistry of their sexually dimorphic brain areas in an ongoing fashion.

Synthesis

The following general scenario summarizes our current understanding of the steroid-sensitive development of vertebrate brain sex (Figure 12.5). The same regulatory genes may control the origin of homologous brain circuits in males and females. Although these circuits first develop monomorphically, their subsequent fate depends on the pattern of secretion of gonadal steroids, the spatial distribution of enzymes which convert these steroids to different forms, and the location and number of intracellular receptors which respond to the presence of hormones.

Sex-specific genes such as SRY and MIS lead to sex-specific gonadal differentiation. The expression of steroid receptors and steroid metabolizing

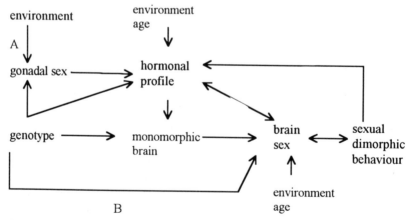

Figure 12.5. Causes of brain sex. This schema indicates possible interactions between the genotype and the physical and social environment for the development of brain sex and sexually dimorphic behaviour. 'A' indicates the environmental sex-determination pathway (see Chapter 20 by Pieau and colleagues). 'B' indicates a cell-autonomous pathway.

enzymes in certain brain regions together with a sex-specific pattern of gonadal hormone secretion causes the initial emergence of sex differences in brain circuits by regulating parameters such as neuron number. A difference in neuron number in one or several areas can lead to differences in connectivity, synapse density, electrophysiological properties and neuronal survival in these areas and in their afferent and efferent connections. The subsequent action of steroids strengthens these sex differences.

The initial steps in brain differentiation rely on patterns of hormone secretion and spatial patterns of receptor and enzyme expression in the brain which appear to be resistant to environmental input. It is unclear what determines these initial differences. In later developmental stages, physical, social and environmental factors (ageing, social ranking, stress, nutrition) that alter sex-steroid production can help to determine the further fate of sexually dimorphic circuits.

Functional consequences of sexually dimorphic brains

While the last 25 years have seen significant progress in documenting male–female brain differences and in studying their developmental origins, there has been surprisingly little progress in making direct links between vertebrate brain sex-differences and behaviour. This section discusses why work on functional consequences has lagged behind work on brain differences, and makes some suggestions about how to remedy this situation.

Relevant patterns of behavioural sex differences

Sex differences in behaviour come in two forms. 'Sex-specific' behaviours are those that are shown exclusively by one sex of a species. Examples are male intromission, female live-birth and egg-laying, and many male-specific mate attraction displays such as frog mate-calling, singing in bird species such as the zebra finch, and pheromone release by female insects. Much parental behaviour can also be shown in a sex-specific fashion. However, sex-specific behaviours are numerically a minority. Most sexual behaviours are 'sex-typic', meaning that they can be produced by either sex but are shown more frequently in one sex. This can be because the sexes differ in their predispositions to exhibit a particular kind of behaviour in any circumstance. It may be because, although they are equally likely to give the behaviour, it occurs in particular situations which happen more frequently to one sex. A final possibility is that the two sexes give the same behaviour under very different circumstances.

Sex-typic behaviours are hard to work with, because behavioural diffferences can be situation specific. For example, female zebra finches display copulation solicitation behaviour before they mate with a male. This behaviour consists of a distinctive crouch wherein the cloaca is presented to the male, together with a rapid vibration of the wings and specialized vocalizations. Yet the same behaviour can be shown by males after copulation and may have a rather different social function from that of female copulation solicitation. We do not yet know how this kind of subtle situational difference should be reflected in the structure of the nervous system.

Behavioural diversity within a sex is also a problem for relating sex differences in the brain to behaviour. In some species, not all male individuals use the same mating strategy. In vocalizing fish such as the plainfin midshipman, one type of male looks like the female, and shows female-like vocal behaviour. These 'sneakers' gain access to females by inserting themselves into a mating situation and by depositing sperm surreptitiously. Thus, they copulate like territorial males without showing the complete sequence of male courtship and other reproductive behaviour.

Relationships between structure and function

In general, sex differences in endocrine function and behaviour correlate with dimorphic brain structures or endocrine patterns (Table 12.2). For example, the sexually dimorphic SDN correlates with a sexual dimorphism in rat copulatory behaviour. The sexual dimorphism in size and neuronal number in the forebrain song nuclei of zebra finches and canaries correlates with sex differences in their song. Sex differences in the scent-marking behaviour of gerbils correlate with sexual dimorphism in their medial preoptic area, and sex differences in the endocrine architecture of the ventromedial hypothalamus correlate with the lack of lordosis behaviour in the male rat. But beyond this general level of correlation, it can be difficult to match brain differences among individual males and females to precise behavioural consequences.

There are also known cases of dimorphic sexual behaviour that are not correlated with brain differences. For instance, in the all-female lizard *Cnemidophorus uniparens*, females can behave either like the males or the females of related heterosexual species, where differences have been found

Table 12.2 *Function of sexually dimorphic structures of the central nervous system (CNS)*

Sexually dimorphic structures of CNS	Sexually dimorphic behaviour/function	Species
Medial preoptic nucleus	Scent marking, copulation, ultrasonic vocalization	Gerbil m > f
Sexually dimorphic nucleus of the preoptic area	Sexual arousal? Copulation	Rat m > f, montane vole m > f
Medial preoptic nucleus	Copulation	Quail m > f
Bed nucleus of the stria terminalis	Copulation?	Rat, gerbil m > f
Anteroventral-periventricular nucleus	Ovulation?	Montane vole, guinea pig f > m
Locus coeruleus	?	Rat f > m
Cerebral cortex: primary, sensorimotor, monocular area, binocular area	?	Rat m > f
Hippocampus	?	Rat m > f
Hippocampus	Spatial behaviour	Montane vole m > f
Forebrain song areas	Singing	Zebra finch, canary m > f
Spinal nucleus of the bulbocavernosus	Penile reflex	Rat, mouse, dog, human m > f
Larynx-control area	Vocalization	*Xenopus laevis* m > f
Glomeruli of the olfactory lobe	Mate recognition	*Manduca sexta* m > f
Male-specific visual neurons	Localization of rapidly moving objects	Calliphorid flies m > f

Abbreviations: ?, uncertain functions; f, female; m, male; m > f, an area is smaller or missing in the female; f > m, an area is smaller or missing in the male.

in the phenotype of certain brain regions. Yet unisexual individuals showing male or female behaviour patterns do not manifest the brain differences seen in the heterosexual species (see Chapter 2 by David Crews).

Problems encountered in relating structure and function

There are three basic hurdles that must be overcome in suggesting structure–function relationships between the brain and behaviour. The first problem is the spatial localization of function in the brain. It is common procedure to study the effect of lesions or altered neurochemical properties on behaviour or endocrine profiles. These approaches are too indirect, for at best they can show only that the affected brain area somehow participates in a particular behavioural function.

A second problem is encountered in actually trying to document any behavioural alterations which are correlated temporally with lesions or with neurochemical manipulations of brain areas. Most studies are performed under laboratory conditions that may not match the native environment of the animal, which makes interpreting them difficult. Experimenters usually test very few facets of a behaviour in very few social contexts. But the social context in which behaviours are observed is very important for animals that normally live in groups or in pairs. For example, the red-cheeked cordon blue (*Uraeginthus bengalus*) is an African songbird that lives in lifelong pairbonds. Both males and females use their song to remain in contact, and they sing very frequently when isolated, but the female does not sing much while with her mate. When tested alone, male and female rhesus monkeys (*Macaca mulatta*) respond strongly to distress signals from an isolated infant. However, if the same male and female are housed together, only the female responds to the infant. Thus, singing in cordon blues and parental care in rhesus monkeys appear to be sexually dimorphic behaviours when males and females are tested together, but appear to be monomorphic when males and females are tested separately.

A third problem is the meaningful measurement of the size of brain structures or the number of neurons they contain. The determination of the borders of the relevant brain area is a crucial step in this process. Commonly, subjective borders are drawn in histologically stained brain sections (e.g. Nissl stain, see below) that separate the intensely stained portions of brain tissue from the lightly stained regions which surround them. This procedure can lead to erroneous results as shown for the song-control nucleus HVC of the canary and the song nucleus magnocellular nucleus of the anterior neostriatum (MAN) of the zebra finch (Figure 12.6). The use of connectivity or of neurochemical markers to identify HVC and MAN can lead to very different volume determinations of these brain areas from those measured with subjective line-drawing. The subjective procedure led investigators to conclude that the canary HVC undergoes seasonal alterations in size, when in fact it does not. Rather, it undergoes seasonal change in its phenotypic appearance, as seen in a particular type of tissue staining commonly used to visualize neural cell bodies (called the Nissl stain). When delineated by the

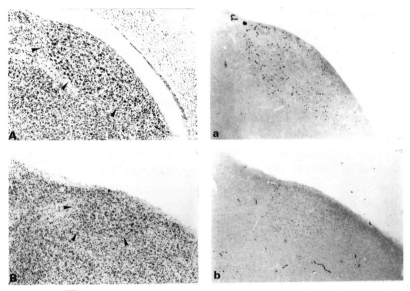

Figure 12.6. Problems in measuring the size of a brain area. Photomicrographs of the HVC of adult male canaries during the breeding period (A, a) and during the non-breeding period (B, b) taken from the same part of the brain nucleus. First, sections were stained immunocytochemically with an oestrogen-receptor antibody (a, b) and then counterstained with a common histological stain (Nissl) (A, B). Oestrogen-receptor containing cells are stained darkly. (a, b) Arrowheads indicate the HVC. During the breeding period, the Nissl-defined HVC (A) and the antibody-stained HVC (a) are similar in size. In contrast, during the non-breeding period the Nissl-defined HVC (B) is difficult to distinguish, whereas the antibody-stained HVC (b) is of unchanged size. (Taken from Gahr (1990).)

distribution of neurochemical markers and the connections that its cells make with other regions, the HVC shows no seasonal difference in size. Furthermore, the HVC is less sexually dimorphic in volume and neuron number when these newer techniques are used.

Relationships between brain area volume and behaviour

Sexually dimorphic brain area volumes are the most commonly reported sexual dimorphism, and, frequently, they are correlated with sex-specific behaviour and function. Since the work of Nottebohm on the zebra finch and the canary, and that of Gorski on the rat, it is commonly accepted that the size of an area correlates with the degree of differentiation of the behaviour controlled by that area. Birds with larger song areas (HVC and RA) are thought to have more complex songs or larger song repertoires. Female rats have a smaller SDN than males which correlates with their lack of copulatory behaviour. In the case of the songbirds, this view is correct only for those species in which the females do not sing at all, such as the zebra finch or the orange bishop bird (*Euplectes franciscanus*). In masculinized female zebra finches that sing like males, the male song areas are still up to $3\frac{1}{2}$ times larger than the female song areas, despite the fact that there is now little difference in male and female song behaviour. In songbird species in which the female normally sings, such as the canary, the cordon blue and the red-winged blackbird (*Agelaius phoenicus*), the degree of size difference between male and female song areas varies widely. In all these cases the areas measured in Nissl-stained material are larger in males compared to females. In duetting species such as the African robin chat (*Cossypha caffra*), female song areas are smaller than those of the male. Yet in most duetting species, males and females can sing the entire duet when isolated from their partners. Even within a sex, the size of song nuclei does not always correlate with the size of the song repertoire. In the rufous-sided towhee (*Pipilo erythrophthalmus*), song nuclei undergo seasonal changes in size (as measured in Nissl-stained sections) without seasonal changes in song repertoire. Finally, in the male orange bishop bird, the volume of song areas correlates not with the repertoire size but with plumage colour.

Female canaries that sing like males after testosterone treatment have song areas that are only about half the size of males with similar songs (Figure 12.7). The number of neurons in the HVC of such females (about 20 000) is only about half the number of a typical male (about 40 000). If females can show male-like song behaviour with only half the number of HVC neurons, then the female brain could control singing differently from the male brain, or the male HVC could have additional functions besides song control. Variation in this aspect of the HVC phenotype may simply be without consequence for singing behaviour.

The medial preoptic area of rats and gerbils is sexually dimorphic in size or contains a sexually dimorphic subarea (SDN) (Table 12.2). The connectivity of the medial preoptic area and various neurochemical properties are sexually dimorphic. The medial preoptic area appears to be involved in the control of different masculine tasks in different rodents, such as sexual arousal in rats, and scent-marking, ultrasonic courtship vocalization, and copulation in gerbils. Even in the most frequently studied sexually dimorphic area of the preoptic area, the rat SDN, there is still some controversy about what particular behavioural functions it controls. Lesions of the SDN of sexually experienced rats do not

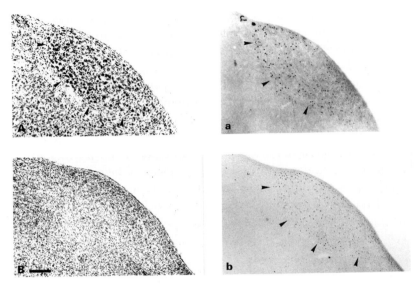

Figure 12.7. A sexually dimorphic brain area. The HVC of adult canaries is twice as large in males (A, a) as in females (B, b). (A, B) HVC as seen with a common histological stain (Nissl); (a, b) The same sections stained immunocyto-chemically with an oestrogen-receptor antibody. Both techniques reveal the same brain area (arrows). Although the female whose HVC is pictured here sang a male-like song, her HVC has a smaller cross-sectional area than the male HVC. (Scale bar is 100 μm. Taken from Gahr, M., Gütteniger, K.-R. and Kroodsina, O. E. (1993) Oestrogen receptors in the avian brain: survey reveals general distribution and forebrain areas unique to songbirds. *Journal of Comparative Neurology*, **327**, 112–23.)

disturb sexual behaviour. In inexperienced rats, lesions affect copulatory behaviour primarily in suboptimal testing conditions with less attractive females. SDN lesions do not affect plasma testosterone, LH and prolactin levels in response to sexual stimulation in experienced or inexperienced rats. Therefore, the SDN of rats is not essential for the display of copulatory behaviour, although a certain role for sexual arousal is suggested. It is unclear whether differences in SDN play any role in mediating behavioural differences among individual animals.

The recent findings of LeVay, and Gorski and coworkers, of brain areas that differ in volume between heterosexual men and women, and homosexual men, require special comment. Both groups claim to have found differences in the size of brain regions which correlate with differences in sexual orientation: the third interstitial nucleus of the anterior hypothalamus (LeVay) was larger in homosexual men and women and the anterior commissure (Gorski) was largest in heterosexual men. Similar to most animal examples, these results are based on a subjective examination using simple histological staining. In the light of the above-mentioned problems that this technique has encountered in similar work on animals, a more careful analysis using connectivity or neurochemical markers to identify the boundaries of a brain area is necessary before a correlation between sexual orientation and the volume of brain areas could be accepted. Furthermore, preoptic–hypothalamic areas that have been reported to be sexually dimorphic between men and women in previous studies were not sexually dimorphic in the recent study of LeVay. This indicates either measuring problems or high inter individual variability in the morphology of the preoptic–hypothalamic area. If, indeed, the volume of brain areas correlates with sexual orientation, causes and consequences of these polymorphisms could then be studied.

Toward the future: better ways of establishing brain–behaviour links

I believe that the overall size of a brain area or its neuron number are poor parameters to use in trying to link CNS structural differences with behavioural functions. Instead, correlations should be sought with the phenotypic characteristics of identifiable sets of cells that are important for differential development or differential function. As shown for the canary song system, measuring the overall volume of brain structures such as the song nucleus HVC has yielded little useful information and, instead, has led to erroneous inferences concerning neural parameters that may contribute to sex-specific behaviour.

There is no clear correlation between the volume of a structure and its underlying cellular properties. For example, no sex differences are found in the size of the splenium of the corpus callosum, but female rats have more axons in this area than male rats. More detailed structural and neurochemical sex differences are known for some steroid-sensitive sexually dimorphic brain areas; these include protein synthesis, neuropeptide contents, dendritic arborization, neuronal size, and synaptic properties (such as the density of synaptic vesicles in the HVC and RA of songbirds). Although these characteristics are probably more instructive than gross regional size and neuronal number (because they are closer to the electrophysiological properties of cells thought to be important for their function in behavioural circuits), the behavioural impact of sex differences in these categories is simply not known at present. The problem of structure–function relationships is not specific to research into sexual dimorphism, but can be applied generally to that of brain physiology.

Future research on brain sex as a cause of behavioural sex in vertebrates should analyze consequences of structural dimorphisms in cellular and network properties rather than collecting more and more 'macroscopic' sexual dimorphisms that correlate with sex differences in behaviour. New technical approaches such as small trans-sexual tissue-transplants or morphologic and genetic manipulations of defined cell populations combined with electrophysiological and behavioural studies could give us the crucial insights we are currently seeking. For instance, one could see whether the distribution of steroid receptors and metabolizing enzymes which mediate many early differences in vertebrate brain sex are regulated in a cell-autonomous fashion by making intersexual brain chimaeras prior to the occurrence of sexual differentiation of the gonad (see Chapter 11 by Evan Balaban). More refined work on how sexually dimorphic cells actually function differently within their neural circuits (using a combination of neuropharmacological and neurophysiological techniques) should also be most enlightening.

The integration of behaviour, endocrinology, neuroscience systems and cell and molecular biology makes the study of brain sex one of the most exciting and active areas in the quest to understand and to explain sex differences. With the current shift in emphasis of this field away from just describing how brain sex-differences are correlated with behaviour toward understanding the development of these differences and their exact relationship to neural function, the study of brain sex should be able to answer many of our most basic questions about what makes the sexes truly different, and how they come to be that way.

Further reading

Belote, J. M. and Baker B. S. (1987) Sexual behaviour: its genetic control during development and adulthood in *Drosophila melanogaster. Proceedings of the National Academy of Sciences, USA*, **84**, 8026–30.

Gahr, M. (1990) Delineation of a brain nucleus: comparisons of cytochemical, hodological, and cytoarchitectural views of the song control nucleus HVC of the adult canary. *Journal of Comparative Neurology*, **294**, 30–6.

Gorski, R. A., Gordon, J. H., Shryne, J. E. and Southam, A. M. (1978) Evidence for a morphological sex difference within the medial preoptic area of the rat brain. *Brain Research*, **148**, 333–46.

Kelley, D. B. (1988) Sexually dimorphic behaviours. *Annual Review of Neuroscience*, **11**, 225–51.

Konishi, M. (1989) Birdsong for neurobiologists. *Neuron*, **3**, 541–9.

LeVay, S. (1991) A difference in hypothalamic structure between heterosexual and homosexual men. *Science*, **253**, 1034–7.

Matsumoto, A. (1991) Synaptogenic action of sex steroids in developing and adult neuroendocrine brain. *Psychoneuroendocrinology*, **16**, 25–40.

Nottebohm, F. and Arnold, A. P. (1976) Sexual dimorphism in vocal control areas of the song bird brain. *Science*, **194**, 211–13.

Possidente, D. R. and Murphey R. K. (1989) Genetic control of sexually dimorphic axon morphology in *Drosophila* sensory neurons. *Developmental Biology*, **132**, 448–57.

Purves, D. (1988) *Body and Brain*. Harvard University Press, Cambridge, MA.

Raisman, F. and Field, P. M. (1973) Sexual dimorphism in the neuropil of the preoptic area of the rat and its dependence on neonatal androgen. *Brain Research*, **54**, 1–29.

Rand, M. N. S. and Breedlove, M. (1988) Progress report on a hormonally sensitive neuromuscular system. *Psychobiology*, **16**, 395–405.

Sex differences in behaviour

13

Hormone–behavior interactions and mating systems in male and female birds

JOHN C. WINGFIELD

At one time or another, most of us have marvelled at the dawn chorus of bird song. Closer investigation reveals not only great diversity of sounds, but also spectacular variations in sexual dimorphism of plumage (Figure 13.1). Dimorphism in sexual tracts is always marked. The male has a testis and vas deferens leading to a simple copulatory organ (cloacal protuberance), whereas the female of the vast majority of species has a single ovary (usually the left), and a corresponding single but highly specialized oviduct that transports and processes the egg prior to oviposition. There are no known instances of hermaphrodites or parthenogens in free-living populations of birds. Sexual behavior is also highly dimorphic in most birds with males and females expressing distinct courtship and copulatory behavior. In a few cases, sex roles are reversed as in polyandrous species in which females actively court males and are more brightly colored. In contrast, patterns of reproductive aggression (territorial behavior, competition for mates or other resources related to reproduction), plumage and associated structures such as wattles, combs, plumes, etc. are extremely complex not only between sexes, but also within sexes and even among individuals within a population! It is this diversity of integumentary structures and reproductive aggression that will be the focus here.

A steroid hormone, testosterone, plays a pivotal role not only in regulating the diversity of morphological structures, but also in modulating the social structure of free-living bird populations. Secretion of testosterone is regulated indirectly by the central nervous system (CNS) (Figure 13.2). Environmental signals, both physical and social, are perceived by the CNS, transduced, and integrated with internal information to regulate release of a neuropeptide, gonadotropin-releasing hormone (GnRH) from neurons in the basal hypothalamus. This peptide is secreted into small

Figure 13.1. Examples of sexual dimorphism in birds. The western gull, *Larus occidentalis* (upper left panel), is monogamous, monomorphic in plumage and, to a great extent also, in behavior. Both males and females are territorial and incubate and feed their young. The mallard, *Anas platyrhynchos* (mid left panel), is also monogamous but there is great dimorphism in plumage and behavior. Males provide no parental care. The redwinged blackbird, *Agelaius phoeniceus*, is a typical polygynous species in which females (upper right panel) are smaller than males and also cryptic in plumage. Males (mid right panel) are strikingly colored (black with red and yellow wing patches) and show virtually no parental behavior. The bottom panels are the female (left) and male (right) northern phalarope, *Phalaropus lobatus*, a typical polyandrous species. Females are larger than males, more brightly colored and males provide all parental care. (All photographs by author.)

blood vessels that form a specialized portal system to endocrine cells in the anterior pituitary. Here GnRH stimulates synthesis and release of gonadotropins (Figure 13.2), particularly luteinizing hormone (LH), into the peripheral blood stream. Endocrine cells within the testis and ovary

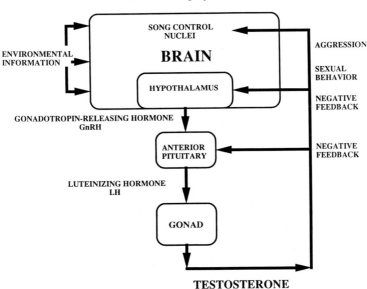

TESTOSTERONE

Figure 13.2. A schematic diagram showing the known links between perception of environmental stimuli and control of gonadotropin secretion and testosterone release and the central effects of testosterone in birds. (Redrawn from Wingfield, J. C. and Moore, M. C. Hormonal, social and environmental factors in the reproductive biology of free-living male birds. In *Psychobiology of Reproductive Behaviour: an Evolutionary Perspective*. D. Crews, ed., © 1987, p. 152. Reprinted by permission of Prentice Hall, Englewood Cliffs, New Jersey.)

are induced by LH to secrete sex steroid hormones such as estradiol and testosterone (Figure 13.2). External environmental signals can be transduced into internal chemical signals via this cascade of events. Sex steroid hormones have a wide spectrum of effects – morphological, physiological and behavioral – and, generally, orchestrate gamete maturation and transport, development of secondary sex characteristics and sexual behavior. Testosterone also has marked effects on aggression in both males and females (Figure 13.2) through organizational (i.e. the hormone controls development of neuronal circuits but is not important in the adult) and activational (i.e. the presence of testosterone is always required to increase the frequency with which a particular behavioral trait is expressed) influences on neurons at various levels throughout the CNS.

Simple experiments show that if the source of testosterone, the testis, is removed, then aggression in reproductive contexts is generally suppressed. In birds, castrates sing less and show fewer aggressive displays related to dominance status or territorial behavior. If the gonad is transplanted back into the castrate, or if testosterone is injected, then aggressive behaviors are restored. This experimental approach has stood the test of time for at least a hundred years. Only in occasional instances has the castration/replacement paradigm failed to show an effect on aggression. However, in some of these latter cases it could be argued that the context of aggression was not truly reproductive. Then, about 20 years ago, the technique of radioimmunoassay allowed us to measure circulating levels of sex steroids in the blood. Since testosterone is transported in the blood stream from the testis to the brain (where obviously it has effects on behavior), it is logical to assume that changes in circulating testosterone should be correlated highly with reproductive aggression. It is here that the relationship of testosterone and aggression becomes confused. In about 50% of investigations, there are no correlations of testosterone with either territorial aggression, or dominance status in social groups. In some cases, several studies on the same species do not agree! However, more recent work suggests that, although testosterone does have activational effects on aggressive behavior in birds, the relationship with circulating testosterone is not always direct.

Testosterone does not affect behavior alone. Many investigations have shown an effect of behavioral interactions on testosterone secretion, suggesting a complex interrelationship of testosterone and aggression. This has led to a possible hormonal basis for differences in mating systems and breeding strategies in birds.

Testosterone, mating systems and breeding strategies in males

Male birds compete with one another over breeding territories, access to females, and for other resources for raising young. The aggressive behaviors expressed as males compete in a reproductive context are thought to be regulated specifically by testosterone. Conversely, failure of many investigations to show a direct correlation of circulating testosterone levels and aggression led to suggestions that testosterone might not be involved. This issue was clarified in an experiment in which castrated male white-crowned sparrows, *Zonotrichia leucophrys gambelii*, were given implants of testosterone, or empty implants as controls, and total aggressive behavior was observed (i.e. songs, threats, etc. – all males were caged singly so that they could see one another but physical contact was not allowed). It was found that both testosterone-treated and control males showed similar increases in aggression following implant, even though one group had circulating levels of testosterone almost two orders of magnitude higher than the other. However, since these birds had been housed together for many months prior to the experiment, it was possible that social relationships had been established for a long time. There is evidence that the rate of aggression among individuals with established dominance–subordinance relationships tends to be lower than when individuals are first introduced and are competing for status. To test this, a novel male white-crowned sparrow, which none of the birds had encountered before, was introduced to each group. Levels of aggression increased immediately in response to this apparent 'intruder', but the testosterone-treated birds showed a significantly greater response than controls. These data suggested that if social relationships are disrupted then the relationships of testosterone and aggression are restored. In socially stable situations, some form of 'inertia' appears to result in low overt aggression and, thus, an apparent lack of relationship with testosterone.

Re-examination of the literature reveals that those investigations of socially stable groups tend to show no relationship between testosterone and aggression. On the other hand, a positive relationship is found during periods of social disruption. These data provide circumstantial evidence for the possibility that testosterone and aggression are related most directly in social instability. This led to design of experiments called the 'simulated territorial intrusion' technique in which a free-living male bird is 'challenged' by a conspecific 'intruder' (a live decoy male in a cage accompanied by play back of tape-recorded songs through an adjacent speaker). A blood sample can be collected (for measurement of testosterone) after the resident has tried to attack and to repel the intruder. Control samples are collected from birds caught in live traps baited with seeds. These birds are foraging rather than being involved in territorial disputes. This method has several advantages. Firstly, one can easily collect behavioral data as the resident male attacks the simulated intruder. Some of these behavioral postures are illustrated in Figure 13.3. The upper panel shows a song sparrow, *Melospiza melodia*, threatening another individual (the caged decoy) with a 'point' in which the beak and head are oriented toward the intruder, the wings are drooped and the tail is elevated. This posture may escalate to a wing wave and trill (lower panel of Figure 13.3) that just precedes a flight or attack. Secondly, if the birds are allowed to interact for about 30 minutes, then it is possible to catch the responding bird in a strategically placed mist net, to remove the bird quickly (Figure 13.4) and to collect a small blood sample from a wing vein. Then testosterone level in the blood can be correlated with behavior during the simulated territorial intrusion.

Figure 13.3. Aggressive postures of a song sparrow, *Melospiza melodia*, responding to a simulated territorial intrusion. The upper panel shows a typical 'point'. Note the elevated tail, drooped wings, and point with the head and beak. The lower panel shows a wing wave (often accompanied with a trill vocalization). This is a higher intensity threat and usually precedes an attack or flight. These displays can be expressed by both males and females. (Photographs by author; from Wingfield *et al.* (1987) with permission of Sigma Xi.)

Using the simulated territorial intrusion technique it became clear that free-living male song sparrows implanted with testosterone show greater frequencies and intensity of aggressive responses to the intruder than do controls (which have a lower circulating level of testosterone). Additional experiments also showed that free-living males have elevated plasma levels of testosterone during these experimental intrusions. These data provide support for the hypothesis that disruption of social stability results in consistent correlations of testosterone and aggression.

Male–male interactions are not the only type of social interaction that can influence secretion of testosterone. It has been shown in many species, from fish to humans, that exposure of males to sexually receptive females also results in an increase in testosterone, possibly linked to increased

Figure 13.4. A song sparrow, *Melospiza melodia*, is removed quickly from a mist net. This technique is used frequently by researchers to catch small birds easily and gently. In this case the sparrow was captured after responding to a simulated territorial intrusion. Immediately after capture a small blood sample can be collected from a wing vein for measurements of circulating hormone levels. The bird can then be weighed, banded and released for further study. Note that this individual already has been banded (colored plastic rings) for recognition in the field. (Photograph by author; from Wingfield *et al.* (1987) with permission of Sigma Xi.)

aggression over mates (including 'mate-guarding' behavior). Note that these increases in testosterone often occur when gonadal maturation is complete and when sexual behavior is already expressed fully. Thus, it is probable that surges of testosterone secretion induced by social cues are involved solely with expression of reproductive aggression. This phenomenon has been called the 'challenge hypothesis'. Note that testosterone levels increase after the aggressive interaction has begun and, as such, do not activate the behavior, but may serve to sustain a high level of aggression over prolonged periods (i.e. several hours of an agonistic interaction over territory).

The challenge hypothesis led to speculations that, since the pattern of circulating levels of testosterone over a breeding season are influenced by social cues, they may be related to mating systems and breeding strategies.

In polygynous species, males spend the breeding season fighting other males and trying to attract as many sexually mature females to their territories as possible. Other males are monogamous and, once they have obtained a territory and mate, show little further aggression unless challenged (as described above). It follows that polygynous males with high rates of male–male aggression and many sexually mature females on their territories may have prolonged high levels of testosterone because of continuous social stimulation. Monogamous males, however, would only have high levels of testosterone early in the season when territories are established and when pairing occurs. Comparisons of reproductive cycles of several species sampled in the field confirm these predictions. Even more striking is the discovery that if males of monogamous species are given implants of testosterone to maintain high circulating levels, then these males become polygynous.

If the pattern of testosterone in blood can have such potent influences on the mating system, and since polygyny potentially could increase a male's reproductive success, why is monogamy so prevalent in birds and polygyny present in only about 10% of species? Clearly, there must be some cost to having high levels of testosterone for prolonged periods in normally monogamous birds. Testosterone-treated males tend to receive more wounds than controls, but this does not always result in debilitation or reduced survival over winter. More importantly, and paradoxically, reproductive success is reduced in testosterone-treated males despite having two or even three females on territory. Experimental manipulation of testosterone levels suggests that suppression of parental care may be a major reason for this reduced fitness. Males of most monogamous species provide parental care and may incubate the eggs and/or feed young. When testosterone implants were given to male house sparrows, *Passer domesticus*, they showed a dramatic reduction in parental behavior, resulting in lower reproductive success. High levels of testosterone reduce male parental care in favor of increased territorial and mate-guarding aggression. Thus, the stimulus for increased testosterone secretion during social interactions must be tempered by the degree of parental care shown by males.

Using these apparently antagonistic stimuli we can outline a possible hormonal basis of mating systems. On the one hand, male–male aggression over territories and mates tends to increase secretion of testosterone, while expression of parental behavior tends to result in lower levels (by mechanisms that are still not completely understood). It is then possible to make predictions about the temporal pattern of testosterone

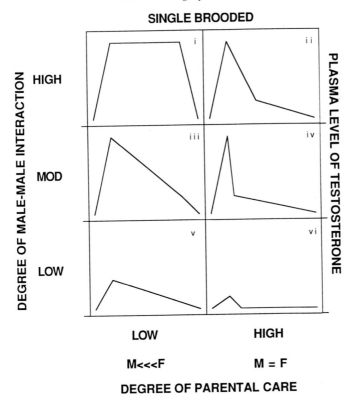

SINGLE BROODED

DEGREE OF MALE-MALE INTERACTION

HIGH — MOD — LOW

PLASMA LEVEL OF TESTOSTERONE

LOW — HIGH

M<<<F — M = F

DEGREE OF PARENTAL CARE

Figure 13.5. Hypothetical seasonal patterns of testosterone levels in blood of birds with varying degrees of male–male aggression and male parental care. Note that the lines within boxes (i–vi) represent the hypothetical pattern of circulating testosterone level through a breeding season for that appropriate ratio of male–male aggression and male parental care. For the Y-axis: LOW, little or no male–male aggression; MOD, (moderate) increased male–male aggression early in the season when territories are established and when females first become receptive; HIGH, intense male–male aggression throughout the breeding season. For the X-axis: LOW, no parental care, or the male may simply escort the young; HIGH, male expresses incubation behavior and/or feeds the young directly. These hypothetical patterns of testosterone secretion are representative of single brooded (i.e. they raise a single brood of young in each breeding season) species. (From Wingfield *et al.* (1990) with permission; © 1990 by The University of Chicago. All rights reserved.)

in relation to degree of male–male aggression and extent to which males show parental care (Figure 13.5). If male–male aggression is high and parental care is low then testosterone secretion should be elevated through most of the breeding season, although the amplitude may vary (Figure 13.5i, iii and v). However, as soon as parental care increases, then testosterone levels must decline (Figure 13.5ii, iv and vi). Thus, by comparing degree of male–male aggression and expression of male parental care it is possible to generate hypothetical patterns of testosterone secretion during a breeding season. Of the 35 or so species that have been studied thus far under natural, or near natural conditions, all show testosterone patterns as predicted from Figure 13.5. As yet we have no species for which both male–male aggression and parental care are low, but if such a species exists, then panel v of Figure 13.5 is the predicted cycle of testosterone.

The question still remains whether the pattern of testosterone is generated entirely by environmental and social cues, or whether some cycles are programmed genetically. It is suggested above that in those species where male–male aggression is high and male parental care is low, testosterone levels may be high because of continual stimulation from social cues. Thus, the pattern of testosterone secretion may be determined by the extent of social stimulation superimposed on a breeding baseline of hormone secretion stimulated by, for example, day length or endogenous rhythms. Before continuing, it is necessary to remind ourselves that testosterone has many biological effects, as described earlier. This may mean that there are 'levels' of testosterone secretion that activate morphological, physiological and behavioral processes differently.

It has been postulated that there is a baseline level of testosterone (Figure 13.6, level a) for the non-breeding season, and a higher breeding baseline (Figure 13.6, level b) that maintains the gonad and secondary sex characteristics in a functional state and activates aggressive, parental and sexual behavior during the reproductive period. Then, social cues can stimulate secretion of testosterone to even higher levels (Figure 13.6, level c) that appear to be involved solely in the maintenance of a high level of reproductive aggression over mates and territories (i.e. the challenge hypothesis). These surges of circulating testosterone above level b are transitory and occur only when appropriate social cues stimulate them. If the stimulus passes, then testosterone levels decline toward level b (this can occur very rapidly, since the half life of testosterone in circulation is a few minutes only). Therefore, the extent to which secretion of testosterone can be elevated above level b toward level c (i.e. (level c − level a)/(level b − level a)) could indicate sensitivity to social cues that regulate secretion of testosterone and, thus, mechanisms that underlie mating system and breeding strategy.

Then we can compare hormonal responsiveness to social interactions ((c − a)/(b − a)) with the ratio of male–male aggression to parental care

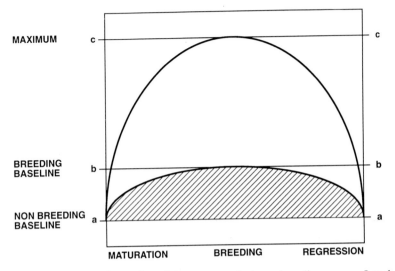

Figure 13.6. Levels of secretion of testosterone during a breeding season. Level a, non-breeding baseline (i.e. lowest levels of the year, but not zero, since some testosterone secretion is required to provide a negative feedback signal for GnRH and LH secretion). Level b, the concentration of testosterone required for complete spermatogenesis, full development of secondary sex characteristics and activation of sexual and aggressive behavior. Cycles of testosterone secretion between levels a and b are regulated by environmental cues such as photoperiod, temperature, etc. Level c, highest levels of testosterone attained. Such increases in testosterone are usually transitory but can occur at any time (i.e. within the upper curved line) and are stimulated by social cues such as male–male aggression and behavior of receptive females. (From Wingfield *et al.*, 1990 with permission; © 1990 by The University of Chicago. All rights reserved.)

calculated from Figure 13.5. To do this we can give an arbitrary score of 1, 2 and 3 to low, moderate and high male–male aggression respectively; and a score of 1 and 2 to low and high parental care respectively, generating a ratio, d. Several predictions result: for example, when d is high (i.e. high male–male aggression and low parental care – typical of many polygynous species) then the ability to secrete testosterone above level b should also be high. Surprisingly, when about 20 species for which we have the relevant information are compared, the opposite trend is found (Figure 13.7). Polygynous species cluster at one end of the curve with low ability to secrete testosterone in relation to social cues (i.e. the ratio of level c to level b is close to 1). In contrast, monogamous species cluster at the other end (the ratio of level c to level b is high). Interestingly, polyandrous species fall within the range for monogamous males. This makes sense, since males of these species provide most, if not

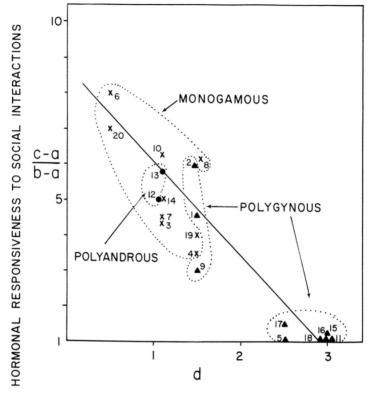

Figure 13.7. The relationship between hormonal responsiveness to social interactions ((c − a)/(b − a); see text) and the ratio of male–male aggression expressed
as a ratio of the degree of parental care (d). The ratio, d, is calculated by applying
arbitrary scores to the level of male–male aggression and that of parental care.
(i) Aggression: 1, low; 2, moderate; 3, high. (ii) Parental care: 1, low; 2, high. Species
are as follows: 1, *Sturnella neglecta*; 2, *Pooecetes gramineus*; 3, *Sturnus vulgaris*; 4,
Melospiza melodia; 5, *Molothrus ater*; 6, *Larus occidentalis wymani*; 7, *Zonotrichia
leucophrys*; 8, *Passer domesticus*; 9, *Ficedula hypoleuca*; 10, *Turdus merula*; 11,
Meleagris galloparvo; 12, *Actitis macularia*; 13, *Phalaropus tricolor*; 14, *Diomedea
exulans*; 15, *Phasianus colchicus*; 16, *Coturnix coturnix*; 17, *Agelaius phoeniceus*;
18, *Lagopus lagopus*; 19, *Streptopelia risoria*; 20, *Aptenodytes forsteri*. (Crosses are
monogamous species; filled circles are polyandrous species; filled triangles are
polygynous species. From Wingfield *et al.*, 1990; © by The University of Chicago.
All rights reserved.)

all, parental care and may mate only with one female. Note also that the
polygynous species that fall close to monogamous species show some
degree of parental care.

   The comparisons made in Figure 13.7 suggest a strong hormonal basis to mating systems and breeding strategies in male birds. Even more interesting is the suggestion that polygynous males, or those species in which parental care is low, actually do *not* show hormonal responses to social cues (i.e. levels b and c are similar). In these cases, other environmental cues (e.g. the annual change in day length) may stimulate gonadal development early in spring and the cycle progresses more as a genetically determined program. Perhaps lack of parental care provides no selection for social modulation of testosterone secretion. If expression of parental behavior is high, then testosterone secretion must be held in check to avoid conflicts with high rates of aggression. However, should a male be challenged for his territory by an intruding male, then the ability to increase testosterone secretion to combat the intruder, or to mate-guard a sexually receptive female during a renesting attempt, is crucial. Hence, plasticity in interrelationships of testosterone and aggression is complex in monogamous species. This finding has implications for future studies on brain mechanisms. Neural pathways for social signals influencing testosterone secretion could be studied in species which show high levels of hormonal responsiveness to social interactions, although species with low levels may provide interesting additional controls.

Testosterone, mating systems and breeding strategies in females

It has long been thought that females are generally not aggressive and that testosterone is the 'male sex hormone'. Both views have turned out to be inaccurate. Female vertebrates can have significant quantities of testosterone circulating in their blood, and may be extremely aggressive, although this aggression in reproductive contexts is directed almost always at other females. Generally, it is thought that females compete with other females: to defend resources such as food or nest sites for breeding; for exclusive access to male investment (e.g. parental care); to space nests and to avoid predation; and to reduce intraspecific brood parasitism. The aggressive postures expressed are often similar to those of males. Females of many species can sing, although the extent to which they do so varies considerably. Otherwise, threats and similar postures are essentially identical to those of males (e.g. Figure 13.3).

Females also show aggression in many other contexts, but the hormonal control of female–female conflict during the breeding season has been neglected. Maternal aggression (defense of young from a predator) has received more attention but provides a different context from the ones discussed here. It has been known for decades that testosterone can

Sex differences in behaviour

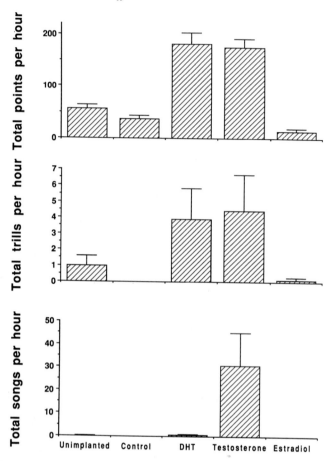

Figure 13.8. Effects of subcutaneous implants of 5-α dihydrotestosterone (DHT), testosterone and estradiol on aggressive behavior of female song sparrows, *Melospiza melodia*. Females were held one per cage on short days (i.e. endogenous levels of sex steroids were basal). Point behavior and trills (including wing waves) are shown in Figure 13.3. These behaviors are common to both males and females. Female song sparrows also sing under natural conditions, but much less frequently than males. (Bars are means and vertical lines are standard errors. The unimplanted group was an extra control group which was unmanipu lated, whereas birds from the control group were captured and surgery was performed.)

activate aggression, including singing (or analogous vocalizations) in females. Furthermore, it is clear that testosterone, and not estradiol, regulates development of a distinct nuptial plumage in those species in which females have the same plumage as males (e.g. Laridae, see Figure 13.1). In Figure 13.8, it can be seen that both testosterone and

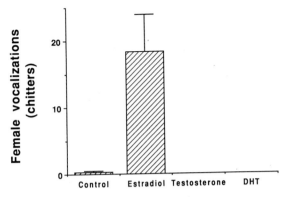

Figure 13.9. Effects of subcutaneous implants of 5-α dihydrotestosterone (DHT), testosterone and estradiol on 'chitter' vocalizations of female song sparrows, *Melospiza melodia*. Females were held one per cage on short days (i.e. endogenous levels of sex steroids were basal). The 'chitter' vocalization is given by the female mostly in the breeding season and is thought to incite the male to attack an intruder (mate-guarding), and also in the context of copulation solicitation. (Bars are means and vertical lines are standard errors.)

its metabolite, 5-α dihydrotestosterone (DHT), activate threatening postures (points and trills with wing wave, see Figure 13.3), but only testosterone activates singing behaviour in female song sparrows. Estradiol is without effect (Figure 13.8). Conversely, estradiol, but not testosterone or DHT, activates a female-specific sexual behavior ('chitters' – used both in an aggressive context to incite the male to attack an intruder, and also for copulation solicitation, Figure 13.9).

Clearly, testosterone can activate aggression in females as it does in males, but do females have patterns of circulating testosterone that parallel expression of reproductive aggression under natural conditions (i.e. is this an actual function of testosterone in females or are the laboratory implantation studies pharmacological rather than physiological)? In species in which females have similar plumage and rates of aggression to those of males, or are even more brightly colored (e.g. polyandrous sandpipers), then levels of testosterone in blood are correlated with aggression. In other species where females are aggressive, but at lower frequencies than in males, cycles of testosterone levels may be very slight or absent. In yet other species, females show high levels of aggression toward conspecific females, but there is no correlation with circulating testosterone. These data suggest that, although some females activate aggression with high testosterone levels, others do not. Perhaps additional mechanisms exist to activate reproductive aggression in females.

It has been postulated, with some supporting evidence, that LH may actually activate aggression in females directly and independently of testosterone. Correlations of aggression in females with LH, but not testosterone, lend further support to this hypothesis. However, experimental evidence is weak and further study is required before this issue can be resolved.

The relationship of testosterone to female aggression deserves closer inspection. Do females that are similar to males in plumage and behavior tend to have higher levels of testosterone relative to males than in species with greater dimorphism? To test this, a dimorphism index is required that standardizes differences between the sexes for all species. This index is outlined in Table 13.1, and involves a mean score for differences in body size, plumage and other integumentary structures, and behavior. We can then compare this dimorphism index with the ratio of peak plasma levels of testosterone in males versus females (to control for differences in absolute levels of testosterone among species that may have no direct bearing on expression of behavior). Fortunately, about 30 species have been investigated in sufficient detail to allow such a comparison. A remarkable relationship is revealed in Figure 13.10. When males and females are similar (dimorphism index is low) then indeed the levels of testosterone in females are high relative to those of males. However, this is not a linear relationship. As the dimorphism index increases, the variability in ratios of testosterone also increases. Some females have high levels of testosterone, others do not (Figure 13.10). Is there any relationship here to mating systems and breeding strategies?

In Figure 13.11, the same relationship has been plotted separately for (a) monogamous and (b) polygamous species. The results are even more striking. In monogamous species there is an almost linear relationship between the dimorphism index and the ratio of circulating testosterone in males versus females with a single exception. The point at the lower right-hand corner of Figure 13.11a represents values for the mallard (Figure 13.1). Interestingly, this species is monogamous, but males provide no parental care. In all the other species, males express substantial parental behavior. In contrast, there is no relationship between the dimorphism index and testosterone ratio in polygamous species. Note also in polyandrous species (in which all males show high parental care) there is no obvious relationship (Figure 13.11b). These data suggest that, at least in monogamous birds, females have higher levels of testosterone (in comparison with males) involved in expression of reproductive aggression, especially when the dimorphism index is low. Females often

Table 13.1 *Dimorphism index*

Body size
1 = 80–100% overlap
2 = 5–80% overlap
3 = No overlap

Plumage
1 = Monomorphic nuptial plumage
2 = Moderate dimorphism (e.g. eye, beak, leg color, size of feather plumes, etc.)
3 = Great dimorphism (one sex has bright nuptial plumage, other sex is cryptic)

Territorial aggression
1 = Both sexes defend territory or compete equally
2 = One sex defends territory or competes more than other sex
3 = One sex defends territory or competes and other sex does not

The index is the mean of the summed scores from body size, plumage and territorial aggression. (See also Figure 13.10.)

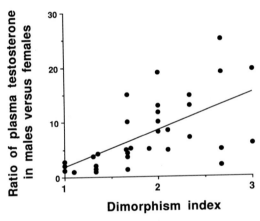

Figure 13.10. The relationship of the ratio of peak circulating testosterone levels in male versus female birds to the dimorphism index (generated as described in Table 13.1).

compete intensely over territory and for access to male investment in terms of paternal care. Conversely, in polygynous species, male parental care is less widespread, although females still compete for other resources such as territories and food, and also to prevent intraspecific brood parasitism. Whether or not this reflects sexual selection on a role for testosterone in females that compete for male reproductive investment (i.e. expression of parental care) requires further study. It is also clear that

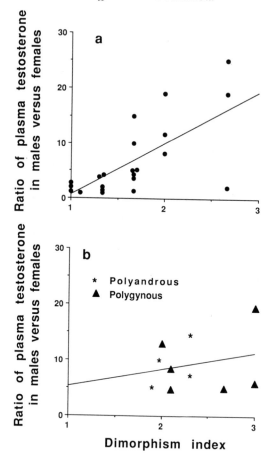

Figure 13.11. Relationships between the dimorphism index (see Table 13.1) and the ratio of peak circulating testosterone levels in males versus females for (a) monogamous species and (b) polygamous species (asterisks are polyandrous species and filled triangles are polygynous species).

control of aggression in females of polygamous species, and for whom dimorphism is great, involves mechanisms independent of testosterone. These mechanisms remain obscure, but the relationships presented in Figures 13.10 and 13.11 may provide a template for further investigation.

Effects of social cues on gonadal development and sex hormones in females

There is extensive evidence that behavior of males can accelerate reproductive development and behavior in females. For example, male song, or its equivalent, and courtship promote ovarian development. Play back of tape-recorded songs of the canary, *Serinus canarius*, can increase the rate of nest-building behavior in females. Moreover, the effect on nest building is even greater if songs of a male with a large repertoire are played. In the quail, *Coturnix coturnix*, play back of male crows also stimulates ovarian development, but it is most effective at night. Normally, this species crows at night rather than during the day. Thus, females may choose among males, preferring to respond to some, but not others, or are receptive to male courtship only during an appropriate time of day. In white-crowned and song sparrows, females may show slight, but none-the-less significant, enhancement of photoperiodically induced ovarian development by play back of tape-recorded male song. However, under natural conditions there is no evidence that implants of testosterone into male song sparrows early in the season advance ovarian development, date of egg laying, or circulating levels of reproductive hormones (Figure 13.12). Thus, it should be noted that, in some instances, effects of male behavior on reproductive development in females are slight and possibly of little relevance under natural conditions.

In some species, discrimination among males is even more marked. Male courtship and vocalizations may be without effect unless the female has chosen that male as a mate. In canvasback ducks, *Aythya valisinaria*, the female does not respond behaviorally to courtship from males unless she has formed a pair bond with that male. Females forced to pair with males randomly, or when their self-chosen mates are switched, do not lay eggs and they respond aggressively to males. When reunited with their original mates, females then lay eggs (Figure 13.13). The consequences of this can be severe. In five of seven cases in which the self-chosen male mate was switched, the new male actually died.

Not all female birds respond readily to male courtship. Red grouse, *Lagopus lagopus scoticus*, females show no response of reproductive hormones to the presence of males. Also in free-living song sparrows, if the males' reproductive season is extended by giving subcutaneous implants of testosterone (that maintain territorial and courtship behavior, and delay molt for up to two months longer than usual), then females mated to these males terminate breeding, molt and abandon the territory at the same time as control females (Table 13.2). In contrast, if the breeding season of females is prolonged experimentally by giving similar implants but filled with estradiol (again this treatment prolongs sexual receptivity and delays molt for up to two months), then males also remain on territory and delay molt for four to six weeks longer than males mated to females

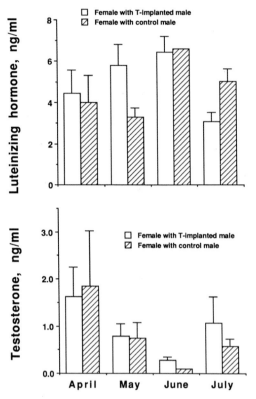

Figure 13.12. Effects of implanting free-living male song sparrows, *Melospiza melodia*, with testosterone on plasma levels of luteinizing hormone (LH) and testosterone in their female mates. Implanted males became polygynous, had larger territories and were more aggressive than control males. Females were forced to compete with other females on the polygynous males' territories, but this had no effect on LH and testosterone compared with females mated to control males. (Bars are means and vertical lines are standard errors. From J. C. Wingfield, unpublished.)

given empty implants as controls (Table 13.3). Males with estradiol-implanted females also have higher circulating levels of testosterone. Clearly, males are responding to the sexual behavior of their mates, but females are not fooled by the artificially altered behavior of the males.

There is experimental evidence for group effects on gonadal development. The house sparrow, *Passer domesticus*, is a highly social species that is semi-colonial when breeding and has complex group social displays all year round. When house sparrows were kept in flocks (three males and three females) with a nest box, testicular development was enhanced

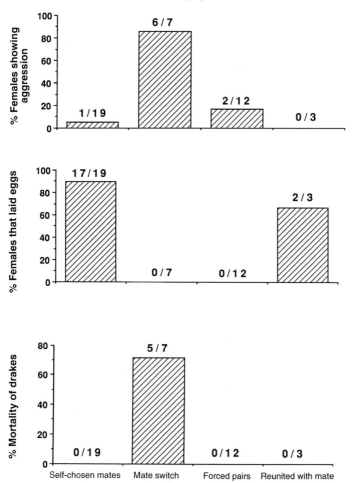

Figure 13.13. Effects of natural and forced pairing, and switching of a self-chosen mate for a novel male on behavior of female canvasback ducks, *Aythya valisinaria* (top two panels), and mortality of males (lower panel). (From Bluhm (1985) with permission of author and Springer-Verlag).

over that of males held one pair per similar sized aviary (with nest box) and another group of males which were held one per pair in a small cage with no nest box (Figure 13.14a, b). However, females showed no such enhancement of ovarian development when housed in a flock (Figure 13.14c, d). Again the trend is for males to show marked responses to females, but females are much more discriminating in their responses to male behavior. Note, however, that if data are reorganized according to whether individuals within the flock group had access to a nest

Table 13.2 *Effect of sexually active male song sparrows on the territorial behavior and onset of molt in females*

Female treatment	Aug	Sep	Oct	Initiation of molt
	Number on territory			
Testosterone implant				
Female (n = 9)	(9–1)[a]	0	0	24 Aug–10 Sep
Male (n = 9)	9	8	8	Sep–Oct
Control implant				
Female (n = 11)	(11–3)[b]	0	0	10 Aug–25 Aug
Male (n = 11)	11	1	0	24 July–25 Aug

[a] Nine females on territory in early August, but only one remained by late August.
[b] Eleven females on territory in early August, but only three remained by late August.
Source: From Runfeldt and Wingfield (1985).

Table 13.3 *Effect of sexually receptive female song sparrows on the territorial behavior and onset of molt in males*

Female treatment	Aug	Sep	Oct	Initiation of molt
	Number on territory			
Estradiol implant				
Female (n = 8)	8	8	8	Sep–Oct
Male (n = 8)	8	8	6	24 Aug–10 Sep
Control implant				
Female (n = 9)	(9–3)[a]	0	0	10 Aug–25 Aug
Male (n = 9)	6	1	0	24 July–26 Aug

[a] Nine females on territory in early August, but only three remained by late August.
Source: From Runfeldt and Wingfield (1985).

box, this did not seem to influence gonadal development (Figure 13.14b, d), suggesting an effect of true social facilitation within the group.

Hormonal responses during female–female interactions are much less well-known. In polyandrous species such as the spotted sandpiper, *Actitis macularia,* and phalaropes, *Phalaropus* sp., females have high levels of testosterone when they are competing for territories and/or mates, but

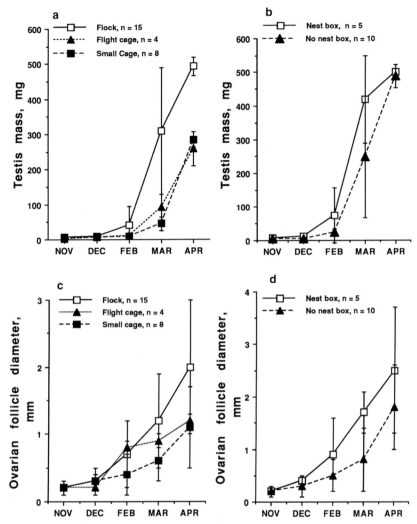

Figure 13.14. (a), (b) Effects of social environment on testicular development in flocks of house sparrows, *Passer domesticus*. Birds were held in five flocks of six (three males and three females) in a flight aviary with a nest box. Others were held in pairs either in a similar sized flight aviary with nest box (*n* = 4), or in a small 'preview' cage without a nest box (*n* = 8). (c), (d) Effects of social environment and presence of a nest box on ovarian development in female house sparrows. (b, d) Only those birds in flocks were compared. (Points are means and vertical lines are standard errors). Sample sizes on the figure refer to the number of birds in each experimental group for which data were available. For example, for the flock group there were totals of 15 males and 15 females. However, only certain birds in the flock had access to nest boxes, owing to social dominance relationships. Thus, one pair of birds in each flock (i.e. *n* = 5) had a nest box and 10 males and 10 females did not have nest boxes (i.e. *n* = 10). (From R. E. Hegner and J. C. Wingfield, unpublished.)

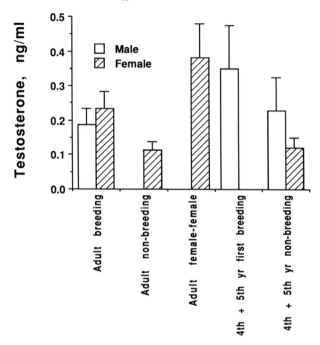

Figure 13.15. A comparison of testosterone levels in male and female, breeding and non-breeding, and subadult (fourth and fifth year) western gulls, *Larus occidentalis wymani*. The birds were sampled under natural conditions in a colony on Santa Barbara Island, southern California. (From J. C. Wingfield, G. L. Hunt and A. Newman, unpublished.)

not higher than those of males. In female western gulls, *Larus occidentalis wymani*, on Santa Barbara Island off southern California, there is a three-fold excess of adult females over males. These females compete fiercely for males and in some cases, form female–female pairs, court one another – even mount and copulate – resulting in both females laying eggs in their shared nest. They are territorial and exclude other females. Plasma levels of testosterone are higher in breeding females, regardless of whether they are heterosexually or homosexually paired, compared with unpaired females that are not territorial (Figure 13.15). This difference is less obvious in breeding and nonbreeding males (including subadults). However, the latter result may be due to the fact that, since there is an excess of females, males do not compete as much as in other populations where females are in shorter supply. Therefore, the testosterone levels in male western gulls may represent breeding baseline levels rather than a reflection of any influences of social stimulation (Figure 13.15).

The data for all species together suggest that it is probable that some of the apparent variation in sensitivity of females to the social cues affecting reproduction may be due to mate choice. The female may respond to a male's courtship behavior selectively, responding only to her self-chosen mate and not to males at random. Clearly the hormonal responses, if any, to female–female interactions need further investigation, and the mechanisms remain obscure.

Mechanisms by which behavioral cues influence gonadal development and sex steroid secretion in males and females

Behavioral interactions, both intersexual and intrasexual, are complex and involve several sensory modalities, e.g. visual, tactile, auditory, chemical. Each modality is perceived by its own specialized sensory receptors and this information is then integrated and transduced by higher centers of the brain. This poses some problems concerning mechanisms by which behavioral cues are transduced into GnRH and gonadotropin secretion that, in turn, regulate gonadal development, sex steroid secretion and behavior. In males, experiments in which the behavioral stimulus is dissected, so that auditory, visual and tactile stimuli can be separated, suggest that a combination of visual and auditory information is important for the hormonal response to male–male aggression and to sexually receptive females. In the song sparrow, males show reduced responses of testosterone secretion to play back of tape-recorded conspecific songs (i.e. auditory stimulus alone), or to a devocalized decoy male (i.e. visual stimulus alone). A greater response was observed in males exposed to a devocalized decoy and play back of tape-recorded song. In all cases, contact was not allowed, suggesting that tactile stimuli are not important. Chemical cues (e.g. pheromones) are not thought to be major components of communication in birds although this is a conclusion reached by default rather than by rigorous experimentation.

The male's response to females can be elicited just by the male seeing a receptive female. Again, tactile stimuli do not appear to be important, since the effect can be induced without contact. For females, however, the problem is that many species respond to male behavior only in restricted contexts. This suggests that, in addition to perception of the stimuli, there must be some sophisticated additional processing of social cues in relation to whether the mate is self-chosen, the time of day, the song repertoire, etc.

Recent ingenious work by M.-F. Cheng, and her associates at Rutgers University, has shown that in the ring dove, *Streptopelia risoria*, the female may actually stimulate ovarian development herself. Male and female ring doves perform a mutual 'bow–coo' display during courtship. If the female is devocalized (by puncturing the interclavicular air sac that regulates air flow over the syrinx – the voice box of birds), then ovarian development in response to a male is diminished greatly. Also, play back of tape-recorded female vocalizations can stimulate ovarian development in isolated females. These remarkable data suggest that, although male courtship may provide context, it is the female's own courtship behavior in return that results in self-stimulation. This has profound implications for the mechanisms by which females of many species discriminate among males. Only if she responds to the mate of her choice, or at the correct time of day, or in preference to a male with a larger repertoire, will the female then self-stimulate. If this kind of mechanism is widespread, then it could explain the apparent variability in published reports concerning sensitivity of females to male courtship.

Summary

As a whole, the class Aves produces a great diversity of sounds, and shows tremendous variation in sexual dimorphism of plumage and behavior patterns. Although sex differences in sexual maturation and sexual behavior are more or less constant throughout all avian species, the pattern of hormone secretion in relation to reproductive aggression, especially territorial behavior, is highly variable. In males and females, testosterone activates territorial aggression in the majority of species studied to date. However, patterns of testosterone secretion are extremely diverse not only between males and females, but also within sexes among species and even among individuals within a population. The pattern of testosterone secretion is determined in part by the degree of aggression (over territory and mates) and parental care. High levels of testosterone facilitate aggressive behavior but inhibit parental care. Male–male aggression appears to stimulate high levels of testosterone secretion, but these levels must decline if parental care is also to be expressed. Thus, the interaction of male–male aggression and parental care (i.e. mating system and breeding strategy) determines the pattern of hormone secretion. In females, it is clear that testosterone can also activate reproductive aggression, although the degree to which this operates under natural conditions is less clear, and highly variable. This variation corresponds roughly to the degree of sexual dimorphism. When dimorphism is low, then patterns of testosterone secretion in females are more similar to males; when dimorphism is high, then females tend to have much less testosterone than males. Thus, it is possible that the patterns of testosterone secretion and aggression of males and females in relation to mating system and breeding strategy may have similar bases.

Behavioral interactions can, in turn, regulate gonadal development and sex steroid secretion. Generally, males respond to the sexual behavior of females with enhanced reproductive development and delayed gonadal regression at the end of the breeding season. Females can also respond to male courtship with increased ovarian development, but this is much more variable and is influenced by mate choice, territory quality, time of season, etc. Neural pathways by which behavioral information is transduced into neuroendocrine and endocrine secretions also appear to be dimorphic. For males, a combination of visual and auditory information may be critical for endocrine responses to both male–male and male–female interactions. For females, visual cues alone or auditory stimuli alone can be equally effective, but the final mechanism may involve self-stimulation through the female's own behavior in response to a male. Although the hormonal basis of mating systems and breeding strategies may have similar foundations in males and females, the mechanisms by which behavioral interactions feed back to influence hormone secretion appear to be highly dimorphic.

Acknowledgements

Preparation of this manuscript and several experiments and investigations described above were supported by a series of grants from the National Science Foundation to J. C. W.

Further reading

Balthazart, J. (1983) Hormonal correlates of behavior. In *Avian Biology*, Vol. 7, pp. 221–365. Eds. D. S. Farner, J. R. King and K. C. Parkes. Academic Press, New York.

Bluhm, C. K. (1985) Social factors regulating avian endocrinology and reproduction. In *The Endocrine System and the Environment*, pp. 247–64. Eds. B. K. Follett, S. Ishii, and A. Chandola. Japan Scientific Press, Tokyo, and Springer-Verlag, Berlin.

Cheng, M.-F. (1983) Behavioral 'self-feedback' control of endocrine states. In *Hormones and Behavior in Higher Vertebrates*, pp. 408–21. Eds. J. Balthazart and R. Gilles. Springer-Verlag, Berlin.

Harding, C. F. (1983) Hormonal influences on avian aggressive behavior. In *Hormones and Aggressive Behavior*, pp. 435–67. Ed. B. Svare. Plenum Press, New York.

Lofts, B. and Murton, R. K. (1973) Reproduction in birds. In *Avian Biology*, Vol. 7, pp. 1–107. Eds. D. S. Farner and J. R. King. Academic Press, New York.

Murton, R. K. and Westwood, N. J. (1977) *Avian Breeding Cycles*. Clarendon Press, Oxford.

Runfeldt, S. and Wingfield, J. C. (1985) Experimentally prolonged sexual activity in female sparrows delays termination of reproductive activity in their untreated mates. *Animal Behaviour*, **33**, 403–10.

Silverin, B. (1988) Endocrine aspects of avian mating systems. In *Acta XIX Congressus Internationalis Ornithologici*, pp. 1676–84. Ed. H. Ouellet. University of Ottawa Press, Ottawa.

Wingfield, J. C. (1991) Mating systems and hormone–behavior interactions. In *Acta XX Congressus Internationalis Ornitholigici*, pp. 2055–62. Ed. B. D. Bell. New Zealand Ornithological Trust Board, Wellington.

Wingfield, J. C. and Farner, D. S. (1993) Reproductive endocrinology of wild species. In *Avian Biology*, Vol. 9, pp. 163–327. Eds. D. S. Farner, J. R. King and K. C. Parkes. Academic Press, New York.

Wingfield, J. C. and Marler, P. R. (1988) Endocrine basis of communication: reproduction and aggression. In *The Physiology of Reproduction*, pp. 1647–77. Eds E. Knobil and J. D. Neill. Raven Press, New York.

Wingfield, J. C. and Moore, M. C. (1987) Hormonal, social and environmental factors in the reproductive biology of free-living male birds. In *Psychobiology of Reproductive Behavior: An Evolutionary Perspective*, pp. 149–75. Ed. D. Crews. Prentice-Hall, New Jersey.

Wingfield, J. C. and Ramenofsky, R. (1985) Hormonal and environmental control of aggression in birds. In *Neurobiology*, pp. 92–104. Ed. R. Gilles and J. Balthazart. Springer-Verlag, Berlin.

Wingfield, J. C., Ball, G. F., Dufty, A. M. Jr, Hegner, R. E., and Ramenofsky, R. (1987) Testosterone and aggression in birds. *American Scientist*, **75**, 602–8.

Wingfield, J. C., Hegner, R. E., Dufty, A. M. Jr and Ball, G. F. (1990) The 'challenge hypothesis': theoretical implications for patterns of testosterone secretion, mating systems and breeding strategies. *American Naturalist*, **136**, 829–45.

Witschi, E. (1961) Sex and secondary sex characters. In *Biology and Comparative Physiology of Birds*, pp. 115–68. Ed. A. J. Marshall. Academic Press, New York

14

Sex differences in the behaviour of African elephants

JOYCE H. POOLE

Among mammalian species, females contribute more energy to the production and rearing of offspring than do males. Since female reproductive success appears to be limited primarily by energetic and nutritional constraints, it has been hypothesized that female grouping patterns in many species are influenced to a large extent by the distribution of food. Thus, females should remain in their natal groups and should cooperate with kin when kin-based alliances increase access to food patches or increase the survival chances of their offspring. In contrast, since the reproductive success of males is limited primarily by the availability of females, male grouping patterns are hypothesized to be dependent upon the distribution of females.

This pattern is illustrated neatly by African elephants where females and their offspring live in tightly knit family units composed of one to five adult females, and males live a more solitary existence with few social bonds. Related females form coalitions against other non-related females, and larger families with older matriarchs are able to dominate smaller families with younger matriarchs, thus competing more successfully for access to scarce resources. Kin-based alliances also have a significant impact on calf survival. Large families with more female caretakers have higher calf survival than do smaller families. As a consequence of a 21.5-month gestation and the long period of calf dependence, adult female elephants come into oestrus only once every four years. Presented with this situation it does not make reproductive sense for a male elephant to live permanently with one family and to attempt to hold a harem. A large, high-ranking male moving from group to group in search of oestrous females could do significantly better than a male guarding one particular family unit. As a result, through the course of sexual selection, a premium has been set on characteristics that assist male elephants to locate receptive females and, once they find them, to compete successfully with other males for matings. Male elephants can hear the infrasonic calls, and probably can detect the smell, of an oestrous female 5–10 km away. Their prolonged growth in body size and weaponry, and their extremely aggressive behaviour during the rutting period of musth, allows the larger, older males to compete successfully for access to these females.

How does this basic difference affect the lifestyles and behaviour of male and female elephants? In this paper, sexual dimorphism in African elephants is described and the behavioural consequences are discussed. The data are drawn largely from a long-term study of social and reproductive behaviour of the African elephant population in Amboseli National Park, Kenya, undertaken by Cynthia Moss, Phyllis Lee and the author.

The study population

Amboseli National Park, in Kenya, covers an area of 390 km^2 and consists of semi-arid wooded, bushed and open grassland interspersed with a series of swamps. The Park provides the dry-season concentration area for 790 elephants including 50 matriarchal families and 170 adult males. The Amboseli elephants have been monitored since 1972 and now form one of the longest studies of a population of individually known large mammals. At each sighting of a single elephant or group, the date, time, location, habitat and group size are recorded. Also noted is the presence or absence of musth male(s) or oestrous female(s), the identification of family units and individual males, and the occurrence of any matings, guarding behaviour, births or deaths.

A *group* of elephants is defined as any number of elephants of any age or sex moving together in a coordinated manner. A *family unit* consists of a single adult female, or two or more related adult females and their immature offspring. A *bond group* consists of two or more (usually) related *family units*. Members of a family unit or bond group show a high frequency of association over time, act in a coordinated manner, exhibit affiliative behaviour towards one another and are known to be related or are putatively related.

Sexual dimorphism in body and tusk size

In elephants, sexual dimorphism in body size is extreme. Males continue to grow in height and weight throughout most of their lives, eventually reaching twice the weight of adult females (Figure 14.1). By the time

Figure 14.1. Male 45, Patrick, in musth, guards an oestrous female. At the time this picture was taken, Patrick was 35 years old and was not yet fully grown. The adult female had reached her full adult size. Note the difference in body size, tusk size and relative head sizes.

males reach sexual maturity, at about 17 years of age, they are taller than the largest females. The ability of elephants to continue growing beyond the age of sexual maturity, when most mammals cease to grow in height, is related to the unusual delayed epiphyseal fusion of the long bones, which is more pronounced in males than in females. Among female elephants, epiphyseal fusion of the long bones occurs between 15 and 25 years of age, whereas in males epiphyseal fusion takes place between 30 and 45 years of age. Undoubtedly, there has been strong selective pressure for large body size in male elephants.

Also, sexual dimorphism of tusk size is pronounced in African elephants. The tusks of males grow exponentially with age, and by 50 years of age, male tusks are seven times the weight of female tusks. The average tusk weight of a 55-year-old male is 49 kg/tusk while the average tusk weight of a female of the same age is a mere 7 kg/tusk. It is because the tusks of males are very much larger than those of females that they have been so vulnerable to ivory poachers. In many populations of elephants in Africa most of the mature males have been killed, and the sex ratio of breeding adults is skewed highly toward females. The much thicker tusks of males create the 'hour-glass' shaped face of mature males (broad foreheads, narrowing below the eyes and widening again at the base of the tusks), and this feature, in combination with the relatively larger heads of males, distinguishes the two sexes at a glance.

Sex differences in behaviour
Lifestyles

Perhaps the most fundamental behavioural difference between male and female elephants is that adult females and their offspring live in tight-knit, stable family groups, while adult males live more solitary independent lives with few social bonds.

Female elephant society consists of complex multi-tiered relationships extending from the mother–offspring bond out through families, bond groups and clans. The basic social unit is the family which is composed of several related females and their offspring. The bond group is made up of from one to five closely allied families that are usually related. Above the bond group is the clan, defined as families and bond groups which use the same dry-season home-range. The size and structure of groups that a female elephant finds herself in depends upon a number of different social and environmental factors including: the basic size of her family unit; the number of individuals that make up her bond group; the strength of bonds between her own and other families; the habitat; the season; and, in many cases, the level of human threat. In Amboseli, adult females tend to be in groups that are larger than their own family unit and, except under extremely unusual circumstances, a female is never found alone. The size of group in which a female finds herself increases during and following the wet seasons when food is well distributed and plentiful. Interactions between females, both within and between families, are generally amicable except when there is competition over a scarce resource.

Male elephants leave their natal families at about 14 years of age and thereafter live in a highly dynamic world of changing sexual state, rank, associations and behaviour. The structure and size of groups with which an adult male associates, and the type of interactions he has with members of these groups, are determined by his sexual state. During non-musth, males spend time alone or in small groups of other males in particular bull areas, where interactions are relaxed and amicable. During musth,

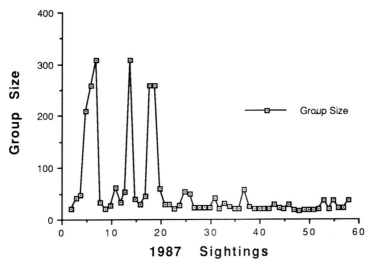

Figure 14.2. The sizes of groups that Echo was seen in during 1987 (see text).

males leave their bull areas and move in search of oestrous females; they may be found alone or in association with groups of females. Musth males interact aggressively with other males.

The sex differences in group size and structure can be illustrated best by following the groups that one female, Echo, from an average family size ($n = 13$), found herself in during the course of 1987 (Figure 14.2) in comparison to the groups that male 22, Dionysius, found himself in during the same period (Figure 14.3). The groups that Echo was found in tended to be larger in the first half of the year, during and following the rains, when Echo and her family were often moving as part of a large aggregation, but they became significantly smaller during the long dry-season. Dionysius began the year by being either alone or in association with several small groups where he was guarding a series of oestrous females. As the long rains began, females gathered into large aggregations of several hundred individuals and he joined them in his continuing search for receptive females. By mid-April, Dionysius had dropped out of musth and spent the rest of the year alone or occasionally with small groups of males.

Sexual behaviour and the phenomenon of musth

Female elephants reach sexual maturity at 10 to 12 years of age and typically produce their first calf two years later. Thereafter, females

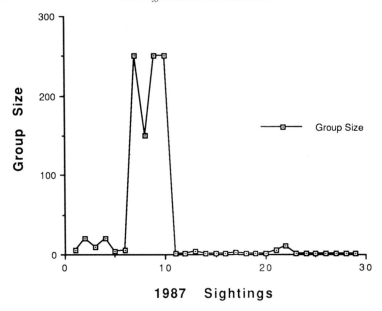

Figure 14.3. The sizes of groups that Dionysius (male 22) was seen in during 1987 (see text).

produce, on average, one calf every four to five years, slowing down as they approach 50 years of age. In Amboseli, females may be observed in oestrus throughout the year, with the highest numbers of females coming into oestrus during and following the rains. By the late dry season, or during drought years, very few females are observed in oestrus. Oestrus lasts between four and six days during which time the female often gives loud (102 dB at 5 m), very low-frequency (fundamental frequency of 18 Hz) calls (Figure 14.4). During the middle two days of oestrus the female is found and then mated and guarded by a high-ranking musth male (Figure 14.5).

Males do not reach sexual maturity until several years later than females, and probably do not father their first offspring until they are between 30 and 35 years of age. Between 20 and 25 years of age, males begin to show distinct sexually active periods, and, by the age of 29 years, males have exhibited their first musth period. The word musth comes from the Urdu, *mast*, meaning intoxicated, and refers to a heightened period of sexual and aggressive activity, or rut. During musth, males secrete a viscous liquid from swollen temporal glands, and they leave a trail of strong smelling urine (Figure 14.6) and call repeatedly in very low

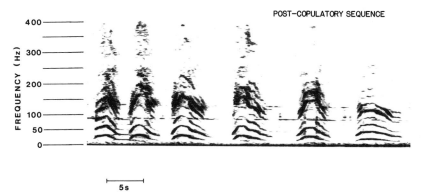

Figure 14.4. Sonogram illustrating the oestrous or post-copulatory call of a female elephant: fundamental frequency of 18 Hz and sound pressure level of 102 dB. (Time bar indicates 5-second intervals. From Poole *et al.* (1988).)

Figure 14.5. Male 45, Patrick, in musth (note swollen and secreting temporal glands) mates with an 11-year-old female.

frequencies (fundamental frequency of 14 Hz; Figure 14.7). Testosterone levels rise over five times their non-musth levels and musth males behave extremely aggressively toward other males, particularly those in musth. Several musth males have been killed in fights with other males in musth.

Figure 14.6. Male 45, Patrick, in musth. Note his swollen and secreting temporal glands and his hind legs that were wet from urine dribbling.

As a consequence of continued growth in height and body weight, larger, older males rank above smaller, younger males; male elephants do not reach their sexual prime until they are over 45 years of age. The oldest male in Amboseli, Iain, is estimated to be 57 years of age and is only just showing signs of a decline in reproductive activity. Due to the ivory poaching that took place around Amboseli during the early- and mid-1970s, Iain has been the highest ranking male in the population for

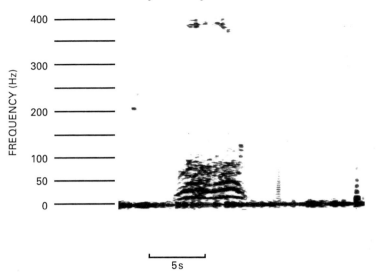

Figure 14.7. Sonogram illustrating the musth rumble: fundamental frequency of 14 Hz. (From Poole *et al.* (1988).)

14 years and is believed to have fathered a large proportion of the babies born during this period.

The occurrence and behavioural patterns of musth have now been followed for 15 years among the Amboseli males. Musth occurs throughout the year, with higher ranking males coming into musth during the period when most females are in oestrus. The musth periods of these older males last several months and occur at a predictable time each year (Figure 14.8). For example, among Amboseli's oldest males, Dionysius (M22) has come into musth in January, February, March and April for 14 years in a row, M13 comes into musth during March, April, May, and M126 comes into musth every year during June, July and August.

By contrast, the musth periods of younger males are short and sporadic (Figure 14.9) and they appear to come into musth relatively opportunistically. I have watched these younger males come into musth within an hour of encountering an oestrous female and be forced out of musth immediately following an attack from a higher-ranking musth male.

As Amboseli's elephant population has recovered from the ivory poaching and hunting that took place in the 1970s, the number of mature males in the population has increased. In 1979, ten males were known to come into musth. Ten years later, over 45 males were known to come into musth. Figures 14.9 and 14.10 illustrate the periodicity and duration

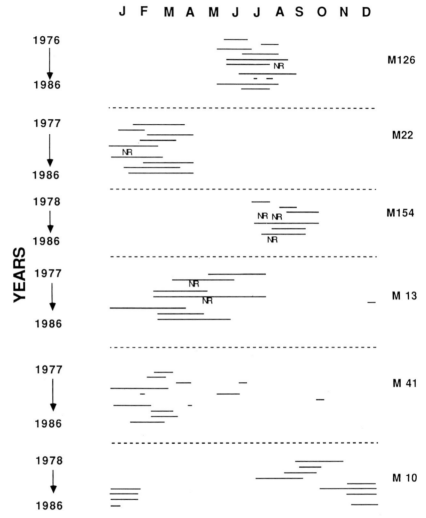

Figure 14.8. Duration and periodicity of musth for six high-ranking males (M126, M22, M154, M13, M41, M10) from 1976 until 1986. (NR, no record. From Poole (1987).)

of musth among males in the Amboseli population in 1980 and again in 1989. As young males have got older, so too have their musth periods grown longer. The data seem to indicate that, although the oldest males still command the best time of year, their musth periods are becoming shorter.

Although males in their late teens and early twenties are sexually active

Born	Male	Jan	Feb	Mar	Apr	May	Jun	Jul	Aug	Sep	Oct	Nov	Dec
1935	13			————	————	————							
1935	41		————	————									
1939	126						————	————	————				
1940	22		————	————									
1941	28				—		————	————	————	————			
1943	44						—						
1944	99				————	————							
1944	117				————	————							
1945	7	—											—
1945	10							————	————				
1945	45							—					
1945	78	-										————	————
1945	80			————									
1946	73									————	————		
1946	51												
1946	107				- —								
1947	119				—								—
1949	154							————	————				
1950	57				-	- -							
1950	114	—											
1952	150												-

Figure 14.9. The timing and duration of musth periods during 1980 ($n = 21$). Males are listed in chronological order starting with the oldest. The year of birth is estimated as ± 3 years.

and attempt to mate, they cannot compete with the larger, aggressive musth males for access to females on the peak days of oestrus. In addition, females prefer to mate with older musth males; they will not stand for the younger non-musth males and, if approached, will solicit guarding behaviour from musth males.

Vocal communication

African elephants have a large vocal repertoire consisting of several different types of sounds: rumbles, trumpets, roars, screams, snorts and bellows. The most frequently used calls are the rumbles. There are over 20 different types of rumbles, almost all of which contain components below the range of human hearing, some of these being totally infrasonic. Audio recordings of elephant vocalizations were made on a Nagra IVSJ recorder equipped with two Seinnheiser MKH110 microphones that was capable of recording these very low frequencies. By analysing the number of vocalizations recorded on tape, we discovered that we were hearing only 30% of what the elephants were saying. Most of the sounds that we were unable to hear were probably calls made by distant individuals, which

Born	Male	Jan	Feb	Mar	Apr	May	Jun	Jul	Aug	Sep	Oct	Nov	Dec
1935	13												
1935	41	Not seen in musth											
1939	126												
1940	22												
1941	28	Dead											
1943	44	Dead											
1944	99	Dead											
1944	117	Dead											
1945	7												
1945	10												
1945	45												
1945	78												
1945	80												
1946	73	Dead											
1946	51												
1946	107	Dead											
1947	119												
1948	175	Not seen in musth											
1949	154												
1950	57												
1950	114												
1950	147												
1951	91												
1952	150												
1952	79												
1953	97	Not seen in musth											
1953	125	Not seen in musth											
1953	131												
1953	132												
1953	145												
1953	174												
1955	137												
1956	5												
1956	100												
1957	12												
1957	86												
1957	93												
1957	135												
1958	129												
1959	156	Not seen in musth											
1959	105	Not seen in musth											
1959	140												
1959	207	Not seen in musth											
1960	146												
1960	201												
1961	159												
1961	104												
1961	191												
1962	32												
1962	115												
1963	48	Not seen in musth											
1965	196												
1965	212												
1966	193												

Figure 14.10. The timing and duration of musth periods for all males seen in musth during 1989 ($n = 40$). Males who were seen in musth in 1980 but have since died ($n = 6$), or males who come into musth but were not seen in musth in 1989 ($n = 8$) are also listed. Males are listed in chronological order starting with the oldest male. The year of birth is estimated as ± 3 years. The diagram illustrates the increasing number of males coming into musth each year as the population recovers from the poaching which, during the early to mid 1970s, seriously reduced the number of mature males.

the elephants, with their very sensitive low-frequency hearing, were certainly capable of detecting. However, it is very likely that many of the undetected calls were made by elephants within close range and were just too low in frequency for our human ears to hear.

The very different lives of male and female elephants are reflected in both the number and types of vocalizations used by each sex. Table 14.1 presents a list of the calls that I have described thus far.

The degree of sex differences in elephant vocalizations is pronounced and worth reviewing. While females use many different vocalizations in active communication between and within family groups, males use many fewer vocalizations, relying more on passive communication to locate groups of females. Of the 26 vocalizations made by adult elephants, 19 are made only by females, three are made by adults of both sexes and only four are made solely by males. An additional six calls are made only by subadults. The infrasonic alarm has not been detected, but is still suspected. Of the 22 calls used by females, 9 are calls usually given in chorus with other family members, while 13 are usually made by an elephant calling on its own. Males do not call in chorus. The majority of female calls are related to group coordination, infant care and social excitement, while the few calls made by males are related to male–male dominance interactions or reproduction.

By comparing the very significant differences in vocal repertoire between males and females, one gains a clearer understanding of the very marked differences between the function of male and female groups. The survival of females and their offspring depends on the cohesion and coordination of the family unit and the bond group, and on their ability to compete with other groups for access to scarce resources. The majority of vocalizations made by adult females are related to family/bond-group dynamics, cohesion and coordination ($n = 12$) or reproduction and communication with musth males ($n = 3$). Many of the very low-frequency calls used by females have high enough sound pressure levels to communicate with elephants 5–10 km away.

Discussion

Among species that show sexual dimorphism in body size, larger males frequently have a higher rate of mating success than smaller males. Since most mammals reach full adult size soon after reproductive maturity, mating success typically peaks during middle age and declines late in life. African elephant males reach twice the weight of females, making them

Table 14.1 *Elephant vocalizations*

		AM	AF	J	C	I
Sexual excitement						
Oestrous rumble	L	—	x	—	—	—
Female chorus	S/Lᵃ	—	x	x	x?	?
Genital testing	S	—	x	xf	—	—
Social excitement						
Greeting rumble	S/Lᵃ	—	x	x	x	x?
'Social' rumble	S/Lᵃ	—	x	x	x	x?
Roar	S/Lᵃ	—	x	x	x?	—
Mating pandemonium	S/Lᵃ	—	x	x	x	?
Play trumpet	S	x	x	x	x	—
Social trumpet	S/L	—	x	x	x?	—
Group dynamics and coordination						
Attack rumble	Sᵃ	—	x	x	?	?
Let's go rumble	S	—	x	x?	—	—
Contact call	L	—	x	x	x?	—
Contact answer	Lᵃ	—	x	x	x	—?
Coalition rumble	Sᵃ	—	x	—?	—	—
Mystery rumble – female	S	—	x	?	?	—
'Discussion' rumble	S	—	x	?	?	—
Distress						
Lost call	L	—	—	x	x	x
Suckle rumble	S	—	—	—	x	x
Suckle cry	S	—	—	—	x	x
Distress call (SOS)	S	—	—	—?	x	x
Reassurance rumble	Sᵃ	—	x	x	x	—
Calf response	S	—	—	—	x	x
Suckle distress scream	S	—	—	—	x	x
Social fear						
Scream	S	—	x	x	x	—
Bellow	S	x	x	x	x	x?
Groan	S	x	—	—	—	—
Fear, surprise of strangeness						
Trumpet blast	S	x	x	x	x	—
Snort	S	?	x	x	x	—
Alarm? Infrasonic?	S/L?	?	?	?	?	?
Dominance						
Female–female respect	S	—	x	x	x	—?
Musth rumble	S/L	x	—	—	—	—
Male–male respect?	S	x	—	—	—	—
Unknown						
Mystery rumble – male	S	x	—	—	—	—

ᵃ Often or always given as a group call. AF, adult female; AM, adult male; C, calf, f, female juveniles only; I, infant; J, juvenile; L, apparent long-distance function; S, apparent short-distance function; x, occurs; —, does not occur; ?, unknown.

one of the most sexually dimorphic of land mammals in terms of body size. Large size confers a significant advantage to male elephants in terms of fighting ability and dominance. Unlike other mammals, however, due to the continued growth of males, and the relationship between body size and rank, reproductive success increases rapidly late in life. Being in musth confers an additional reproductive advantage to older males in two ways: (1) their extremely aggressive behaviour results in musth males ranking above all non-musth males; and (2) oestrous females prefer musth males to non-musth males and express this preference by soliciting guarding behaviour from them and standing for, and often initiating, mating.

In studying sexual dimorphism, emphasis has often been placed on describing morphological differences between the sexes and explaining the evolutionary reasons for sexual selection. Studies of sex differences have also examined the ecological and behavioural consequences of sexual selection: differing habitat preferences of males and females as a consequence of body size; differing life-history patterns and social systems of males and females. There has, however, been very little study of dimorphism in the vocal communication of mammals other than humans. This may be either because most mammals have relatively small vocal repertoires, or because there is large overlap in the calls made by the two sexes. If the latter is the case, why are elephants so very different?

African elephants have a large vocal repertoire, the number and variety of calls rivalling chimpanzees and some of the other more vocal mammals. Chimpanzee vocal communication is unusually complex and, like the elephants, appears to be related to the fission–fusion pattern of their social organization. Compared to the elephants, among chimpanzees and other large mammals (e.g. zebras, horses, dogs, lions, hyenas, vervet monkeys and howler monkeys) the two sexes overlap broadly in their vocal repertoires. One reason may be that, in most of these species, males and females live together in stable groups. It is likely that the reason for the very different repertoires of male and female elephants is, to a large extent, related to the very different lifestyles that they lead. As we learn more about the vocal communication of elephants and other animals, we will, no doubt, gain a better understanding of the causes and consequences of sex differences.

Acknowledgements

I am grateful for financial support from the National Geographic Society. For permission to conduct research in Amboseli National Park, I thank the Office of the President, the National Research Council, the Kenya Wildlife Service, and the Amboseli Wardens. For helpful comments on the draft I am grateful to Andy Dobson. Norah Njiraini, Soila Sayialel and Cynthia Moss spent long hours with me listening to elephants and I thank them for their support.

Further reading

Haynes, G. (1991) *Mammoths, Mastodons and Elephants: Biology, Behaviour and the Fossil Record.* Cambridge University Press, Cambridge.

Lee, P. C. (1987) Allomothering among African elephants. *Animal Behaviour,* **35,** 278–91.

Moss, C. (1988) *Elephant Memories: Thirteen Years in the Life of an Elephant Family.* William Morrow and Co., New York.

Moss, C. S. and Poole, J. H. (1983) Relationships and social structure of African elephants. In *Primate Social Relationships: An Integrated Approach,* pp. 315–25. Ed. R. A. Hinde. Blackwell Scientific, Oxford.

Poole, J. (1987) Rutting behaviour in African elephants: the phenomenon of musth. *Behaviour,* **102,** 283–316.

Poole, J. (1989) Announcing intent: the aggressive state of musth in African elephants. *Animal Behavior,* **37,** 140–52.

Poole, J. (1989) Mate guarding, reproductive success and female choice in African elephants. *Animal Behavior,* **37,** 842–9.

Poole, J., Payne, K., Langbauer, W. and Moss, C. (1988). The social contexts of some very low frequency calls of African elephants. *Behavioural Ecology and Sociobiology,* **22,** 385–92.

15

The costs of sex

T. H. CLUTTON-BROCK

Breeding systems, reproductive success and competition for mates

In many long-lived birds as well as in a few mammals, a male is mated monogamously to the same female for life. In these species, the number of adult males and females at pairing is approximately equal and both sexes have to compete for mates to approximately the same extent. Since males are mated permanently to particular females, a male's breeding success depends largely on the breeding success of his mate and variation in lifetime reproductive success (the closest estimate of fitness that is usually available) is of similar magnitude in the two sexes. For example, both in kittiwakes *Rissa tridactyla* and in Bewick's swans *Cygnus columbianus*, individual differences in lifetime breeding success (measured by the number of young fledged) are similar in males and females (Figure 15.1a), although variation in male breeding success may be larger than these estimates suggest if some males engage regularly in extrapair copulations. The factors affecting breeding success are also usually similar for males and females and the breeding success of males and females changes in the same way with age (Figure 15.2a).

In contrast, in animals that breed polygynously, a minority of males monopolize mating access to females. As a result, variation in lifetime breeding success is usually greater in males than in females. For example, in red deer *Cervus elaphus* (Figure 15.1b), observational measures of breeding success suggest that the most successful stags produce around 30 surviving offspring in their lifetimes, while the must successful hinds rarely produce more than nine surviving offspring. Recent studies using DNA fingerprinting show that, although observational measures of male success give reliable estimates of relative mating success in red deer, they underestimate the number of young that are fathered by successful

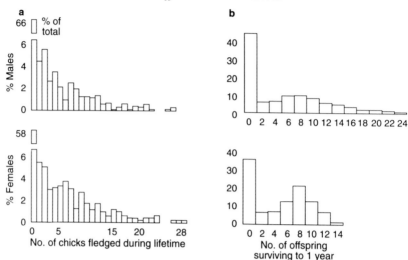

Figure 15.1. Individual differences in lifetime reproductive success in male and female kittiwakes and red deer. Histograms show the proportion of all individuals fledged (a. kittiwakes) or born (b. red deer) that produced different numbers of offspring during their lifetimes.

males and overestimate the number of young that are fathered by unsuccessful ones.

Where particular males are disproportionately successful in acquiring mates, the ratio of receptive females to sexually active males (the 'Operational Sex-Ratio') typically shows a strong bias towards males. As a result, males have to compete intensely for mates. Although females may sometimes compete for access to particular males in poly-gynous species, they seldom have any difficulty in finding mates.

Reproductive competition and breeding lifespans

Where males are forced to compete intensely for females, competition usually involves displays, struggles or fights that are physically demanding (Figure 15.3a–c). As a result, a male's breeding success is typically related closely to his fighting ability.

One important cost of the intense competition for mating access that occurs among males in polygynous species is that successful breeding is commonly delayed until males have reached full adult size, although sexual maturity may be attained several years previously. For example, in red deer, males rarely mate successfully before reaching adult weight

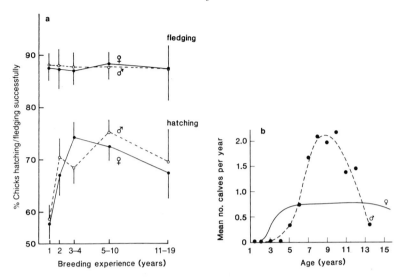

Figure 15.2. Effects of age on breeding success in (a) kittiwakes and (b) red deer.

at seven or eight years of age, while bull elephant seals *Mirounga angustirostris* rarely breed before the age of nine years. Similar delays in the onset of successful breeding occur in many polygynous birds. In addition, breeding success and survival commonly decline rapidly after males have passed their prime. Red deer stags, for example, seldom breed successfully after the age of 12 years and bull elephant seals seldom breed after they have passed 13 years of age. In contrast, in both these species, the effective breeding lifespans of females are usually around twice as long as those of males (see Figure 15.2b).

   The degree of compression in their breeding lifespans suffered by males of polygynous species varies with the form of intermale competition. In primates living in multimale troops, commonly, males assist each other in competition for females and a male's status and breeding success may be influenced strongly by the ranks of the other males with which he has developed reciprocal cooperative relationships. Here, male breeding success appears to be affected by increasing age less than in harem-forming mammals, like red deer or elephant seals, where a male's mating success depends on his fighting ability. Indeed, in some polygynous human societies, the breeding success of males increases in old age as a result of the accumulation of wealth and wives, and men have longer reproductive lifespans than women.

Figure 15.3. (a) Mature red deer stags can hold harems of 20 or more females and can father more than 10 offspring each year of their reproductive lifespans. (b) Mature stags fight for access to females. Larger stags generally win.

Figure 15.3. (*continued*) (c) Fighting is costly, and can have severe consequences for the individual's immediate breeding success and subsequent survival. Consequently, stags assess rivals in roaring contests before committing themselves to a fight. (d) Early development is related closely to adult size and breeding success. Of these two 15-month-old stags, photographed in 1972, the better-grown individual on the left was a consistently more successful breeder than the smaller stag on the right.

Determinants of breeding success in males and females and the evolution of sexual size dimorphism

Where competition for access to mating partners is more intense among males than among females, traits like dominance rank or body size, which affect success in competitive interactions, are likely to exert a stronger influence on breeding success in males than in females. For example, in red deer, body size and dominance rank contribute to breeding success in both sexes, but both measures have a stronger effect on breeding success in the males. This is partly because variation in breeding success is greater among males than among females and partly because competition for mates is more likely to be determined by fighting than by competition for food. Similar differences in the effects of body size on breeding success in the two sexes probably occur in many other polygynous species where male fighting involves pushing.

However, relationships between the size or rank of males and their breeding success should not be expected in all species. In Camargue stallions, for example, which fight by biting and kicking rather than by pushing, male breeding success is not consistently related to body size. Similarly, where males form competitive coalitions, as in some cercopithecine primates that live in multimale groups, a male's breeding success depends more on his social connections than on his size (see above).

Where large body size has a greater effect on the breeding success of males than on that of females, males are likely to evolve to be larger than females and this is presumably why pronounced sexual size dimorphism is a common feature of polygynous vertebrates. In contrast, where body size has little effect on male success or has similar consequences for males and females, little or no sexual size dimorphism would be expected. For example, in equids, where male size has little influence on fighting or breeding success (see above) there is little or no sexual size dimorphism in adult size. In polygynous species where females are the larger sex, such as spotted hyenas, *Crocuta crocuta*, larger female size has evolved presumably because size exerts a stronger effect on the fitness of females than that of males – although this has yet to be demonstrated. This prediction is supported by evidence that in polyandrous birds where individual females monopolize breeding access to multiple males (including phalarope, painted snipe and button quail), females are typically larger and brighter than males.

Sex differences in early growth

Where a male's breeding success increases with his body size, early growth rates are likely to exert an important effect on adult breeding success (see Figure 15.3d). Although compensatory growth is common in domestic mammals maintained on high levels of nutrition, in naturally regulated populations any initial setback in growth is likely to affect subsequent growth rates and, eventually, adult size. For example, in red deer, stag calves born light or late or subjected to food shortage during their first winter are smaller than average when adults and show low breeding success. Similarly, hind calves born below average weight turn into small adults and, in their turn, produce light calves that have a low chance of survival.

The close relationship between early development and adult size has led to selection favouring heavier birth weights and faster postnatal growth rates in males. In size-dimorphic mammals, males are typically born heavier than females and grow faster during the early weeks of life. In size-dimorphic birds, hatching weights are usually similar in the two sexes but, subsequently, males grow faster, and sexual size dimorphism is typically as large in fledglings as in adults.

Costs of sexual size dimorphism

Energy costs of sexual size dimorphism to adults

The large differences in body size found in many polygynous mammals have important energetic consequences. While metabolic rate increases with body size as around body-weight$^{0.75}$, commonly, the rate of food intake may increase less rapidly with body weight. For example, in grazing animals feeding on short swards, the rate of food intake may be proportional to the breadth of the incisor arcade, which increases as body-weight$^{0.33}$, or may even show little increase within the normal range of body size. In addition, in at least some size-dimorphic mammals, males have higher metabolic rates and greater energy demands per unit body weight than females.

On account of their large body size and greater nutritional requirements, commonly, males have to spend more time collecting food than females, especially when food availability is low. For example, in late winter when food is short, red deer stags on Rum spend more time feeding in total, feed relatively more at night when temperatures are low and, compared to hinds, feed less on highly preferred communities where the standing crop is low. These differences have important consequences for habitat use by males and females: in many size-dimorphic ungulates, males are segregated, partly or totally, from females during the non-breeding season and, commonly, are found feeding in habitats where food is more abundant but of lower nutritional quality than in areas used heavily by females.

Energy costs of sexual size dimorphism to juveniles

Sex differences in juvenile growth are generally associated with differences in energy intake. For example, in the sexually dimorphic great-tailed grackle *Quiscalus mexicanus*, measurements of oxygen and food consumption under controlled conditions show that males, which are around 50% heavier than females by the twelfth day after hatching, require around 20% more food. Male red-winged blackbirds *Agelaius phoeniceus* are around 30% heavier than females at fledging and require approximately 27% more energy. In fur seals *Arctocephalus*, measurement of milk intake shows that male pups take around 30% more milk than females (although they ingest less milk relative to their body mass) while in the less dimorphic grey seal *Halichoerus grypus*, the energy costs of rearing sons are only around 10% higher than those of rearing daughters. In a substantial number of other mammals, including red deer, goats, American bison and African elephants, measures of milk intake are not available but males suck more frequently or for longer than females.

Although the energy requirements of males exceed those of females in many dimorphic vertebrates, this is not always the case. In the highly dimorphic southern elephant seal *Mirounga leonina*, there appears to be no difference in the costs of rearing sons and daughters, although males are born heavier than females, while in some ungulates showing unusually fast juvenile growth rates, including pronghorn antelope *Antilocapra americana* and Soay sheep *Ovis aries*, there appear to be no differences in suckling time between the sexes.

Survival costs to juveniles

Where resources are short, the faster growth rates and higher metabolic requirements of male juveniles commonly have substantial costs. Food shortage depresses the growth rates of males more than those of females in many dimorphic species and, in some cases where males are usually the larger sex, males may be absolutely smaller than females at weaning, where food availability is low. Commonly, male juveniles show higher levels of mortality than females where environmental conditions are harsh. Experimental studies of wood rats *Neotama floridana* and domestic pigs have shown that reduction in food availability reduces the survival of males more than that of females, while, in several ruminants, the survival of male juveniles relative to female juveniles falls as population density increases. As would be expected, these differences are most pronounced among those offspring whose access to food resources is affected most strongly by food availability, including the offspring of subordinate mothers and the offspring of mothers belonging to groups of above average size.

Sex differences in juvenile survival are most pronounced in species showing high degrees of size dimorphism in adults. In mammals, it is frequently difficult to separate the effects of sex differences in nutritional requirements or mortality from the effects of sex differences in dispersal but, in birds that rear their young in nests, sex differences in mortality commonly occur before the young fledge. Across a variety of bird species, the extent to which the sex ratio (% males) changes between hatching and fledging increases with the degree of size dimorphism in adults (Figure 15.4).

The faster growth rates and higher energetic requirements of males may even affect the relative survival of the two sexes before birth. Research on a variety of different mammals has shown that the sex ratio of fetuses declines during the course of gestation. In addition, studies of several different rodents show that litter size and birth sex-ratio (percentage of male offspring) fall when pregnant females are subjected to food shortage or other forms of environmental stress. Perhaps the most remarkable results come from studies of golden hamsters, *Mesocricetus auratus*. As in other rodents, food restriction of females during pregnancy leads to a significant reduction in litter size and to a decline in the sex ratio at birth. Moreover, restricting food given to virgin females during their first 50 days of life causes them to produce smaller litters and female-biased sex-ratios during adult life, even when, subsequently, they are replaced on *ad lib* diets (see Figure 15.5a). The effects of food restriction during development even span generations: the daughters of food-restricted females, which were themselves reared on *ad lib* diets, produced smaller litters and relatively fewer sons than the daughters of control females that were not food restricted (Figure 15.5b).

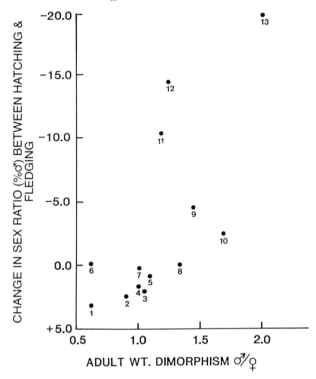

Figure 15.4. Changes in the sex ratio (% males) between hatching and fledging in bird species showing different degrees of adult weight dimorphism (male weight/female weight): (1) peregrine falcon; (2) American kestrel; (3) eastern bluebird; (4) starling; (5) snow goose; (6) European sparrow-hawk; (7) red-cockaded woodpecker; (8) blue grouse; (9) red-winged blackbird; (10) yellow-headed blackbird; (11) rook; (12) common grackle; (13) capercaillie.

Costs to mothers

The faster growth rates of males and the higher energetic requirements associated with them have important costs to mothers that raise sons. Female red deer that have reared sons are more likely to die the next winter than females that have reared daughters; if they survive, they are less likely to calve the next year; and if they do breed again the next year, they conceive and calve later in the season. In several other ungulates, raising males affects the mother's breeding performance, weight, parasite load or survival more than raising females. Although the greater costs of raising sons appear to be most obvious in sexually dimorphic ungulates, they also occur in many other mammals. For example, in Mongolian

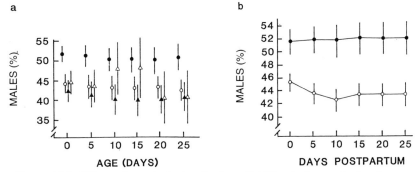

Figure 15.5. Effects of manipulating food availability during early development on birth sex-ratios in golden hamsters, *Mesocricetus auratus*. (From Huck *et al.*, 1986, 1987.) (a) Sex ratios from birth to 28 days of age of litters produced in adulthood by females raised on *ad lib* diets (filled circles), or food-restricted during their first 25 days of life (filled triangles), their second 25 days of life (open circles), or their first 50 days of life (open triangles) and subsequently maintained on *ad lib* diets. (Huck *et al.*, 1986.) (b) Sex ratios from birth to 25 days of litters produced in adulthood by daughters of females who were themselves raised on *ad lib* diets (filled circles) versus daughters of females food-restricted during the first 50 days of life (open circles). (Huck *et al.*, 1987.).

gerbils *Meriones unguiculatus*, females that have raised all-male litters show longer periods of vaginal closure (preventing copulation) after birth and, subsequently, produce smaller litters than females that have raised all-female litters.

The heavier energetic costs of raising sons do not affect all females similarly. In red deer, for example, they are confined to subordinate females, while dominant mothers that rear sons do not differ in subsequent survival or fecundity from those that have raised daughters. Moreover, other factors can affect (or even reverse) the higher costs of rearing sons. In some cercopithecine primates, unrelated females attack one another's female offspring selectively, perhaps because they represent potential competitors. As a result, mothers of daughters allow their offspring more frequent access to the nipple and this has direct effects on the mother's secretion of gonadotrophin and, consequently, on the length of lactational amenorrhoea. In rhesus macaques, mothers that have reared daughters do not conceive again as quickly as mothers that have reared sons. As in red deer, differences in the costs of rearing male and female progeny are confined to subordinate mothers.

It has been suggested that the higher mortality of male juveniles may occur because mothers that cannot afford the higher energetic costs necessary to rear a potentially successful male terminate investment in their sons prematurely, allowing them to die. There is, however, no evidence that mothers suckle their daughters more than their sons when food is scarce. In addition, sex differences in juvenile mortality occur when juveniles are subjected to nutritional stress in the absence of their parents. Instead, it seems likely that these differences are by-products of sexual selection favouring rapid growth in males.

Parental strategies and sex-ratio manipulation

Where sons cost more to rear than daughters, the most successful males are generally the offspring of mothers of superior genotypic or phenotypic quality that can invest heavily in their offspring. Inferior mothers may be unable to afford to invest heavily enough in their offspring to produce males large enough to have a reasonable chance of breeding successfully. For example, in red deer, most of the successful males are the offspring of dominant mothers, while the sons of subordinate mothers have little chance of surviving to breed successfully.

Under these circumstances, females would increase their fitness by varying the sex ratio of the offspring they produce. Mothers that can afford to invest heavily in their offspring might be expected to produce sons that would have a high chance of breeding successfully, while females that cannot invest as heavily in their offspring should concentrate on producing daughters. There is plentiful evidence that some invertebrates manipulate the sex ratio of their progeny in this fashion – but can vertebrates, whose sex is determined chromosomally, do the same?

As I have described already, food shortage or other forms of environmental stress is commonly associated with higher mortality of males *in utero* and with birth sex-ratios that are biased (sometimes heavily) towards females (see Figure 15.5). These relationships have commonly been cited as examples of adaptive control of the birth sex-ratio by females, but an alternative interpretation is that they represent the costs of sexual selection favouring more rapid growth in males.

However, in some mammals, mothers that can afford to invest heavily in their offspring produce a significantly male-biased sex-ratio at birth. For example, in red deer, socially dominant hinds, which consistently show superior reproductive performance, produce consistently more sons than daughters, while subordinates produce more daughters than sons (Figure 15.6). Similar effects have now been demonstrated in several other mammals. In wild populations of common opossums (*Didelphis*

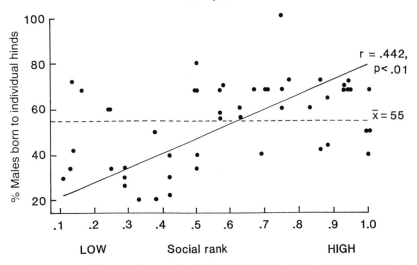

Figure 15.6. Birth sex-ratios produced by individual red deer hinds differing in social rank over their lifespans. Measures of maternal rank were based on the ratio of animals that the subject threatened or displaced to animals that threatened or displaced it, weighted by the identity of the animals displaced. Values of this ratio range from 0.1 (low ranking) to 1.0 (high ranking).

virginiana), experimental feeding of young females led to a significant increase in the number of males produced while control females produced an approximately equal sex-ratio.

An important test of these predictions is to determine whether the relationship between maternal quality and the sex ratio is reversed when females are more costly to rear than males. In social insects where females are bigger and more costly to rear than males and the sex ratio is not determined chromosomally, females commonly produce a large excess of daughters where resources are abundant. There is some suggestive evidence that similar trends may occur in cercopithecine primates where, for social reasons, the costs to subordinate females of rearing daughters exceed those of rearing sons (see above). Some recent studies of yellow baboons *Papio cynocephalus* and rhesus macaques *Macaca mulatta* have shown that, consistently, dominant mothers produce more daughters than sons, while subordinates produce more sons than daughters (Figure 15.7).

Discussion

The contrasting factors affecting reproductive success in males and females and their associated costs thus have far-reaching consequences for many

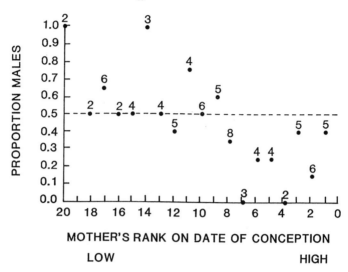

Figure 15.7. Birth sex-ratios in relation to maternal rank in wild yellow baboons in the Amboseli National Park, Kenya. Numbers of offspring produced per female are shown above each square. (From Altmann, J. (1980) *Baboon Mothers and Infants.* Harvard University Press, Cambridge; Altmann, J., Hausfater, G. and Altmann, S. R. (1988) Determinants of success in savannah baboons, *Papio cynocephalus.* In *Reproductive Success*, pp. 403–18. Ed. T. H. Clutton-Brock. University of Chicago Press, Chicago.)

other aspects of vertebrate biology. Where the reproductive success of males depends on their fighting ability and body size, selection commonly favours larger body size, higher metabolic rates, and development of weaponry. In many species, large adult size is dependent on rapid early growth rates, and, consequently, selection favours faster growth in males than in females.

Since the potential benefits of rapid growth and increased body size are evidently large in polygynous vertebrates, only a strong opposing selection pressure will prevent sexual selection continuing to increase male size. It should not surprise us that male size and growth has evolved to a point at which it has substantial costs to survival under adverse environmental conditions. In polygynous species which show little sexual size dimorphism, like horses and zebras, we can expect to find either that body size has similar benefits to males and females or that the costs of increased size to males are unusually high.

Since sex differences in growth occur before birth, it is not surprising that male fetuses, like male juveniles, are more likely to be affected by adverse environments. These sex differences in viability provide a likely explanation of the tendency for birth sex-ratios to show a female bias in some mammals where mothers are subjected to environmental stress.

Although it might be in their evolutionary interests to do so, it is not yet clear whether parents treat their sons and daughters in different ways. In several size-dimorphic birds and mammals, there is firm evidence that raising sons has greater costs to the mother (both in terms of energy and fitness) than raising daughters, but this may be because juvenile males attempt to suck or to feed more frequently than females. There is, as yet, no firm evidence that parents are more responsive to the demands of sons. Similarly, while parents might increase their fitness by abandoning sons that they cannot afford to rear successfully, there is no firm evidence to show that higher mortality occurs in males when resources are short because mothers discriminate against their sons. A more likely explanation is that both the higher costs of sons and their higher mortality are by-products of selection for rapid growth and large size in males.

Perhaps the most contentious issue of all is the extent to which female vertebrates can manipulate the sex ratio of their young before birth. In haplodiploid insects, where the sex of offspring can be manipulated easily by the parent, mothers commonly vary the sex ratio of their progeny from greater than 90% male to greater than 90% female in relation to variation in resource availability (see Chapter 4). However, firm evidence of adaptive sex-ratio variation in vertebrates is scarce. There is little firm evidence of consistent trends in hatching sex-ratios among birds. In mammals, a considerable number of trends has been reported but many have not been tested adequately. In addition, significant sex-ratio trends are more likely to be published than non-significant ones. As a result, published examples of sex-ratio trends represent a biased sample of investigations and there is a danger that a higher proportion of results may represent cases where the null hypothesis has been rejected wrongfully. Finally, many of the most convincing sex-ratio trends represent cases where environmental stress is associated both with a decline in litter size and with a decline in the birth sex-ratio, suggesting that they may be caused by differences in the viability of male and female fetuses rather than by parental manipulation (see above).

Despite these problems, there is a growing number of examples where females in superior condition or with access to plentiful resources produce a significant excess of males, which are unlikely to be statistical artefacts and coincide with the predictions of adaptive theories. As yet, the mechanisms underlying these effects are unknown. The physiological basis of established sex-ratio trends is an area that might well repay detailed investigation by reproductive physiologists in the future.

Further reading

Charnov, E. L. (1982) *The Theory of Sex Allocation.* Princeton University Press, Princeton, N.J.

Clutton-Brock, T. H. (ed.) (1988) *Reproductive Success.* University of Chicago Press, Chicago.

Clutton-Brock, T. H. (1991) *The Evolution of Parental Care.* Princeton University Press, Princeton, N.J.

Clutton-Brock, T. H. and Iason, G. (1986) Sex ratio variation in mammals. *Quarterly Review of Biology*, **61**, 339–74.

Emlen, S. T. and Oring, L. W. (1977) Ecology, sexual selection and the evolution of mating systems. *Science*, **197**, 215–23.

Glucksman, A. (1974) Sexual dimorphism in mammals. *Biological Reviews*, **49**, 423–75.

Huck, U. W., Labov, J. D. and Lisk, R. D. (1986) Food restricting young hamsters (*Mesocricetus auratus*) affects sex ratio and growth of subsequent offspring. *Biology of Reproduction*, **35**, 592–8.

Huck, U. W., Labov, J. D. and Lisk, R. D. (1987) Food-restricting first generation juvenile female hamsters (*Mesocricetus auratus*) affects sex ratio and growth of third generation offspring. *Biology of Reproduction*, **37**, 612–17.

Trivers, R. L. (1972) Parental investment and sexual selection. In *Sexual Selection and the Descent of Man*, pp. 136–79. Ed. B. Campbell. Aldine, Chicago.

Trivers, R. L. and Willard, D. E. (1973) Natural selection of parental ability to vary the sex ratio of offspring. *Science*, **179**, 90–92.

16

Pheromones: the fruit fly's perfumed garden

JEAN-FRANÇOIS FERVEUR, MATTHEW COBB,
YUZURU OGUMA and JEAN-MARC JALLON

At the end of the nineteenth century, the French entomologist Henri Fabre showed that moths of different sexes could communicate over long distances. However, it was not until the end of the 1950s that Karlson and Butenandt were able to show the chemical nature of these signals. From hundreds of thousands of female silk moth (*Bombyx mori*) glands, they extracted a few milligrams of a molecule which could attract conspecific males from long distances. This molecule, which they named bombykol, is a primary alcohol with a 16-carbon primary aliphatic chain containing two double bonds (10,12 hexadecadienol). Karlson and Butenandt invented the term 'pheromone' to describe those chemical signals produced by an organism which modify the behaviour or physiology of another organism. At about the same time, Jeanne Pain and her coworkers identified a substance produced by queen bees which leads to the degeneration of the ovaries of worker bees. Since then, the list of examples of pheromones has grown larger each year. Hundreds of pheromonal molecules have been identified through the close collaboration of biologists and chemists, following the example of Karlson and Butenandt.

Pheromones have been identified principally in insects, and especially in lepidopterans. The relevant molecules are predominantly those containing medium-length carbon chains, often unsaturated with alcohol, acetate or aldehyde groups. Recently, vertebrates have also been the subject of intensive study and a number of their pheromones have been identified. Pheromones have been subdivided into 'releasers' and 'modifiers', depending on whether they act instantaneously or over some time period, respectively. Behavioural effects of pheromones are often focused on reproduction by inducing particular behaviours, by modifying the reproductive state of the recipient or by enabling individuals to identify other members of the same species and, thus, to distinguish members of

363

other species. Pheromones also can elicit aggregation or defence against predators.

In light of the discovery of chemical signals which control the interaction between insects and their host plants, the concept of 'pheromone' has been restricted to interactions between members of the same species or of closely related species. The term 'allelochemicals' was developed to describe chemical signals between unrelated species such as plants and animals. In a final round of linguistic refinement, both pheromones and allelochemicals have been grouped together under the title 'semiochemicals'.

Following the initial discovery of pheromones, newly isolated attractive molecules were put to use in an attempt to control pests and to protect crops. These initial field trials failed to live up to expectations, which provided an impetus for further, more detailed laboratory work. During the 1970s, it became apparent that chemical communication between butterflies and moths was far more complicated than had been thought initially. In particular the major pheromonal components that had been isolated initially seemed to be more effective if mixed with more minor components that were normally present to form pheromone blends. These complex mixtures usually consist of fixed proportions of the various components for a given population.

During this period, other pheromonal substances were identified which showed far lower levels of volatility and, therefore, which could not act at long distances. In 1971, David Carlson and coworkers showed that a cuticular hydrocarbon, Z9-tricosene, is a pheromone in the housefly *Musca domestica*, attracting males to females. Subsequent studies revealed that this substance also induced male courtship behaviour. In the tsetse fly, cuticular pheromones are totally non-volatile; physical contact between individuals has to take place for the message to be detected. There is thus a difference between volatile, airborne pheromones and non-volatile, contact ones. Volatile pheromones are always produced in specialized glands, usually in minute amounts, whilst contact pheromones are more abundant and are often synthesized in cells distributed throughout the subcuticular tissues.

A review of this nature could not begin to do justice to the incredible diversity of pheromonal functions that have been identified so far. Rather than tiring the reader by reciting endless case-studies, we have chosen to concentrate on work done in one taxonomic group, the fruitfly *Drosophila melanogaster* and its closely related sibling species. Because of the genetic tools available in this organism, it is the species of choice for studying the behavioural, genetic, biochemical and evolutionary aspects of pheromones. We would like to use it as a window through which the intricacies of sex differences in chemical communication can be viewed.

Pheromones in *Drosophila*

There are hundreds of *Drosophila* species; most of which are poorly studied and are only distantly related to *D. melanogaster*, the geneticist's friend. Seven other species are closely related to *D. melanogaster*, and together these eight species form the *melanogaster* subgroup. Substantial ecological diversity exists among these species: two of them (*D. melanogaster* and *D. simulans*) are distributed throughout the world; the other six are endemic to Africa. Four of these African species are found on the continent (*D. yakuba* and *D. teissieri*, which are more closely related to each other than the other two continental species, *D. orena* and *D. erecta*). The final two species are found only on islands in the Indian ocean: *D. mauritiana* is confined to the island of Mauritius whilst *D. sechellia* has been found only on the Seychelles archipelago. These species have been the subject of intensive phylogenetic studies, examining their morphology, karytoypes and DNA polymorphisms. Important differences have been found for secondary sexual characteristics, in particular for the form of the penis and the male genital arch.

Behavioural studies of *D. melanogaster* male courtship were initiated by Margaret Bastock and Aubrey Manning in the 1950s; Kevin Connolly and Robert Cook published the first study of female behaviour in 1973. In the late 1960s, Henry Bennet-Clarke and Arthur Ewing showed that *D. melanogaster* males produce an acoustic signal or 'song' during courtship which serves both to identify the male as a conspecific and to induce the female to copulate (see Chapter 11 by Evan Balaban). Comparative studies of quantitative and qualitative behavioural differences between the species were limited to *D. simulans* initially, but, following the discovery of the other members of the subgroup in the 1970s and early 1980s, studies were broadened to include them as well. The courtship behaviour of members of the *melanogaster* subgroup is particularly interesting because it is relatively slow, lasting for several minutes. In comparison, houseflies have a very brief courtship sequence which generally takes place on the wing, and some species of tsetse flies have virtually instantaneous copulation with no courtship. The behaviour of male and female *Drosophila* can be studied relatively easily in a large number of pairs, using detailed behavioural measures. In principle, the stimuli involved in each step of the courtship sequence can be identified.

The pheromonal repertoire of D. melanogaster

For many years, chemical signals were thought to play a role in *Drosophila* sexual behaviour. Yet evidence was relatively weak and indirect, and it is only in the last 12 years that pheromones have been identified in the *melanogaster* subgroup and their role has been revealed. Much of this work has come from our laboratory. The pheromonal repertoire of these insects is particularly rich: there are excitatory and inhibitory pheromones in both males and females, as well as naturally occurring polymorphisms which have helped us to distinguish a number of fundamental aspects of *Drosophila* chemical communication.

Initially, we discovered a cuticular pheromone in mature *D. melanogaster* females which induces courtship behaviour in conspecific males. This pheromone is a necessary, but not a sufficient precondition, for some elements of male courtship behaviour which appear to be produced by the synergy of the pheromone and visual signals. The pheromone was purified and studied in a bioassay, using the amount of wing vibration (singing) displayed by the tester male as a measure of pheromonal stimulus efficacy.

In order to isolate a sex-specific pheromone, we compared cuticular extracts from males and females using gas chromatography. In *D. melanogaster*, this revealed the existence of long-chain hydrocarbons (23–29 carbons) in both sexes and the presence of carbon chains with two double-bonds only in females. Use of different types of chromatography enabled us to separate a fraction containing heptacosadiene (27 carbons, two double-bonds) which induced high levels of male wing vibration. Initially, this work used a frequently studied laboratory strain, Canton-S. The main cuticular hydrocarbon in the females of this strain is 7,11 HD (*cis cis* 7,11 heptacosadiene), present at levels of 500 ng per mg of body weight, a relatively high concentration.

Synthesis of 7,11 HD and its use in bioassays provided the final evidence that this hydrocarbon is the principal component in a female 'courtship' pheromone. Comparison of dose–response curves of male wing vibration following stimulation with synthetic or natural HDs suggests that synergistic interactions may be taking place between 7,11 HD and two other forms with the same constituent molecules arranged

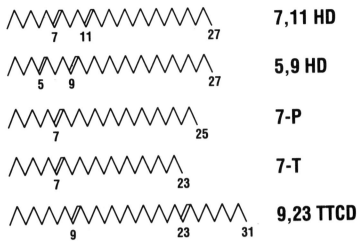

7,11 HD

5,9 HD

7-P

7-T

9,23 TTCD

Figure 16.1. Schematic representation of principal *Drosophila* cuticular hydro-carbons. Zig-zags represent carbon chains. Double lines represent double-bonds. 7,11 HD, 7,11 heptacosadiene (27 carbons), the *D. melanogaster* female pheromone. 5,9 HD, 5,9 heptacosadiene (27 carbons), a position isomer of 7,11 HD, present in high levels in females of some *D. melanogaster* strains. This substance is not known to have any behavioural activity. 7-P, 7-pentacosene (25 carbons), found in high levels in *D. melanogaster* males from near the equator. 7-T, 7-tricosene (23 carbons), the main hydrocarbon in most *D. melanogaster* males. 9,23 TTCD, 9,23 tritriacontadiene (31 carbons), found in the cuticles of young *D. melanogaster* flies of both sexes and in *D. erecta* females.

in different geometries (the 5,9 and 9,13 position isomers) which are also present in low levels in Canton-S females (Figure 16.1).

Male pheromones in **D. melanogaster**

The situation in males is somewhat different: 7,11 HD is completely absent, whilst other hydrocarbon molecules with only one double-bond, called monoenes, are present in large quantities. One form of these molecules with a double-bond in position 7, called *cis* 7-tricosene (7-T), is particularly abundant in Canton-S males, and appears to be the main class of hydro-carbons in the cuticles of most *D. melanogaster* males. The high levels of these molecules in male cuticles, but not in the cuticles of homotypic females, and the behavioural effects of this hydrocarbon on both sexes have led us to speculate about its role in courtship. We believe that 7-T may play the dual role suggested for many sexually selected male charac-teristics (see Chapter 3 by John Reynolds and Paul Harvey) by simulta-neously stimulating females and inhibiting courtship from rival males.

Male flies possess another apparent chemical signal, *cis* vaccenyl acetate, but the behavioural role of this substance is complex. It is not a cuticular component but, instead, is produced from endocrine tissue in the region of the male's ejaculatory bulb, and is transferred from males to females during copulation. It has been shown to play the role of an aggregation pheromone, showing higher levels of attractiveness when mixed with food odours. *cis* Vaccenyl acetate also appears to be able to inhibit male courtship behaviour, and, thus, may be responsible for the decline in attractiveness shown by females who have copulated, relative to virgin female flies. The pheromonal odour is perhaps diffused by the extrusion of the female's ovipositor, which is a classic response of a fertilized female to male courtship. These two different behavioural effects may be produced by different concentrations of the pheromone; aggregation effects have been observed with pheromonal concentrations of the order of tens of nanograms, whilst courtship is inhibited by *cis* vaccenyl acetate concentrations in the range of hundreds of nanograms.

The accessory glands of *D. melanogaster* males synthesize another peptide that is transferred to females during copulation. Then it represses female sexual receptivity and stimulates oviposition. This peptide is a good example of a pheromone of the modifier type. It consists of 36 amino acids resulting from the expression of a single gene located in a known position on the third chromosome. The gene is also active in closely related sibling species, but not in phylogenetically more distant flies.

Development and maturation of sex differences in pheromones

When flies eclose from their pupal case, none of the adult cuticular pheromones are present. The cuticular hydrocarbons of young flies are very different from their adult counterparts. There are no sex differences, the carbon chains are much longer (up to 37 carbons) and a wide range of molecules called dienes are present with different double-bond positions (e.g. 9,23 in the young flies compared to 7,11 in adult females). As the neuroendocrine system of an individual matures, this youthful cocktail is replaced by the adult sex-specific molecules. *cis* Vaccenyl acetate is also first synthesized in males during this period.

In *Musca domestica*, the production of the female pheromone (Z9-tricosene) is blocked following early ovariectomy. Such clearcut evidence for ovarian control of pheromone synthesis has not been found in *D. melanogaster*. Evidence does exist, however, for some role played by the hormone ecdysone in pheromone maturation in this species. Ecdysone is found in higher levels in mature females than in mature males, and its site of synthesis is not limited to the ovaries. Clearly, it plays a role in pheromone synthesis, because the mutation *ecdysoneless* shows a 50% reduction in 7,11 HD levels in females, together with an increase in the level of 7-heptacosene. This suggests that ecdysone may also play a role in the synthesis of the molecules containing a second double-bond that are found in females.

Young *D. melanogaster* of both sexes induce high levels of courtship in Canton-S males, although the adult males tend to give incomplete courtship sequences and break off courtship altogether in response to young flies. It is probable that certain hydrocarbon chains called dienes, characteristic of young flies of both sexes, are responsible for this confusion. In *apterous* mutants, which have very low levels of juvenile hormone, this period of misdirected courtship continues for a much longer period of time. Thus, juvenile hormone seems to affect the switch from the undifferentiated cuticular hydrocarbons typical of young flies to the sex-specific spectrum of hydrocarbons which appear with maturation.

Interspecific recognition

D. melanogaster is a cosmopolitan (widely distributed) species which, over much of its range, is sympatric with its sibling species *D. simulans*. Reproductive isolation between the two species is generally strong, although hybrids can be obtained in the laboratory and are found occasionally in the wild. Although qualitative and quantitative differences are apparent in the male courtship behaviour of the two species, and differences in male courtship song have also been reported, we were interested in discovering whether chemical signals play any role in interspecific discrimination, especially of females by males.

D. simulans females have no cuticular dienes; 7,11 HD, the main cuticular pheromone in *D. melanogaster* females, is completely absent. This enables the males to distinguish conspecific females from members of the other species. The cuticular hydrocarbon profiles of *D. simulans* are also strikingly different from those of *D. melanogaster*: there is no qualitative sexual dimorphism for cuticular hydrocarbons in this species. Furthermore, the cuticular profile of both sexes of *D. simulans* is highly similar to that of *D. melanogaster* Canton-S males, with *cis* 7-T being the principal hydrocarbon. Bioassay studies of the behavioural effect of synthetic 7-T on *D. simulans* males showed that this substance induces male wing display in this species in a dose-dependent fashion. A substance present in the cuticle of both sexes thus stimulates sexual behaviour in at least one of them (as yet, we have no data on the effect of this substance on *D. simulans* females).

This situation should lead to frequent homosexual interactions between males. Such interactions are easily observable in this species, although they are frequently broken off and rarely progress beyond the male wing-display and courtship song. This suggests that inhibitory signals are also being produced and that, as in *D. melanogaster*, other sensory modalities are almost certainly involved in mediating the full range of courtship behaviour. Some of these inhibitory signals are relatively easy to discern. A courted male will often fly away, kick at the courter, or flick his wings, producing a 'rejection song'. All these behaviours tend to break off courtship. *cis* Vaccenyl acetate is also found in *D. simulans* males; this substance could be released as a way of terminating male–male courtship (Figure 16.2).

Combined chromatographic and behavioural studies have been extended

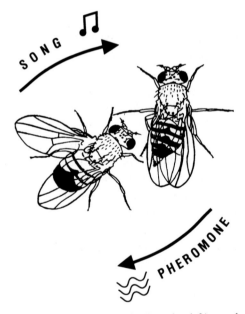

Figure 16.2. *Drosophila melanogaster* male (on the left) produces a courtship 'song' by vibrating its wing in response to the female's pheromone. In the situation depicted here, the female is extruding her ovipositor, perhaps diffusing an inhibitory pheromone. (Adapted from Burnet, B. and Connolly, K. (1974) Activity and sexual behaviour in *Drosophila melanogaster*: In *The Genetics of Behaviour*, pp. 201–58. Ed. J. H. F. van Abeelen. Elsevier Science Publishers, Amsterdam.)

recently to the remaining members of the *melanogaster* subgroup. Of the eight members of the subgroup, three (*D. melanogaster*, *D. sechellia* and *D. erecta*) show a pattern of sex differences in cuticular hydrocarbons similar to that of *D. melanogaster*, with females having high levels of dienes which are absent from males.

The predominant cuticular hydrocarbon of *D. sechellia* females is 7,11 HD, as for *D. melanogaster* females. The males of this species show high levels of 6-tricosene (6-T) on their cuticles. This substance may play a role in male-to-female signalling. *D. erecta* females show high levels of a long chain diene, i.e. 9,23 tritriacontadiene (9,23 TTCD), which is also found on the cuticles of young *D. melanogaster* flies of both sexes. Males of these three species are attracted both to *D. erecta* females and to young flies of both sexes.

The remaining five species (*D. simulans*, *D. mauritiana*, *D. yakuba*, *D. teissieri* and *D. orena*) display no sexual dimorphism in their cuticular hydrocarbons; both males and females show high levels of 7-T. This situation should lead to considerable confusion between sexes and species, with males being unable to distinguish flies on the basis of their cuticular hydrocarbons. Our behavioural studies revealed this to be the case, with high levels of interspecific heterosexual and homosexual courtship taking place between members of the non-dimorphic group of species. On the other hand, males of these species are generally indifferent to young flies.

Thus, misdirected courtships towards individuals who are not potential sexual partners take place in all the species of the *melanogaster* subgroup. It seems probable that in the wild these misdirected courtship bouts are rapidly terminated via the behaviours described above.

These results all reinforce the conclusion that *Drosophila* pheromones are but part of a complex multichannel communication system, with different messages being exchanged by the sexual partners at different moments. Messages can be confused easily and need to be confirmed via other channels of communication. This becomes even more apparent when we examine pheromonal variations within species.

Pheromonal polymorphisms

The first pheromonal polymorphism was described in the European cornborer, a moth which has a pheromone blend consisting of a mixture of two geometric isomers of 11-tetradecenyl acetate. The proportion of these two isomers was found to vary between the north and the south of the United States of America. Males from each morph preferred their homotypic pheromone blend; hybrid females showed intermediate blends, which hybrid males preferred accordingly. This result, together with studies of acoustic communication in other animals, indicates that there must be a tight genetic coupling between a signal and a receptor for any coadapted communication system to function.

We have surveyed scores of strains of *D. melanogaster* and *D. simulans* from all over the world, and have discovered several different types of pheromonal polymorphism based on differences in chain length and/or in double-bond position. One of the clearest variations affects the relative abundance of *cis* 7-T and *cis* 7-pentacosene in the cuticles of *D. melanogaster* males and *D. simulans* flies of both sexes (the sum of the two products is virtually constant). The major effect on double-bond position alters the proportions of 7,11 HD and 5,9 HD in *D. melanogaster* females. (Note that the relative positions of the two double-bonds remain the same in the two morphs.) Of these two isomers, only the former is behaviourally active, absolute amounts varying from 50–500 ng per female, according to the strain.

The biogeographic variations of these two polymorphisms are very different. In *D. melanogaster* males the variation of 7-T/7-P apparently follows a world-wide north–south cline: 7-T levels decline and 7-P levels increase towards the equator. In *D. simulans*, the variation is much more restricted: males and females with high levels of 7-P and low levels of 7-T have been found only in strains from around the Gulf of Benin (West Africa). Female *D. melanogaster* flies from sub-Saharan Africa show high levels of 5,9 HD and low levels of 7,11 HD. In virtually every other strain we have studied, the situation is reversed (see Figure 16.3). The only exceptions are the Caribbean and Guyana (South America) where we have found strains in which the females show typically African hydrocarbon profiles. Human-associated dispersal of African flies to these localities almost certainly explains these results. The clear chemical differences and geographical demarcation of the pheromonal morphs suggest that these two variants constitute different chemical races within the same species.

This variation in the levels of the *D. melanogaster* female pheromone could lead to a tendency towards isolation between the two chemical races. Studies of mating frequency and mating kinetics of individuals of the different morphs showed no sexual isolation, with copulation occurring in all cases. However, males from low 7,11 HD strains direct very intense courtship towards females from high 7,11 HD strains. This is probably because they are more sensitive to 7,11 HD than are males

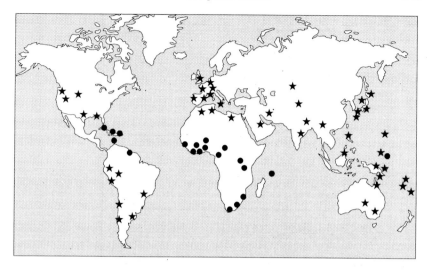

Figure 16.3. Distribution of *Drosophila melanogaster* strains in which 7,11 hepta-cosadiene (star) or 5,9 heptacosadiene (circle) are the main female cuticular hydrocarbons. Both isomers are present in all strains.

from strains with high levels. These results indicate that long-term evolutionary changes in mating patterns could be produced as a result of the differences in pheromone level we have detected.

Genetic control of cuticular hydrocarbon production

The existence of marked quantitative intraspecific pheromone poly-morphisms gave us a way of analysing the genetic control underlying pheromone production. We began our investigation with a study of 7-T/7-P variation in *D. simulans*, because this system is inherently simpler than that of *D. melanogaster* – no sexual dimorphism and only one double-bond. Experiments were carried out by crossing flies from a high-7-T/low-7-P strain (Seychelles) with those from a low-7-T/high-7-P strain (Cameroon). The individual cuticular hydrocarbon profiles of male and female descendants of reciprocal crosses between those two strains, and crosses of hybrids to both parental strains, were analysed. Genetic localization studies using morphological markers found the ratio of 7-T to 7-P is controlled by a single gene (called *Ngbo*) which is located on the second chromosome. Two alleles of this gene are known to exist, one from the Seychelles population (S, producing high levels of 7-T), and one from the Cameroon population (C, producing high levels of 7-P). These two alleles are codominant; the quantity of 7-P in a given individual being proportionate to the number of copies of the C allele (0, 1 or 2).

We have described a second locus which affects the production of 7-monoenes without altering their ratio. This gene was identified following the creation of mutants using a chemical mutagen. The mutation reduces the adult quantities of 7-T and 7-P by about 60%. This gene, called *kété*, is situated on the X chromosome and has deleterious pleiotropic effects, leading to female sterility, as fertilized eggs fail to hatch.

The genetic control of double-bond position polymorphism in *D. melanogaster* females is less well understood. Initial studies showed clearly that the gene(s) controlling this character are not situated on the sex chromosomes. Substitution of the II and III autosomes between strains showing high levels of 7,11 HD (Canton-S) or of 5,9 HD (Taï) showed that characters on chromosome III alone were responsible. Localization of the genetic control of this variation by using morphological markers has revealed the existence of a locus called *Thaïs*.

The biosynthesis of these cuticular substances, all of which have long chains and most of which have a double-bond in position 7, has been studied extensively following topical application of radioactive precursors onto male and female flies. The results show clearly that hydrocarbon biosynthesis continues *de novo*. The hydrocarbons are all produced from saturated linear fatty acids of medium chain length (C14–C18) produced by the condensation of several units of 2–3 carbons in length. This process appears to be under the control of a small number of enzymatic steps: elongation, decarboxylation and desaturation. These results agree with known details about pheromone biosynthesis in moths and in other species of flies.

Sex determination and the production of sex-specific molecules

Many pheromones are sex specific. In the case of *Drosophila*, even where pheromones are present in one sex only, both sexes share elements of the biosynthetic pathway. Thus, the fruitfly constitutes an ideal preparation for the study of the role of the genetic programme which determines sexual dimorphisms in the production of pheromones.

The sex of a mammal is determined by the presence or absence of the Y chromosome. This is not the case in *Drosophila*. Calvin Bridges showed that in the fruitfly, sex is determined chromosomally by the ratio of the number of X chromosomes to the number of autosomes. The fly is female if this value is 1, and male if the value is 0.5. To complicate matters, autosomal genes also exist which can modify the sex of the individual.

Transformer (*tra*), a gene on chromosome III, and transformer 2 (*tra2*), a gene on chromosome II, both have mutations which transform female XX homozygotes into phenotypic males. Intersex (*isx*), a gene on chromosome III, has a similar effect, whilst doublesex (*dsx*), which is also on chromosome III, transforms XX or XY flies into intersexes. These genes function in a cascade and, in their mutant state, have epistatic interactions which function hierarchically.

These genes affect the fly at all levels, from external morphology through behaviour (see Chapter 12 by Manfred Gahr for their effects on the nervous system). They also affect the 'sex' of the cuticular hydrocarbons: chromosomal females which are homozygous for the *tra* or *tra2* mutation have no dienes, but show high levels of various 7-monoenes, like chromosomal males. The *dsx* mutation has a more specific effect: it 'masculinizes' typically female hydrocarbons, stopping the production of dienes and increasing the production of 7-T. Thus, this gene affects the final steps of diene biosynthesis, in particular the introduction of the second double-bond in position 11. This may indicate that the *dsx* gene product affects a specific desaturase enzyme.

Thus, it seems that the genetic control of pheromone production may take place at several different points in a hierarchical pathway. Each point has controls that are more or less specific to it. The executive level of control involves genes such as *Ngbo* or *Thaïs* which control specific steps of biosynthesis; the regulator level includes *kété*, *nerd* and *fruitless* which act not only on chemical messages but also on other mechanisms of sexual behaviour or physiology. Finally, the selector genes consist of those genes involved directly in sex determination.

Production and detection of pheromones

In the Lepidoptera, pheromonal mixtures are produced by specialized, sex-specific glands. In some insects, in particular the social insects, several glands may exist, each of which emits a given mixture. Flies do not have any pheromone-producing glands, but biosynthesis appears to take place in one or more subtegumentary tissues in the abdomen. This has been shown unambiguously for *Musca* by *in vitro* studies of pheromone biosynthesis in slices of fly abdomen. In *Drosophila*, indirect evidence has been provided from studies of gynandromorphs or mosaics, individuals which contain a mixture of male and female tissues (Figure 16.4).

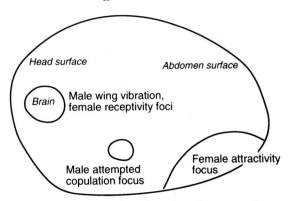

Figure 16.4. Fate map of *Drosophila* embryo based on gynandromorphic studies. 'Focus' refers to a part of the embryo which has to be of the relevant sex in order for the behaviour or the characteristic to be expressed. (For further details, see text.)

These sexual mosaics result from the loss of a special, unstable X-chromosome from certain cells during the first cellular divisions of female zygotes. This produces a situation in which half the cells of the resultant individual have the full sex-chromosome complement (and are thus female), whilst the other half are male hemizygotic cells and express any recessive genes which are carried on the remaining X-chromosome. By using recessive markers, male tissues can be detected and the role of given organs or regions ('foci') in determining sex-specific behaviours can be estimated. We measured the courtship responses of male tester *D. melanogaster* to mosaic individuals, and found a focus for the synthesis of female cuticular pheromones, located in the ventro-posterior region of the abdomen. This result is similar to that found in comparable studies of houseflies.

In Lepidopteran species, pheromones are detected by the antennae, which often show remarkable sexual dimorphisms. For example, male moths generally have highly developed antennae which are covered with specialized olfactory sensilla. The antennal nerve projects from the peripheral detection organ to the antennal lobes, a part of the insect brain which receives a majority of its input from nerves originating in the antennae. Sexual dimorphism is also present at this level (see Chapter 12 by Manfred Gahr). Integration of the different elements of the pheromone blend takes place in a macroglomerulus which is absent in females. In flies, contact pheromones are not detected by the antennae but rather by the prothoracic legs; the mode of detection is thus more akin to gustation than to olfaction. The tarsi show a clear sexual dimorphism: there are around twice as many sensory hairs on male prothoracic legs as on those of females. Neurons from these receptors converge on the thoracic ganglion, which shows a sexual dimorphism.

In *D. melanogaster*, the thoracic ganglion is part of the 'heterosexual courtship behaviour' focus, together with part of the hindbrain. On the basis of gynandromorph studies, this brain region has also been implicated in the control of homosexual courtship behaviour. The area involved may be the mushroom bodies, named for their shape, which are known to play a role in courtship signal integration. These organs receive projections from the antennal lobes, which can be stimulated by the detection of *cis* vaccenyl acetate or by acoustic signals detected by vibration receptors in the antennae known as Johnson's organs. There are also contacts between the mushroom bodies and the thoracic ganglion: all parts of the nervous system which have been implicated in courtship communicate directly with each other.

Conclusion

The new tools of molecular biology enable us to alter the expression of a given gene or to limit its expression to a given tissue. In the near future we should be able to understand exactly which groups of cells and organs are involved in the different steps of pheromone production, detection and response. We will be able to investigate how biosynthetic mechanisms common to both sexes interconnect with sex-specific enzymatic processes to produce 7,11 HD in female *D. melanogaster*. But no doubt the biggest and most exciting challenges will come from future studies of how the insect brain integrates information from different sensory channels and directs it towards different behavioural programmes. We believe that the 'perfumed garden' of fly courtship will open its gates onto a vast unexplored continent rich in surprising new discoveries.

Further reading

Bell, W. J. and Cardé, R. T. (1984) *Chemical Ecology of Insects.* Chapman and Hall, London

Blomquist, G. J., Dillwith, J. W. and Adams, T. S. (1987) Biosynthesis and endocrine regulation of sex pheromone production in Diptera. In *Pheromone Biochemistry*, pp. 217–43. Eds. G. D. Prestwich and G. J. Blomquist. Academic Press, London.

Chen, P. S., Stumm-Zollinger, E., Aigaki, T., Balmer, J., Bienz, M. and Böhlen, P. (1988) A male accessory gland peptide that regulates reproductive behavior of female *D. melanogaster*. *Cell*, **54**, 251–98.

Ferveur, J.-F. (1991) Genetic control of pheromones in *Drosophila simulans*. 1. *Ngobo*, a locus on the second chromosome. *Genetics*, **128**, 293–301.

Ferveur, J.-F. and Jallon, J.-M. (1993) Genetic control of pheromones in *Drosophila simulans*. 2. *Kété*, a locus on the X chromosome. *Genetics*, **133**, 561–7.

Ferveur, J.-F. and Jallon, J.-M. (1994) *Nerd*, a locus of chromosome III, affects male reproductive behavior in *Drosophila melanogaster*. *Naturwissenschaften*, in press.

Ferveur, J.-F., Cobb, M., and Jallon, J.-M. (1989) Complex chemical messages in *Drosophila*. In *Neurobiology of Sensory Systems*, pp. 387–409. Eds. R. Naresh Singh and N. J. Strausfeld. Plenum Press, New York.

Gailey, D. and Hall, J. C. (1989) Behavior and cytogenetics of *fruitless* in *D. melanogaster*. *Genetics*, **121**, 773–85.

Hotta, Y. and Benzer, S. (1976) Courtship in *Drosophila* mosaics: Sex-specific foci for sequential action patterns. *Proceedings of the National Academy of Sciences USA*, **73**, 4154–8.

Jallon, J.-M. (1984) A few chemical words exchanged by *Drosophila* during courtship. *Behavior Genetics*, **14**, 441–78.

Karlson, P. and Butenandt, P. (1959) Pheromones in insects. *Annual Review of Entomology*, **4**, 49–58.

Klun, J. A. and Maini, S. (1979) *Environmental Entomology*, **8**, 423–6.

Matsumoto, S. G. and Hildebrand, J. G. (1981) Olfactory mechanisms in the moth *Manduca sexta*: response characteristics and morphology of central neurons in the antennal lobes. *Proceedings of the Royal Society of London. Series B*, **213**, 249–77.

Steinman-Zwicky, I., Amrein, H., and Nöthiger, R. (1990) Genetic control of sex determination in *Drosophila*. *Advances in Genetics*, **27**, 189–237.

Genetic and environmental control of gonadal sex

17

Evolution of mammalian sex-chromosomes

MARY F. LYON

The two most common chromosomal sex-determining systems found among animal groups are the XX:XY system, where females are chromosomally XX and males are XY, and the ZZ:ZW system, in which the females have the heteromorphic ZW chromosomes, with males being chromosomally ZZ. Mammals have the XX:XY system, and so also do some insects, including the fruit fly *Drosophila*. Birds, and some insects such as the butterflies and moths (*Lepidoptera*), have the ZZ:ZW system.

Susumu Ohno pointed out that lower vertebrates, such as fish and amphibia, in many cases do not have morphologically distinguishable sex-chromosomes. Thus, the mammalian system must have been developed during the evolution of vertebrates, and the similar systems seen in insects must have evolved independently. The X and Y chromosomes of present mammals differ markedly in size, and the X carries many genes unconnected with sex, which have no homologues on the Y. In parallel with the evolution of distinct sex-chromosomes, there has evolved a system of dosage compensation, equalizing the effective gene dosage of X-linked genes in XX females and XY males. Dosage compensation systems are found not only in mammals but also in other animals with XX:XY mechanisms or XX:XO systems, including the fruit fly *Drosophila* and the nematode worm *Caenorhabditis elegans*. However, the means of achieving equal gene dosage in the two sexes are different in different animal groups. In mammals, dosage compensation is obtained by inactivation of one X chromosome in the somatic cells of females, so that both sexes effectively have a single dose of X-linked genes. However, in *Drosophila* both X chromosomes of females are equally active, and transcription of the single X chromosome in males is enhanced to equal that of the two X chromosomes of the female. Thus, a similar end result,

381

of equal doses of X-linked genes in males and females, has been achieved in evolution by different means in different animal groups. By contrast, in organisms with ZZ:ZW sex-determining mechanisms, where the female has heteromorphic sex-chromosomes, dosage compensation has not yet been found.

The chromosomal sex-determining mechanism in mammals also differs from that in *Drosophila*. In mammals, the Y chromosome is the key. Any animal with a Y chromosome normally becomes male, even if it is chromosomally abnormal with extra X chromosomes, e.g. XXY, or XXXXY. Conversely, all animals without a Y chromosome are female, even if there is a single X only, i.e. XO. In *Drosophila* and *C. elegans*, sex determination depends on the balance between the number of X chromosomes and the number of sets of chromosomes other than sex chromosomes: the autosomes. If one set of autosomes is denoted by A, animals with a ratio of 2X:2A (i.e. XX) become female, whereas those with a ratio of 1X:2A (i.e. XY or XO) become male. Thus, as with dosage compensation, the various systems of chromosomal sex-determination by XX:XY means have evolved along different paths.

Considerable evidence has now accumulated concerning the evolution of the sex-chromosomes, and of dosage compensation, in mammals. Part of this evidence comes from studies of the different subclasses of mammals: the prototheria or monotremes, the metatheria or marsupials, and the eutherians or placental mammals.

Other chapters in this volume deal with the details of the mechanisms of sex determination and sexual differentiation in these groups. Here we deal with the content of the sex-chromosomes and the mechanisms of dosage compensation by X-chromosome inactivation.

Size and genetic content of chromosomes

In both marsupials and eutherian mammals the Y chromosome is typically a small chromosome, whereas the X is relatively large, constituting about 5% of the genome in eutheria and 3% in marsupials. In monotremes, however, the X and Y chromosomes are more nearly equal in size (Figure 17.1). Thus, monotremes may show an early stage in the differentiation in size of the sex chromosomes.

Concerning the X chromosome, Susumu Ohno formulated the concept that has become known as Ohno's Law. This law states that a gene X-linked in one mammalian species is X-linked in all. The underlying rationale is that, owing to X-chromosome inactivation, the organism

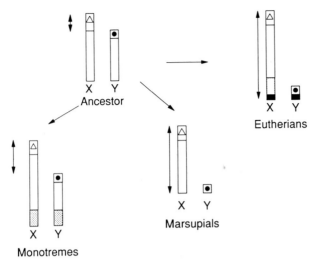

Figure 17.1. Change in size and content of sex-chromosomes during evolution of mammals. In the primitive ancestor the X and Y chromosomes were of equal or nearly equal size and content with the Y carrying a sex-determining gene. In the line to present monotremes there has been some reduction in size of the Y accompanied by inactivation of a corresponding small segment of the X (the extent is indicated by the vertical arrowed line), and some material from an unknown autosome has been added to X and Y. In the marsupials, the Y has become very small and this has been accompanied by extension of X-inactivation to the full length of the X. In eutherians, again the Y has become very small. Material from autosomes has been added to the X, and X-inactivation has spread into this material also. (Black circle, sex-determining gene; open triangle, X-inactivation centre; cross hatching and single hatching, different autosomal additions; black bars, pseudoautosomal region of X–Y homology and crossing over.)

would be adapted to single gene dosage of X-linked genes, interacting with the normal diploid dose of autosomal genes. If an X-linked gene were to be translocated to an autosome the resulting doubling in amount of gene product would upset the balance of X-chromosomal and autosomal gene products and, hence, such a translocation would be detrimental and would be eliminated. A converse effect would apply if an autosomal gene were translocated to the X chromosome.

Much evidence has now accumulated that in eutherian mammals Ohno's Law is indeed obeyed. Studies in about 20 species from various orders have shown that the same genes are X-linked in all. This contrasts with the situation in autosomes, in which there has been much exchange of genes among chromosomes in the course of evolution. The two species in which chromosome mapping is known in most detail are the mouse and man. Each autosome of the mouse carries genes with homologues on at least three different human autosomes. These homologous autosomal genes are arranged in segments varying in length from very small up to about one-third the length of a chromosome. Thus, the X chromosome, as Ohno postulated, indeed differs from autosomes in having all its homologues on the same chromosome.

However, the X chromosome has clearly undergone rearrangements during evolution. The shapes of the mouse and human X chromosomes are very different. On the basis of genetic content, the chromosome can be divided into five segments (Figure 17.2). Within a segment the order of genes has been conserved in evolution, but the segments have been rearranged with respect to each other, by chromosomal inversions or by movement of the centromere.

In marsupials the situation is rather different. Jenny Graves (see Chapter 18) and her colleagues have shown that homologues of genes on the long arm (Xq) of the present day human X chromosome are X-linked in marsupials. However, genes from the human X short arm (Xp) are autosomal in marsupials. In a monotreme, the platypus, again genes from human Xq are X-linked but those from Xp are autosomal. The monotremes and marsupials are thought to have diverged separately from the evolutionary line leading to eutherian mammals. Hence, the simplest interpretation is that genes now found on the human Xp were autosomal in the primitive ancestors of mammals. In the line leading to eutherians, these genes have been transferred from the autosomes to the X chromosomes, presumably at an early stage in the evolution of X-chromosome inactivation (Figure 17.1). Thus, the arrangement of genes on the human X is probably nearer to the original arrangement than in the mouse, where genes from human Xp are distributed along the X in at least three separate segments (Figure 17.2).

In the platypus, as mentioned previously, the X and Y differ little in size (Figure 17.1). The short arms of each look diffferent whereas the long arms of the two chromosomes look cytogenetically similar in their G-band patterns. Some of the genes homologous to those on human Xq are found on the long arm. Thus, the monotreme Y may carry these genes also on its long arm, and the monotremes may be exhibiting an early stage in the evolution of differences in size and content of the sex-chromosomes.

The Y chromosome is postulated to have become smaller in evolution, by loss of genes that are now found on the X chromosome only. Some evidence for this comes from genes now remaining on the Y. In the mouse, the gene for an enzyme, steroid sulphatase, is located in the small

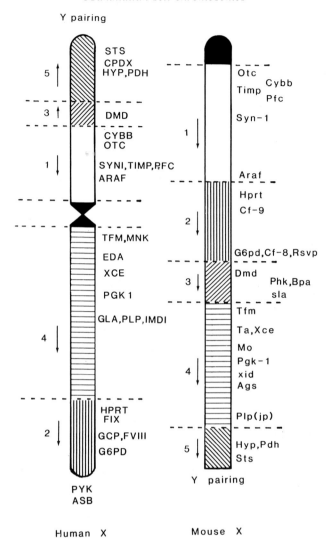

Figure 17.2. Comparison of human and mouse X chromosomes, giving the symbols of homologous genes. By convention, human genes are represented by capital letters, and mouse genes by lower case. Each chromosome is divided into five segments (1 to 5). Within a segment the order of genes is conserved, but the segments are rearranged. The arrows show the order of the genes and that segment 1 is reversed relative to the centromere in the two species. The centromere is solid black near the centre of the human X and at one end of the mouse X. (Reprinted with permission from Searle, A. G., Peters, J., Lyon, M. F., Hall, J. G., Evans, E. P., Edwards, J. H. and Buckle, V. J. (1989) Chromosome maps of man and mouse IV. *Annals of Human Genetics*, **53**, 89–140.)

region of homology between the X and Y, the pairing segment or pseudoautosomal region, and it has an active homologue on the Y. In the human, the corresponding gene is located near but outside the pseudoautosomal region of the X chromosome, and does not have an active homologue on the Y. However, there is a steroid sulphatase 'pseudogene' on the human Y. Thus, it is possible that the mouse and human represent different stages in the evolution of the steroid sulphatase gene, in its loss from the human Y but not the mouse Y. Similarly, the amelogenin gene and the gene for ubiquitin-activating enzyme are located near the pseudoautosomal region of the human and have related sequences but not active genes on the Y, and, thus, could be exhibiting a stage in the loss of genes from the Y in evolution.

X-chromosome inactivation

In 1961, I put forward the hypothesis of X-chromosome inactivation. This idea came while I was studying a mouse X-linked coat colour gene, which in heterozygous females produced a variegated coat pattern, as in the well-known tortoiseshell cat. It seemed to me that such a pattern would result if the two different colour genes on the two X chromosomes of these heterozygous females were each active separately in different cells. It has been fascinating to see this idea borne out by the work of many scientists over the years.

The current view is that X-chromosome inactivation in present eutherian mammals involves spreading of a signal for inactivation along the chromosome from an X-inactivation centre. In the embryo soon after fertilization both X chromosomes are in the active state (Figure 17.3). Early in embryonic development a single X-inactivation centre in each cell receives a signal which blocks it, and the X chromosome with this blocked centre remains active. All other X-inactivation centres in the X chromosomes of the cell (normally one in a female and none in a male) then commence the process of inactivation by spreading of a signal in both directions along the chromosome (Figure 17.4a). Once the initiation of inactivation has occurred in the embryo, it remains stable with each X chromosome retaining its active or inactive state throughout further cell divisions in the life of that animal. An exception occurs in the germ cells, where, in the female, reactivation occurs so that both X chromosomes are active, and in the male the single X chromosome becomes inactive.

In addition to its lack of transcription the inactive X chromosome has

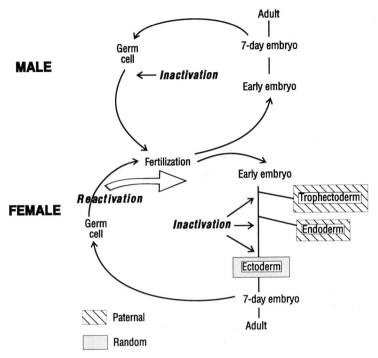

Figure 17.3. Cycle of X-inactivation in eutherians as exemplified by the mouse embryo. In the male (upper section) the single X chromosome remains active, except in the germ cell where it becomes inactive at about the time of meiosis. In the female (lower section) both X chromosomes are active in the early embryo. Then inactivation of one X chromosome occurs, and persists throughout life, except in the germ cell, where reactivation occurs and continues after fertilization to reactivate the inactive X chromosome contributed by the male germ cell, thus beginning the cycle once again.

the properties of late replication of DNA, and of condensation during interphase, to form the sex chromatin body.

No evidence of X-chromosome inactivation has yet been found in vertebrates other than mammals. Within the mammals there are considerable differences in the various subclasses.

X-chromosome inactivation in monotremes

In monotremes, where the X and Y are nearly equal in size, in some tissues a small part of one X chromosome of females shows late replication. This is the region of the X short arm in which it differs morphologically from

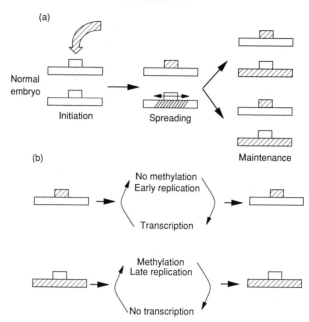

Figure 17.4. Mechanism of X-inactivation. The X-inactivation centre is shown as a box. At initiation, (a) a trans-acting factor acts on one inactivation centre and blocks its activity. A signal for spreading of inactivation is then initiated in both directions from the unblocked centre. In maintenance, (b) the active or inactive state of each X chromosome is maintained through successive cell divisions, by a feedback mechanism.

the Y. The existence of this small late-replicating segment suggests that the monotremes may be showing a rudimentary form of X-chromosome inactivation. Perhaps the X-inactivation centre is present. However, the spreading mechanism is apparently either not present or imperfectly developed. Jenny Graves (see Chapter 18) and Jaclyn Watson have pointed out that it is surprising that, although the homologues of the human Xq genes are found on the monotreme X, some of them are in the long arm, which is apparently not inactivated. This seems to suggest that their retention on the monotreme X during evolution may have been due to chance, rather than to effects of genic imbalance as implied by Ohno's Law.

X-chromosome inactivation in marsupials and eutherians

There are various differences in X-chromosome inactivation between marsupials and eutherians. Briefly, whereas in eutherians the choice of chromosome for inactivation is random, with either the maternally or paternally derived X chromosome being inactive in different cells, in marsupials the paternally derived X is inactivated in all cells. In addition, inactivation in marsupials is less stable than in eutherians, with reactivation of some genes occurring in cultured cells or even *in vivo*. These differences may provide clues to the evolution of the phenomenon and to its mechanism, including the role of the X-inactivation centre and the mechanisms of spreading and stabilization of inactivation.

X-inactivation centre

The evidence for the existence of the inactivation centre comes from the study of X-autosome translocations and of X-chromosome deletions. When an X chromosome is broken by a translocation, only one of the two X-chromosome segments thus formed undergoes inactivation; the other remains active in all cells. Similarly, some X chromosomes with deleted segments remain active in all cells. The interpretation is that inactivation is a positive process initiated from a certain point, the inactivation centre, and that segments lacking the centre cannot undergo inactivation. By comparing different translocations and deletions it has been possible to map the centre to a precise spot on the X chromosome, both in human and mouse. Recently, a gene, termed the *Xist* gene, has been cloned from the region of the centre, which has the unusual property of being transcribed from the inactive, but not the active, X chromosome. From the combination of its position and its unusual expression, this gene seems a strong candidate for a role in the inactivation centre, but its mode of action is not yet clear.

The existence of a centre for control of dosage compensation in mammals is a somewhat unusual phenomenon. In the *Drosophila* and *C. elegans* dosage compensation systems there is no comparable centre, and individual genes behave autonomously. Thus, this provides yet further evidence that, although the end result of dosage compensation is similar in these different groups of animals, it has evolved independently in each case. At present, there is no evidence about the type of structure from which the X-inactivation centre of mammals might have arisen.

The paternal X-inactivation seen in marsupials is presumably mediated through the inactivation centre. It is one example of genomic imprinting, the phenomenon by which chromosomes retain some cell memory of their gametic origin. It has been suggested that paternal X-inactivation may be the primitive form from which random inactivation evolved. Some evidence in favour of this comes from the fact that paternal X-inactivation is also seen in some extraembryonic placental tissues of mice and rats, namely those forming the trophoblast and extraembryonic endoderm. Experimental work on mouse embryos has indicated that the imprint is not an essential part of X-inactivation, but instead affects the choice of X-inactivation centre to become the active one.

It is possible that, by making the probabilities of inactivation of the two X chromosomes unequal, imprinting may confer a selective advantage. Where two homologous chromosomes must behave differently within a cell, as in X-inactivation, there is likely to be a risk of error, with either both chromosomes being inactivated or neither. If one homologue is much more susceptible to inactivation than the other the risk of such an error is probably considerably reduced.

Imprinting in mammals is not confined to the X chromosome but also occurs in autosomal genes, in that these may be expressed differently according to their parental origin. Only a tiny proportion of genes are affected, and some are expressed more strongly from the maternal and others from the paternal homologue. However, the end result is that an embryo can survive only if it has an equal complement of genes from the mother and the father. As shown by Azim Surani and others, for embryos with two maternal chromosome sets, the embryo itself begins development fairly normally, but the embryonic membranes, forming the placenta, are imperfectly formed, resulting in death. For those with two paternal sets the reverse occurs, with the embryonic membranes and placenta developing fairly normally, but the embryo itself failing to grow. Thus, imprinting in autosomes may have a different function from that in the X chromosome.

Eliezer Lifschytz and Dan Lindsley suggested that X-inactivation in somatic cells may have arisen from an earlier inactivation of the differential segment of the single X chromosome in male meiosis. Inactivation of the X chromosome in male meiosis is seen commonly in other XX:XY systems. There is evidence that the presence of unpaired pairing sites in male meiosis is deleterious. As the X and Y begin to differentiate in evolution, and the Y loses genes, there will be unpaired pairing sites on the X. Inactivation may be a means of protection against the potentially deleterious effect of these sites. Thus, X-inactivation may have arisen in male meiosis in an early ancestor of the mammals, when the X and Y chromosomes began to diverge in size, presumably before the stage of present monotremes, and the mechanism may have been reused later to provide dosage compensation in somatic cells of females.

Spreading of X-inactivation

The evidence for spreading of X-chromosome inactivation again comes from X-autosome translocations. From the segment of the X which undergoes inactivation, the signal spreads into the attached autosomal material and the genes become inactivated. Inactivation can spread a considerable distance, over many megabases of DNA, up to one-third or one-quarter of a chromosome in length. Like the inactivation centre, spreading seems to be an unusual system. In this case the unusual feature is the long distance of spread. In *Drosophila*, there are so-called 'position effects' in which the expression of a gene is affected by nearby chromatin, but these operate over much shorter distances.

Some evidence from differences between human and mouse suggests that, in addition to a spreading signal, there is a response of individual genes to the signal. In the human, various genes are known which are not inactivated. Some are in the pairing segment, where inactivation is not expected because no dosage compensation is needed. Others are near the pairing segment. However, in yet other cases non-inactivated genes are flanked on both sides by genes which are inactivated in a typical manner. This means that the spreading signal must have run past these genes and they must in some way have resisted the signal. Human non-inactivated genes include those with the symbols *STS*, *ZFX*, *UBE1* and *RPS4X*. In the mouse the homologous genes undergo inactivation. So far no X-linked genes are known in the mouse which are not inactivated. Apparently, the mouse has more effective dosage compensation than the human, and does not need a double dose of any X-linked genes. Chromosomally, XO mice are normal, except for reduced fertility, whereas nearly all human XOs die as fetuses, with the few survivors malformed with Turner's syndrome. This difference in phenotype of XOs is thought to be due to the requirement in humans for two doses of the non-inactivated genes. In some other myomorph rodents, phenotypically normal XOs have been reported and, indeed, XO is the normal type in females of a few species, such as *Microtus oregoni*, the creeping vole (see Chapter 19 by Karl Fredga). This suggests that the myomorph rodents as a group have evolved a more complete dosage compensation than that of humans. How has this evolution come about? In view of the evidence from X-autosome translocations that autosomal genes can become inactivated, it seems unlikely that specific X-chromosomal sequences are needed for susceptibility to inactivation. In *Drosophila*, there is evidence that genes can have regulatory sequences termed 'boundary elements' which protect them from position effects due to nearby heterochromatin. It may be that in mammals also, such 'boundary elements' exist and that in the evolution of the X chromosome there has been a tendency to lose any such protective sequences. In the X chromosomes of rodents, loss of protective sequences may be more complete than in humans, so that mouse genes have more completely lost any resistance to inactivation.

X-autosome translocations also provide some evidence of evolution of susceptibility to passage of the spreading signal. Although spreading proceeds over considerable distances in autosomal material, it appears not to travel as far as in the X chromosome itself. Autosomal genes at some distance from the translocation breakpoint may not undergo inactivation, and the late replication and condensation which are features of X-inactivation typically extend some way into the autosomal segment but not to its end. This limitation of spread in autosomes implies that there is either some factor in X-chromosomal material which facilitates the spread or some feature of autosomal chromatin which inhibits it. Art Riggs has suggested that there may be 'booster elements' along the X which promote the spread. As with the postulated sequences affecting response of genes to the signal, the sequences which promote the spread need not necessarily be X-chromosome-specific. Rather, the X chromosome may have an unusual array of the interspersed repetitive elements which are found in all mammalian chromatin. Some of these repetitive elements may promote the spread of inactivation and others may retard it. In evolution there could be selection for types of elements on the X chromosome that promote spreading.

The mechanism of spreading of inactivation remains unknown. Various authors, including Art Riggs, have suggested differential methylation of cytosine bases by a methyltransferase which runs along the chromosome. Genes on eutherian X chromosomes indeed show differential methylation. In the 5' promoter regions of some genes there are clusters of dinucleotide pairs with a cytosine (C) base followed by guanine (G), and these clusters are known as CpG islands. The CpG islands on the eutherian inactive X chromosome are typically heavily methylated. However, no such differential methylation has been found in marsupial genes, and this implies that methylation is not a requirement for spreading. The roles of methylation and other properties of the inactive X chromosome are discussed further below.

Stabilization of inactivation

Reactivation of the inactive X chromosome occurs normally in oocytes at around the time that meiotic division begins, in both eutherians and marsupials. In eutherians, X-inactivation is otherwise very stable. In cultured cells, partial reactivation can be induced by treatment with demethylating agents, and is accompanied by loss of methylation of CpG islands. *In vivo*, a minor level of reactivation has been seen in the mouse in autosomal coat colour genes translocated to the X chromosome, and in the X-linked gene for ornithine transcarbamylase, *Otc*. In marsupials, by contrast, reactivation of some genes occurs readily both in cultured cells and *in vivo* in certain tissues. This suggests that, in eutherians, evolution has resulted in a more stable and hence more complete form of dosage compensation than in marsupials. An exception to the stability of X-inactivation in eutherians occurs in the extraembryonic placental membranes, cells from which can undergo reactivation in culture. Thus, the two types of paternal X-inactivation, in marsupials and in extra-embryonic membranes of eutherians, are less stable, and it is possible that there is some interaction between the imprint and the stabilizing mechanism. A distinguishing feature of the inactive X chromosome of eutherians is the methylation of CpG islands. It is now suggested that methylation is part of the stabilization mechanism. It is possible to envisage a feedback loop, in which, at DNA replication, a maintenance methylase methylates already half-methylated sites, whereas unmethylated sites remain unaffected. Art Riggs, and Steve Grant and Verne Chapman, have suggested that late replication of DNA also could result in a feedback loop and could be involved in stabilization, if late replication prevents transcription, and transcription is required for early replica-tion (Figure 17.4b). Thus, the suggestion is that, in marsupials, late replication provides the stabilizing mechanism, whilst eutherians have evolved a more complex system with both late replication and methylation, and hence have greater stability of inactivation (Figure 17.5). There is also the condensation that is typical of the inactive X chromosome to be considered. As a mechanism for this, Art Riggs has suggested type I DNA reeling, in which an enzyme at a chromosomal scaffold attachment site reels in DNA towards itself in both directions. A question yet to be resolved is the relation of the various properties (the condensation, late replication and differential methylation) to each other, and the role of the X-inactivation centre in bringing about the differentiation of the active and inactive X chromosomes.

Extent on X		Tissues	Imprint	Complete or stable	Methylation
⬚	**Monotremes**	Some	?	?	?
⬚	**Marsupials**				
	Embryonic tissues	Some	Yes	No	?
	Adult	Most or all	Yes	No	No
⬚	**Eutheria**				
	Extraembryonic tissues	Most or all	Yes	No	No
	Embryo & adult	All	No	Yes	Yes

Figure 17.5. Comparison of X-inactivation in monotremes, marsupials and eutheria. In monotremes, inactivation affects only part of the X in some tissues. The extraembryonic endoderm tissues of eutherians resemble marsupials in having preferential paternal X-inactivation (imprinting), less stable inactivation, and no differential methylation.

Conclusions

The present state of knowledge about the evolution of the mammalian sex-chromosomes and dosage compensation can be summarized as follows. At some stage in the evolution of mammals from the mammal-like reptiles, before the divergence of the lines leading to monotremes, marsupials and eutherians, the X and Y chromosomes began to differ in content. A small differential segment developed, with the Y chromosome carrying one or more male-determining genes. At this stage, a mechanism developed for inactivation of the differential region of the X chromosome during male meiosis to prevent the deleterious effects of the presence of unpaired pairing sites. This mechanism probably involved an X-inactivation centre and spreading of the inactivation signal over a short distance. Somewhat later in evolution the same mechanism was reused to provide somatic X-inactivation, so as to give dosage compensation by equalizing the effective doses of X-linked genes in the differential segment of the two sexes. There would have been preferential inactivation of the paternal X chromosome. The mechanism would have involved late replication of DNA and probably also condensation of the inactivated segment. This stage in evolution of X-chromosome inactivation would have been similar to the system seen in present monotremes (Figure 17.5).

As evolution proceeded, the X and Y chromosomes became more differentiated with loss of genes from the Y. There was selection for improved spreading of the inactivation signal along the chromosomes and for the response of genes to the spreading signal. Stabilization of inactivation would have been obtained by a feedback loop system involving late replication. This stage would resemble that seen in present marsupials, and the extraembryonic tissues of mice and rats (Figure 17.5).

In the line leading to the eutherians, the X chromosome became larger, due to recruitment of genes from autosomes, early during the evolution of X-inactivation. There was stronger selection in favour of efficient spreading of the signal, and for response of genes to the signal, but, in some present eutherians, response to the signal is not complete. There was further selection for stabilization of inactivation. This involved a change from paternal X-inactivation resulting from imprinting, to a random type of inactivation, accompanied by development of a feedback loop resulting from differential methylation (Figure 17.5).

The function of the X-inactivation centre and the nature of the spreading signal remain unknown. The *Xist* gene is a strong candidate for a role in the inactivation centre, and further advances in knowledge are likely to result from studies of this gene, and of genes on the human and mouse X chromosomes which escape or undergo inactivation.

Further reading

Cooper, D. W., Johnston, P. G., Sharman, G. B. and Vandeberg, J. L. (1977) The control of gene activity on eutherian and metatherian X chromosomes: a comparison. In *Reproduction and Evolution*, pp. 81–7. Eds. J. H. Calaby and C. H. Tyndale-Biscoe. Australian Academy of Sciences, Canberra.

Gartler, S. M. and Riggs, A. D. (1983) Mammalian X-chromosome inactivation. *Annual Review of Genetics*, **22**, 155–90.

Grant, S. G. and Chapman, V. M. (1988) Mechanisms of X-chromosome regulation. *Annual Review of Genetics*, **17**, 199–233.

Graves, J. A. M. and Watson, J. M. (1991) Mammalian sex chromosomes: evolution of organization and function. *Chromosoma*, **101**, 63–8.

Hodgkin, J. (1987) Sex determination and dosage compensation in *Caenorhabditis elegans*. *Annual Review of Genetics*, **21**, 133–54.

Lucchesi, J. C. and Manning, J. E. (1987) Gene dosage compensation in *Drosophila melanogaster*. *Advances in Genetics*, **24**, 371–429.

Lyon, M. F. (1988) The William Allan Memorial Award Address: X-chromosome inactivation and the location and expression of X-linked genes. *American Journal of Human Genetics*, **42**, 8–16.

Lyon, M. F. (1992) Some milestones in the history of X-chromosome inactivation. *Annual Review of Genetics*, **26**, 17–28.

Ohno, S. (1967) *Sex Chromosomes and Sex-Linked Genes.* Springer-Verlag, Berlin.

Riggs, A. D. (1990) Marsupials and mechanisms of X chromosome inactivation. *Australian Journal of Zoology*, **37**, 419–41.

Riggs, A. D. and Pfeifer, G. P. (1992) X-chromosome inactivation and cell memory. *Trends in Genetics*, **8**, 169–74.

Surani, M. A., Kothary, R., Allen, N. D., Singh, P. B., Fundele, R., Ferguson-Smith, A. C. and Barton, S. C. (1990) Genome imprinting and development in the mouse. *Development*, Supplement, 89–98.

18

Mammalian sex-determining genes

JENNIFER A. MARSHALL GRAVES

Of all the differences between male and female mammals, the primary one seems to be the development of the testis in males (see Chapter 8 by Jean Wilson and Chapter 9 by Marilyn Renfree). Early in development the mammalian embryo is ready for anything. Male and female embryos are morphologically indistinguishable; rudiments of the gonad formed as a germinal ridge on the embryonic kidney remain undifferentiated either as testis or ovary, and the embryo is equipped with both male and female internal ducting. The development of the gonad into a testis triggers a cascade of hormone-controlled changes, so once this decision – testis or no testis – has been taken, the sex of the embryo is determined. The first step we can identify is thus the determination of a testis, and the factor which accomplishes this has been called the testis-determining factor, TDF for short.

The development of the gonad into a testis is followed by the production of two potent testicular hormones; testosterone (and its derivative) which stimulates the growth of the male reproductive tract and genitalia, and Müllerian inhibiting substance (MIS), which represses development of the female ducts. In the absence of testis formation, the gonad develops into an ovary, and female development is set in motion. All subsequent differences between the sexes result from these initial decisions. If something goes wrong with the male developmental pathway, the usual result is female development; thus the female is the so-called 'default sex' in mammals.

The pathways for the determination of the male or female body type involve many steps, and each step is controlled by an enzyme or other protein made by a gene. In this sense, then, there are probably hundreds, or thousands, of sex-determining genes in the pathway which determines maleness or femaleness. Potentially, any of these genes could control the

397

whole male- or female-determining pathway if its action were confined to one sex. Other genes in the same pathway need not, then, be sex-specific. Evidently TDF controls a critical step in a male-determining pathway, and, because it is expressed only in males, it has taken on the job of master switch for the whole pathway.

Body sex ('somatic phenotype') and the production of eggs or sperm are two quite different processes. Germ cells develop in a region of the embryo that is distant from the gonads, then move into the testis or ovary, where they differentiate into sperm or eggs. There are mutations in mice which affect sperm development but not the male phenotype, implying that germ-cell development is under independent genetic control. However, these pathways interact because, in the absence of gonadal development, no sperm or eggs can be formed.

Which gene acts as the testis-determining factor? What makes the action of TDF male-specific? How does TDF control the development of the testis? The identification of TDF has been the subject of intense investigation for decades. Not only would a solution to this problem satisfy the centuries-old questions about sex determination, but answers might have immediate practical usefulness in management of domestic animals, as well as diagnosis and possible cure of human conditions which involve abnormal sexual differentiation. Also, the testis-determining pathway is likely to bear similarities to the pathways by which other organs develop – perhaps there is a liver-determining factor or a brain-determining factor with an action similar to TDF! Defining the first step of the sex-determining pathway may reveal new families of genes which control some of the earliest events of organogenesis. The sex-determining pathway is a good model system, since mutations of sex-determining genes do not cause death – just femaleness.

The starting point of traditional genetics is the obvious phenotypic difference, from which we deduce the underlying biochemical differences and ascertain what protein controls these. With this knowledge, we can attempt to isolate the gene itself. However, the differences between the sexes include such a bewildering array of biochemical differences, that it would be hard to choose one which is the cause from all the others which are effects.

The new 'reverse genetics' approach is the complete opposite. First, the position of the gene is determined by genetic mapping techniques, which can be refined to pinpoint the gene extremely accurately to a tiny region of a chromosome. The DNA of this region can be isolated physically and characterized by 'recombinant DNA' techniques, and the sequence of the

DNA which constitutes the gene can be identified from what we know about gene structure and control signals. We can then use our knowledge of the genetic code to predict the amino-acid sequence of the protein product it will make, and use comparisons to sequences of other well-known proteins to predict how the product of the unknown gene works. This is the approach which has been spectacularly successful in the isolation and characterization of the gene which determines sex. Here I will describe how gene-mapping approaches have been used to pinpoint TDF in a region so small it could be dissected at the molecular level. I will then describe the genes which were first proposed to be candidates for the sex-determining master switch, and the evidence, often from surprising quarters, which discounted these claims. I will recount the story of the cloning of the SRY gene, the evidence that it is, indeed, the sex-determining gene, and the clues we have about how it performs this role. I will conclude by comparing mammalian and other vertebrate sex-determining systems, and by speculating about how, and from what, the mammalian sex-determining system evolved, and what we might find if we returned for a follow-up study of mammalian sex-determination 100 million years from now.

The how and Y of sex chromosomes

Where do we look to find the gene which acts as the testis-determining factor (TDF)? The human genome (and genomes of all mammals) contains about 3 billion bases, strung into about 1.6 metres of DNA. This represents between 50 000 and 100 000 genes and an inordinate amount of genetic junk, which is highly repeated, simple DNA sequences with no obvious function. Finding a gene in this enormous genome is a real needle-in-a-haystack problem.

However, with TDF we have an advantage because we already know that part of the genome is specialized to determine sex. The human genome has been sliced up into 23 pieces, which are manoeuvred more easily when the cell divides; these smaller lengths of DNA are complexed with protein and are visible as the condensed chromosomes we see in dividing cells. Each of these 23 chromosomes contains a unique set of genes. Each has a typical length, pattern of light and dark bands after treatment with dyes, and a characteristic shape at mitosis, conferred by the position of the centromere which holds the two replicated copies of the chromosome together.

Humans and other mammals, being diploid, have two complete genomes,

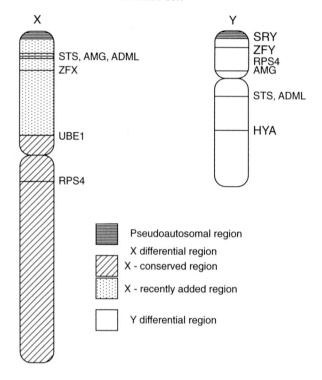

Figure 18.1. The human X and Y chromosomes differ in size and gene content, although they evolved from a homologous pair. The only homology left is within the small pseudoautosomal region (PAR), which contains several genes (not marked), and several genes shared by the human X and Y (STS, AMG, ADML, ZFX/Y and RPS4) in the differential regions. UBE1 is shared with the Y in mouse, but not human. The three candidate sex-determining genes HYA, ZFY and SRY are shown in larger print.

one from the mother and one from the father; this is why chromosomes come in pairs. The one exception is the sex-chromosome pair. In fact, the X and Y chromosomes can hardly be considered to be a pair at all, for the X is large (about 5% of the genome) and gene-rich, whereas the Y is small and composed largely of genetic junk (Figure 18.1). Females have two copies of the X chromosome, whereas males have a single X and a single Y chromosome. The sex-chromosome constitution of all mammals may be represented as XX female:XY male.

The names of the sex chromosomes derive from the puzzle presented by the single large 'X for unknown' chromosome in the male. How could a proper gene balance be achieved if females had two copies of each

X-borne gene, whereas males had only one? The answer is a 'dosage compensation' mechanism which turns off the genes on one X chromosome in females so that both males and females have only one active copy of each X-linked gene.

The Y chromosome is an obvious place to look for a testis-determining gene whose action must be male-specific. Indeed, there is good evidence that it is the presence or absence of the Y chromosome, rather than the number of X chromosomes, which determines maleness in mammals. Like many important genetic finds, this knowledge comes from studies of humans and mice with variations to the normal genes and chromosomes. Individuals with two X chromosomes plus a Y (i.e. an XXY constitution) are male, and those with a single X and no Y (XO) are female. This crucial finding led to the idea that the testis-determining gene must be unique to the Y chromosome, since this is present in males and lacking in females.

However, a unique-Y gene is not the only way in which an X and Y chromosome could effect sex determination. A plausible alternative is that sex is determined by the number of copies of a gene which is present on both X and Y chromosomes. There would be two active copies of this gene in males, but X-inactivation would ensure that there was only one in females. The idea that dosage compensation forms an integral part of the mammalian sex-determining system is attractive by analogy with other sex-determining systems in insects, worms, reptiles and birds, in which dosage differences, or dosage compensation, is involved intimately in sex determination.

The X and Y chromosomes of man and mouse have been mapped in some detail, and reveal a number of unique features associated with their unpartnered state. The X is a large chromosome, and bears an estimated 3000–5000 genes. The 249 genes which have been mapped and characterized, as a result of the Human Genome Project, present an unremarkable mixture of genes which code for household enzymes (like the classic glucose-6-phosphate dehydrogenase) or specialized functions (like blood-clotting factor and visual pigments), as well as interspersed repetitive sequences that do not appear to code for anything.

The Y chromosome is much smaller, and has so much repeated sequence DNA that it literally glows in the dark when stained with fluorescent dyes. Apart from the testis-determining gene, no genes with obvious outward effects have been attributed to this chromosome. Before the advent of DNA technology, the only other function ascribed to the Y was the male-specific histocompatibility antigen, HYA. More recently, DNA sequences have been isolated from the human Y or mouse Y. Some of these have been shown to represent functional genes, while others are inactive sequences that are related closely to genes (degenerate genes called 'pseudogenes'). Interestingly, every single one of these Y-borne active genes as well as some pseudogenes proves to have a relative on the X chromosome; we now know of several such genes shared by the X and Y chromosomes in either human or mouse or both (Figure 18.1).

The sex chromosomes of different eutherian ('placental') mammals are remarkable because of their invariant size and gene content. There is not a single exception to the rule that genes on the X in one species are on the X in all others: you've seen one X, you've seen them all. The sex chromosomes seem to have been protected from rearrangements with other chromosomes, presumably because this might disrupt their function in sex determination and dosage compensation. This lack of variety offers us little information on how the mammalian X and Y evolved. However, comparisons between eutherians and the other two major mammalian groups, marsupials and monotremes, show that only part of the sex chromosomes are conserved, and allow us to reconstruct how bits of other chromosomes must have been added to an ancestral X and Y early in the 200 million year history of mammals (Figure 18.1).

What do the sex chromosomes do about segregating into gametes at meiosis? Two copies of each of the other chromosomes ('autosomes') are present in spermatocytes, and these homologues line up and pair along their length in preparation for an orderly segregation, which ensures that a single copy of each chromosome goes into each (haploid) gamete. Unpaired chromosomes often fail to move and get lost.

If the sex chromosomes do not pair, how do they survive meiosis? In female meiosis there is not a problem, since the two X chromosomes can pair and segregate, ensuring one X per egg. However, since the X and Y are physically and genetically different, it is not so obvious how they can achieve pairing and segregation in male meiosis. What is observed is that the human, and the mouse, X and Y do pair, along a very tiny region at the tips, which allows the X and Y to undergo orderly alignment and segregation at male meiosis. This delivers an X into half the sperm and a Y into the other half and ensures a 1:1 sex ratio. The existence of a pairing region between the X and Y was predicted many years before a homologous region could be demonstrated by molecular methods, and pairing and exchange of the X and Y within this region have been observed genetically. Since this part of the sex chromosomes acts more like one of the autosomes, it has been called the 'pseudoautosomal region' (PAR).

The existence of these X–Y shared genes, within and outside the

PAR, supports an old hypothesis that the Y was originally homologous to the X. This idea grew from comparisons of the sex chromosomes of mammals and other higher vertebrates, which do not subscribe to the XX-female:XY-male system. Birds, reptiles, amphibians and fish show a great variety of chromosomal sex-determining mechanisms, and many species have no differentiated sex chromosomes at all, preferring to leave the matter to environmental forces. In birds and snakes, for instance, it is the female which has a large Z chromosome, and a small W full of repeated sequences, while the male has two Z chromosomes. The mammalian X chromosome has no homology with the bird Z, suggesting that the X and Z chromosomes must have evolved from different chromosome pairs. This means that a common ancestor of mammals, birds and reptiles probably had sex chromosomes that differed only at one or a few sex-determining genes.

Zeroing in on TDF

Back to the needle-in-a-haystack problem of finding TDF. The knowledge that TDF resides on the Y chromosome means that at least the haystack is now smaller, and the chance of finding TDF is better. More detailed gene-mapping to narrow down further the region of the Y that contains TDF has been vital to the search.

The position of TDF on the Y has been refined by a procedure known as 'deletion analysis', in which the sex of humans or mice having only part of a Y chromosome can tell us which bit must logically contain TDF (Figure 18.2). Accidents to the human Y chromosome occur, producing a visibly abnormal Y that lacks a larger or smaller piece. The normal Y chromosome has its centromere near one end, dividing it into a 'short arm' and a 'long arm'. Several individuals have been examined who have the long arm of the Y, or some derivative of it, but lack the short arm; the fact that these patients are female excludes TDF from the long arm and places it on the short arm. Conversely, several males have been observed who possess only the short arm, and lack most or all of the long arm, again mapping the TDF gene to the short arm of the Y.

Even more commonly, abnormal X and Y chromosomes are produced when regions of homology outside the PAR indulge in illegitimate exchange between the X and Y, transferring part of the Y to the X, and part of the X to the Y. This may give rise to apparently XY females who possess most of a Y but obviously lack TDF, or apparently XX males with a small part of the Y, which must bear TDF. The position of TDF

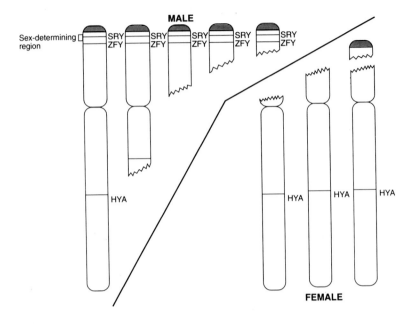

Figure 18.2. Locating the testis-determining factor from the sex of people with deletions of various parts of the Y chromosome. People with the short arm of the Y, or even the region just below the pseudoautosomal region (PAR, i.e. shaded area), are male; people who lack this region are female. This identifies a very small sex-determining region, within which the SRY gene was found.

was refined further by carefully mapping DNA marker sequences on the regions of the Y possessed by these XX males, as well as the bits that were absent in XY females. The region which must contain TDF became progressively smaller. The task of identifying the proverbial needle in this shrinking haystack became less and less impossible. The progress of this narrowing down, and ultimate success of this positional cloning strategy is illustrated in Figure 18.3.

However, one obvious problem was that the characteristics of the putative TDF gene were quite unknown. Perhaps we might miss it! Is TDF a large or a small gene? Is it all in one piece, or, like many genes in higher organisms, split into many pieces by 'intron' sequences? Is it unique to the Y, or, like the other genes on the Y, does it have a relative on the X? Are we sure that it is a gene at all? Maybe instead of coding for a protein which turns on testis development, TDF sequences act as

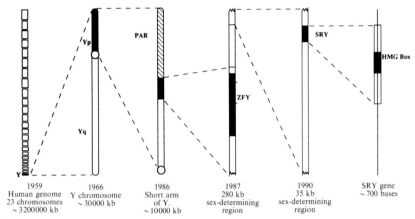

Figure 18.3. Progressive narrowing down of the region of the human genome which must contain the testis-determining factor, from the demonstration that the Y chromosome is male determining to the isolation and characterization of the SRY gene in 1990. (Adapted from McLaren, A. (1991) The making of male mice. *Nature*, **351**, 96.)

a sort of 'sink' to bind a repressor of testis determination. Thus, even with a much smaller haystack, searching for a needle of unknown characteristics was not going to be easy.

Candidates for TDF

Assuming that TDF is a real gene that codes for a real protein, a sensible premolecular biology approach was to search for a protein that was present in males and absent in females. This was done in the 1970s by immunizing female mice with cells taken from males of the same inbred strain. An antibody produced by these females identified a male-specific surface protein which acted as a minor histocompatibility antigen, Hya. HYA could also be detected in males, but not in females, of other species, including humans. Given that HYA was the only gene ascribed to the Y at this stage, it is not surprising that it was thought for a long time to be the TDF. A great number of papers in the 1970s and 1980s reported its properties and speculated on how it could perform this function.

However, gene mapping finally relieved the HYA gene of its sex-determining responsibilities, for it was found that human patients who possessed the long arm of the Y expressed HYA but were female, whereas patients who possessed the short arm lacked HYA but were male (Figure 18.2). Tdf and Hya are closer together on the mouse Y, but

were found eventually to be genetically separable in this species, too
(mouse gene nomenclature differs from human in the use of lower case).
We still do not know what HYA is or does, but, clearly, it does not
determine sex.

The demise of HYA removed from consideration the only genetic
function known on the human Y or mouse Y. Attention turned to repeated
sequences when it was discovered that one class of repeats, first isolated
from a snake (the banded krait *Bungarus fasciatus*), were represented on
the sex chromosomes, not only of mouse, but also snakes and birds and
even the fruit fly *Drosophila melanogaster*. This 'Bkm' sequence (named
for the said snake), being composed of many monotonous repeats of a
four-base (GATA) sequence, could hardly code for any sensible protein;
however, perhaps it could act by amassing DNA binding protein and by
preventing it from binding and controlling the activity of sex-determining
genes. This hypothesis accounted well for sex determination in mouse,
in which Bkm sequences were located in the sex-determining region;
however, Bkm is barely represented on the human Y, so it seemed unlikely
to have a sex-determining role.

The modern approach to finding TDF requires the detailed mapping
of the locus to a region small enough to be isolated physically and to be
searched for suspicious sequences. The short arm of the Y chromosome
still contains about 10 million base pairs, or 10 000 kilobase-pairs (kb); it
is a practical proposition to isolate and characterize 20 or 50 or maybe
a few hundred kilobase-pairs. Therefore, the 'positional cloning' technique
was employed by many groups around the world, collaborators and
rivals, who together narrowed down the search to the upper part of
the short arm of the Y. Pre-eminent among these workers during the
1980s were Peter Goodfellow's research group, at the Imperial Cancer
Research Fund in London, and David Page's research group, at the
Whitehead Institute in Boston.

David Page's group was first to claim success. They used DNA from
two exceptional patients to pin down TDF – a male having only 280 kb
of the Y, and a female with a Y chromosome apparently intact except for
140 kb in the same region. Lining up the bits of Y chromosome mapped
TDF to a 140 kb region, which was small enough to be isolated physically
(Figure 18.3).

To do this, Page broke up the DNA into smaller fragments and inserted
them into small self-replicating bacterial virus genomes, which could
be multiplied by allowing this 'recombinant' virus to infect bacterial cells.
These 'cloned' DNA fragments could then be examined to determine

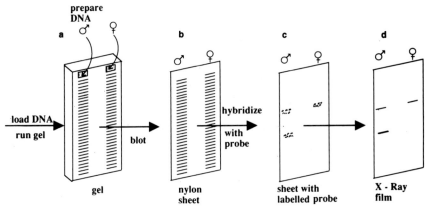

Figure 18.4. Southern blotting analysis of candidate TDF genes. For this technique, DNA is cut into specific fragments by enzymes which recognize short sequences of bases. The fragments are separated by size by using electrophoresis (a), then blotted and bound onto a nylon sheet (b). Radioactive 'probe' DNA is added, which binds only to fragments carrying the same gene (c). This specific binding, which is crucial to all molecular biology, depends on the complementary base pairing which forms the basis of the double helical structure of DNA; adenine bonding only to thymine, and guanine only to cytosine. When the DNA fragments and the probe DNA are both made single stranded, they will 'hybridize' only with complementary sequences. Bands of DNA containing the fragments which are complementary are detected by overlaying the membrane with X-Ray film (d), which will show darkened bands at the sites of radioactive decay within the membrane. This 'Noah's Ark' blot containing DNA from a male and female human, shows a male-specific band, as well as a band shared by males and females. This probe therefore detects a gene on the Y chromosome, as well as a gene on the X or one of the autosomes. (Adapted from A. McLaren (1991) The making of male mice. *Nature*, **351**, 96.)

which of them carried human DNA sequences that could possibly be TDF. Not knowing exactly what to look for, he first used a technique called 'Southern blotting' (the name of this important technique does not derive from a mystic orientation of the gel tray, but honours its inventor, Ed Southern) to determine whether the sequence was specific to males (Figure 18.4). Page used his candidate fragments as a 'probe' to detect homologous sequences in DNA from male and female pairs of several species – human, gorilla, monkey, cattle, mouse, loaded two by two on a 'Noah's Ark' blot. He then determined the base sequence of these DNA fragments and searched for sequences which had the trade marks of active genes: typical start and stop sites, and an 'open reading frame' which could specify a long run of amino acids, devoid of stop signals. The other important criterion was that the sequence should have stayed much the same throughout evolution, since most important genes will not change too much over time.

Page found just one gene that fit these criteria. It detected a male-specific fragment in all mammals investigated, and was highly conserved between species, as shown by the strong binding between fragment and probe DNA. Its base sequence would code for a protein with an amino-acid sequence reminiscent of several known proteins with a 'zinc finger' trademark: loops of amino-acid chain held together with zinc ions.

Therefore, this zinc finger gene (ZFY), like other proteins with the same motif, could bind to specific DNA sequences and perhaps could control the activity of other genes, a sensible sort of function for a sex-determining gene to have.

One unexpected feature of the ZFY gene was that it detected, as well as a male-specific sequence, a homologous sequence (called ZFX) shared by both males and females. This sequence was mapped to the X chromosome, giving rise to the speculation that ZFY and ZFX might work in partnership to determine sex by means of their differential dosages in males and females.

However, the role of ZFY in sex determination was thrown into doubt when it was discovered that the gene was not on the Y chromosome in marsupial or monotreme mammals, but was present on an autosome along with other genes from the same region of the human X. This supported the theory that an autosomal region, containing ZFX/ZFY, was added recently to the X in placental mammals (Figure 18.1). This recently added region seemed a strange place to find TDF! The simplest conclusion was that ZFY was the wrong gene, after all. Subsequently, four XX male patients were analysed who had only a tiny piece of the Y chromosome near the PAR, and lacked the ZFY gene, providing definitive evidence that ZFY was not TDF.

Since this time, another five genes have been discovered that have copies on both X and Y chromosomes in human, mouse or both (Figure 18.1). These genes code for tooth enamel, enzymes and adhesion molecules, functions that have no obvious relevance to sex determination. There is no particularly good reason why these genes must be on the X and Y chromosomes, and they may simply have got stranded on the diverging sex chromosomes. The ZFX/ZFY gene appears to be one of this motley lot of X–Y shared genes, and, although the X- and Y-borne copies seem to have diverged in their pattern of expression in mouse (Zfx is expressed everywhere but Zfy is expressed only in germ cells), apparently they play no direct role in sex determination.

Discovery of the SRY gene

So it was back to the drawing board – or, rather, to the pitchfork and the haystack. Workers in Goodfellow's group used DNA from the four XX male patients with only tiny fragments of the Y chromosome to define a new sex-determining region hard up against the PAR (Figure 18.2). Careful screening of this 35 kb region revealed several sequences that had the trade marks of genes. Of these, only a single sequence was specific to males, and was conserved in other species. This sequence, named SRY (for sex-determining region, Y chromosome) was suggested to be the elusive TDF gene.

Three pieces of evidence support this claim. Firstly, several patients with an outwardly normal Y chromosome who are, nevertheless, female (so must lack TDF activity) proved to have mutations in their SRY genes. These single base changes or small deletions would alter the amino acids, and therefore the properties, of the protein product so that it no longer worked to determine testis formation. Curiously, the XY female whose DNA was used by Page to delineate a sex-determining region around ZFY also provided evidence which supported the sex-determining role of SRY, for her Y proved to have a second deletion which included the SRY gene. A mutant line of mice with XY females was found to have a deletion of the corresponding Sry gene.

The most spectacular evidence that SRY is the testis-determining gene came from experiments in which the gene was injected into mouse embryos (Figure 18.5). The cloned mouse Sry gene was injected into cultured mouse embryos, a mixture of XX and XY. The injected embryos were returned to the uteri of host mothers, and allowed to develop, then their chromosomes were examined, and their DNA was screened to check that the Sry 'transgene' was, indeed, incorporated into their genomes. Half the injected mice were XY, and were expected to be male with or without the transgene. Uninjected XX mice were all female, as expected, but among the injected XX mice were two that were decidedly male in appearance. One of these XX male mice was named 'Randy', in appreciation of his decidedly male mating behaviour. Both XX male mice proved to have the Sry transgene incorporated in their DNA. It was the presence of the Sry transgene which ensured that they developed as males instead of females. The two XX Sry transgenic mice were, however, sterile. This was expected, because sperm development is independent of sexual phenotype, and is thought to require other genes elsewhere on the Y chromosome. Thus, the Sry gene was demonstrated to be both necessary and sufficient for the determination of male body type in humans and mice.

fertilized mouse eggs

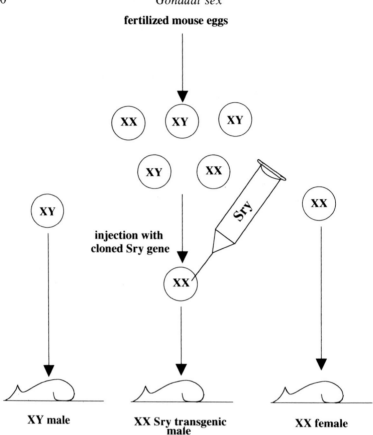

Figure 18.5. Sry transgenesis. Fertilized eggs are flushed from the oviducts of mice. Some have two X chromosomes and, normally, would develop into females. Others are XY and, normally, would develop into males. Some of these embryos are injected with many copies of the cloned and purified mouse Sry gene. This gene will be incorporated into the genomes and will be expressed in some of, but not all, the mice. Some mice were found which were XX but were male; these mice were transgenic for Sry. Thus, Sry is sufficient to produce male development.

Because mapping in marsupials had proved to be the critical test of the credentials of the ZFY gene, it was a matter of some interest whether the marsupial Y chromosome contained a homologue of the SRY gene. This was not at all a straightforward study, because it turned out that there are a number of genes in the mouse and the human genome which share some homology with SRY. Probing DNA from marsupial males and females had to be done under conditions which allow the diverged SRY sequences to hybridize. These more forgiving conditions also showed up a whole family of SRY-like sequences. The first studies, using DNA from cells cultured from male and female animals, revealed several bands shared by both sexes, but no bands which were exclusive to male-derived DNA. However, DNA prepared from liver samples of male animals did show a male-specific band. It turned out that these cultured cells had jettisoned their Y chromosomes; these XO cells provided a splendid (if unexpected) negative control. The DNA fragments represented by the male-specific band were then isolated by cloning into a virus genome, and the base sequence of this marsupial version of the SRY gene was determined. Mapping confirmed that this gene resided on the Y chromosome.

Like other comparisons between very distantly related mammals, this work also revealed quite unexpected features of SRY. One was that there seems to be an X-linked relative of SRY in marsupials. Another unexpected feature was that the base sequence of the SRY gene was not highly conserved, as we might have expected of a gene with such a critical function in reproduction. Both these findings have far-reaching implications for our understanding of how the mammalian sex-determining system works and how it evolved.

Function of the SRY gene

How does the SRY gene determine sex? Following the reverse genetic approach, we can use the cloned gene to investigate the time and tissue in which it is active, as well as the structure and function of its protein product.

Genes act to specify the amino-acid sequence of a particular protein. The base sequence of the DNA that constitutes the gene is first copied into an RNA molecule by a process of complementary base pairing, in which adenine pairs with thymine (uridine) and guanine with cytosine. This messenger RNA leaves the nucleus to be translated into protein in the cytoplasm. Each triplet of bases specifies one of the 20 amino-acids in the corresponding position in the protein product. Of the more than 50 000 genes in the mammalian genome, only a subset are active at any time in any tissue. Genes important in embryonic development are switched on and off according to a complex genetic program.

As for Southern blotting analysis, detection of a particular messenger RNA depends on the specific binding between complementary base pairs. DNA from the mouse Sry gene will line up and bind to complementary Sry–RNA-sequences, so the cloned Sry gene can be used as a probe to detect Sry messenger RNA. Probing for these Sry transcripts in different tissues of adult mice showed that the gene is active only in the testis. Investigating when the gene is expressed in the embryo required even more sensitive techniques because very little RNA can be obtained from the tiny embryos, and the gene may be expressed only transiently. So a new technique has been employed, using short 'primer' DNA sequences synthesized to be complementary to the two ends of the Sry messenger RNA. In the presence of polymerase enzymes which make DNA copies of the sequence between these primers, many DNA copies of the Sry sequence can be amplified and detected under UV as a bright band of DNA. This polymerase chain reaction (PCR) allowed the detection of Sry messenger RNA in XY male mouse embryos 10.5–11.5 days after fertilization. This narrow window of expression just precedes the first changes observed in the appearance of the undifferentiated testis in XY embryos. Expression of Sry seems, therefore, to coincide with testis determination.

The SRY gene has been cloned from human, mouse, rabbit and two marsupial species, and the base sequences of these genes have been determined. We can now use the universal code table you see in every elementary genetics textbook to determine the amino-acid sequence of the protein it codes for. Comparisons of base sequence of the gene and amino-acid sequence of its protein product should offer us clues to its function.

The coding region of the human SRY gene, identified by the usual start and stop signals, is 669 base pairs (bp) long. Many genes are patchworks of coding sequences and interspersed regions of junk ('introns'), which are cut from the RNA molecule before it is shipped out into the cytoplasm. However, the SRY gene and its RNA transcript seem to line up perfectly, so the gene seems to be one of the few which contain no introns.

The SRY base sequence would code for a protein of 223 amino-acids. This amino-acid sequence, by itself, tells us a little about the properties of the protein. Searching DNA and protein databases for similar runs of bases or amino-acids is more informative, for we find that particular runs of amino-acids are used over and over again in proteins with similar functional domains. For instance, the 'zinc finger' motif which was identified in ZFY is shared with many other proteins which use this portion of the protein to bind to DNA.

There was great excitement when the first sequence to be identified as being similar turned out to be a protein which is involved in sex in the lowly yeast. Maybe there really is a universal sex-determining gene! However, it soon became apparent that this was only one of a very large class of proteins, all of which shared a recognizable 79 amino-acid domain.

This group of proteins included the so-called 'high mobility group' proteins (so-called because they are separated by their faster movement through an electrophoretic gel), and the shared region was therefore called the 'HMG box'. These HMG proteins have rather non-specific effects on the activity of other genes. The SRY relatives also include other proteins which modulate the activity of other genes. Some of these related genes are thought to turn on transcription of other genes, and, by analogy, it was expected that SRY, too, functions as a 'transcription factor', turning on testis-differentiating genes.

Studies with the human or mouse SRY gene showed that there were a number of other genes in the genome which were partly homologous to SRY, including a similar HMG box sequence. The SRY gene seemed, therefore, to define a new class of genes. These 'SOX' genes might include other developmentally important genes which, perhaps, regulate other pathways in mammalian organogenesis.

The next step in the characterization was to isolate the SRY protein. This would be almost impossible to do by standard biochemical techniques, given the minute amounts present in tiny embryos over such a short time. What has been done, instead, is to trick bacteria into making a 'fusion protein', part bacterial, part SRY, from the cloned SRY gene, spliced into a special virus-derived genome which allows its expression. Studies with this artificial protein show that, indeed, the SRY protein, like the transcription factors to which it is similar, seems to bind to DNA, recognizing the same or similar target sequences. This binding is absent in protein made by cloned mutant genes which cause sex reversal, suggesting that the binding to DNA has biological meaning. Very recently, it has been found that the SRY product binds to a specific target sequence in DNA, and holds it in a looped configuration, which might well act to shut down or to open up a whole region of DNA for transcription.

Does the SRY protein turn on genes which control the next step in the testis-differentiating pathway (Figure 18.6a)? By analogy with transcription factors which share the HMG box, the SRY protein was thought to enhance the rate at which other genes are expressed. However, one puzzle is the finding, particularly obvious when comparisons are made between the sequence of human, mouse and marsupial SRY, that the sequence is not really very conserved even within the 'conserved' HMG box, and the homology disappears altogether outside this box region. This is surprising for a gene that is expected to control sex determination, which must be a critical factor in the reproductive success of a species. If the SRY protein were a transcription factor, its sequence would be

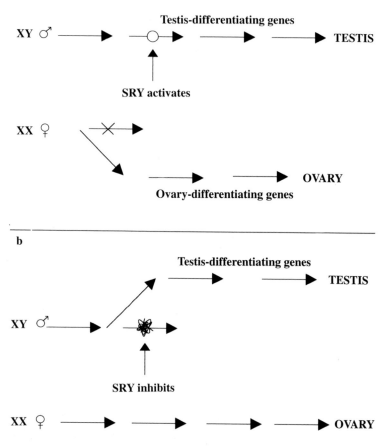

Figure 18.6. Two possible ways in which SRY acts to turn on testis-differentiating genes. (a) SRY acts to stimulate testicular development; (b) SRY acts to repress ovary-differentiating genes, allowing the testis-differentiating genes to act.

expected to be constrained by the necessity to bind, not only to a specific DNA sequence, but also to several other proteins required for expression.

Is it possible, then, that the SRY gene acts, not by turning on male-determining genes, but by turning off female-determining genes (Figure 18.6b)? Such a repressor may not be nearly as dependent on its sequence for its action in blocking transcription of target genes, since a good deal of the protein could function merely to keep other transcription factors at bay.

The next step in the investigation of SRY activity will be identification of genes which contain the sequence to which SRY protein binds specifically. Cloning these SRY-controlled genes and investigating their expression, and the sequence and function of their protein products, should tell us how the Y chromosome ultimately has its effect on sex determination.

Evolution of sex determination and sex-determining genes

How ancient is our sex-determining system? Is SRY the sex-determining gene in other vertebrates? Even in other animals?

There are genes with homology to SRY in birds and reptiles, but no sign that any of them are sex-specific; they are more likely to represent the equivalents of some of the SOX family of related genes in mammals. Also, gene-mapping work makes it clear that the X and Y chromosomes of mammals bear no homology to the Z and W chromosomes of birds and reptiles. This suggests that birds chose one gene in the sex-determining pathway, and mammals chose another; the two chromosomes which bore these different genes were destined to become the ZW pair or the XY pair. Nevertheless, mammals, birds and reptiles shared a common ancestor some 200 million years ago, so that their sex-determining systems must have derived from a common system. What was this common system? How and when did the SRY gene take on its role as the master switch?

There is some evidence that sex determination in birds relies on a balance system in which the number of Z chromosomes per haploid genome determines sex. Dosage differences and dosage compensation are recurring themes in invertebrate sex-determination as well. Is it possible that sex-specific dosage of a gene on the X, or inactivation of a gene shared by the X and Y, does have an underlying effect on mammalian sex-determination? It is difficult to assess the effect of extra copies of the X chromosome in mammals because all but one X chromosome is inactivated in females. However, there are several XY female patients who have a duplication of a part of the short arm of the X, suggesting that an extra copy of a gene in this region may override the SRY gene. Perhaps mammalian, like avian, sex was determined originally by the numbers of X chromosomes, or by the number of active copies of a gene shared by the X and Y, rather than by the presence or absence of a unique gene on the Y.

At what point did the dominant male-determining SRY gene evolve? New genes do not arise from scratch in evolution; they evolve from pre-existing genes with related functions. The observation that (like ZFY and all the other sequences on the Y) SRY has a related sequence on the X suggests that a progenitor gene may have been located on both members of the chromosome pair which became differentiated into the X and Y. Was part of the SOX family of HMG box genes involved in organogenesis (e.g. liver-determining or brain-determining genes)? Perhaps it just happened to be on the chromosome pair that differentiated into the X and Y. Then, as the X and Y diverged, the Y-borne copy of the gene took on a new function in the regulation of gonadal differentiation. Either this may have been the initial event in mammalian sex-chromosome differentiation, or it may have superseded the action of a more ancient dose-dependent gene on the X and/or Y.

What do we expect from a follow-up study of mammalian sex in about 100 million years' time? It is clear from the differences between mammalian species that the process of differentiation between the X and Y chromosomes continues. It is likely that our follow-up study would show the complete elimination of the PAR; the last vestiges of X–Y homology would have vanished. The function of SRY as the master switch of sex determination could also be overridden by other controlling genes, which interact with SRY. At present, certain vole species have no Y chromosome, while XY females are observed frequently among lemmings; perhaps these species have invented new sex-determining systems. Will they catch on? It seems certain that chromosomal sex-determining systems evolved independently many times, and are continually being revised and refined as sex-determining genes fight it out for supremacy.

Summary

In all mammals, the presence of the small Y chromosome determines male development. The Y consists largely of genetic junk, but must logically bear a gene which determines testis development, the first step in the male-determining pathway. The search for this TDF gene in man and mouse has depended on a 'reverse genetic' approach which involves the detailed mapping of the testis-determining function in humans and mice having only part of a Y chromosome. This analysis has pinpointed TDF on part of the short arm of the human Y.

The cloning and characterization of rival candidate genes from this sex-determining region has provided an exciting genetic detective story. The first contender for the TDF role, a gene (HYA) detected immunologically, was found to map in a region of the Y that is distant from the testis-determining function. A second candidate gene (ZFY), which makes a protein with a 'zinc finger' structure, was eliminated first by comparative studies which showed it was not on the sex chromosomes in marsupial and monotreme mammals; subsequently, it was found to be absent in some human males having only the very tip of a Y chromosome. More recently, a new gene (SRY) has been cloned from this part of the human Y. It has excellent credentials for its testis-determining function, as shown by changes within the SRY gene in females with a Y, by the male development of XX mice transgenic for SRY, and by the presence of an SRY homologue on the marsupial Y. This gene makes a protein which binds to DNA and presumably turns other genes in the pathway off or on.

The gene which determines sex in mammals is different from the gene which determines sex in other vertebrates, and may not even be the original mammalian sex-determinant. The SRY gene may have evolved from another gene on the ancestral X and Y chromosomes with some other function in development. A follow-up study in 100 million years' time may well show that the SRY gene, in turn, has been superseded by a new master gene.

Acknowledgements

I thank Joanne La Rose for help with drafting the diagrams.

Further reading

Ballabio, A. and Willard, H. F. (1992) Mammalian X chromosome inactivation and the XIST gene. *Current Opinion in Genetic Development*, **2**, 439–47.

Bull, J. (1983) *Evolution of Sex Determining Mechanisms*. Benjamin Cummings, Menlo Park.

Burgoyne, P. (1992) Y Chromosome function in mammalian development. *Advances in Developmental Biology*, **1**, 1–29.

Graves, J. A. M. (1990) The search for the mammalian testis-determining factor is on again. *Reproduction, Fertility and Development*, **2**, 199–204.

Graves, J. A. M. and Schmidt, M. M. (1993) Mammalian sex chromosomes – design or accident? *Current Opinions in Biology*, **2**, 890–901.

Graves, J. A. M. and Watson, J. M. (1991) Mammalian sex chromosomes: evolution of organization and function. *Chromosoma*, **101**, 63–8.

Graves, J. A. M., Foster, J. W., Hampikian, G. K. and Brennan, F. E. (1993) Marsupial SRY-related genes and the evolution of the testis-determining factor. In *Sex Chromosomes and Sex-Determining Genes*, pp. 325–36. Eds. K. C. Reed and J. A. M. Graves. Harwood Academic Publishers, Switzerland, Australia.

Jones, K. W. and Singh, L. (1985) Snakes and the origin of sex chromosomes. *Trends in Genetics*, **1**, 55–61.

Koopman, P., Gubbay, J., Vivian, N., Goodfellow, P. and Lovell-Badge, R. (1991) Male development of chromosomally female mice transgenic for Sry. *Nature,* **351**, 117–21.

Page, D. C., Mosher, R., Simpson, E. M., Fisher, E. M. C., Mardon, G., Pollack, J., McGillivray, B., de la Chapelle, A. and Brown, L. G. (1987) The sex-determining region of the human Y chromosome encodes a finger protein. *Cell,* **51**, 1091–104.

Sinclair, A. H., Berta, P., Palmer, M. S., Hawken, J. R., Griffiths, B. L., Smith, M. J., Foster, J. W., Frischauf, A.-M., Lovell-Badge, R. and Goodfellow, P. N. (1990) A gene from the human sex-determining region encoding a protein with homology to a conserved DNA-binding motif. *Nature,* **346**, 249–54.

Watson, J. D. and Jordan, E. (1989) The Human Genome Program at the National Institutes of Health. *Genomics,* **5**, 654–6.

19

Bizarre mammalian sex-determining mechanisms

KARL FREDGA

Some males become females which give birth only to daughters. This provocative statement is true for the wood lemming, a small rodent inhabiting the mossy forests of the Eurasian taiga. The wood lemming belongs to an exclusive group of mammals, all rodents, which do not follow the normal chromosomal rules for sex determination, viz. that males are XY and females are XX.

Several species, belonging to different orders, have multiple sex-chromosome systems due to translocations between an autosome and the original X- or Y-chromosome. Among these, the male has one chromosome more or less than the female. However, the sex chromosomes are distributed regularly at male meiosis, two kinds of sperm and one kind of egg are produced, and sex is determined in the orthodox way by Y-linked gene(s). The most interesting group consists of species with complicated, or bizarre, mechanisms for sex determination, and some of these species have fascinated scientists for many years. Already in the 1950s the legendary cytogeneticists Robert Matthey and Michael White were puzzled by the fact that the mole vole *Ellobius lutescens* had 17 chromosomes only and had identical karyotypes in males and females. We still cannot distinguish the XO karyotype of males from that of females, and we do not know how sex is determined in this species.

Investigations of these exceptional species may contribute to our understanding of sex determination in mammals, the cascade of molecular happenings taking place in an embryo when an indifferent gonad develops into an ovary or a testis. Now, it is accepted generally that the Sry gene is the long-sought testis-determining factor in the Y chromosome (see Chapter 18 by Jennifer Graves). Therefore, it is of great interest to know whether Sry is present in these bizarre species or not. As will be shown below, there are fertile XY females testing positively for the Sry gene,

Table 19.1 *Rodents with bizarre mechanisms for sex determination*

	Sex chromosomes			
	Female		Male	
	soma	germ	soma	germ
Order Rodentia				
Family Muridae				
Subfamily Hesperomyinae				
Species				
Akodon azarae + 2 spp.	XX XŸ*	XX XŸ*	XY	XY
Subfamily Arvicolinae (Microtinae)				
Species				
Dicrostonyx torquatus + ? spp.	XX X̊X X̊Y	XX X̊X X̊Y	XY	XY
Myopus schisticolor	XX X̊X X̊Y	XX X̊X X̊X̊	XY	XY
Microtus cabrerae	XX XŸ*?	XX XŸ*?	XY	XY
Microtus oregoni	XO	XX	XY	YO
Ellobius lutescens	XO	XO	XO	XO
Ellobius tancrei + 2 spp.	XX	XX	XX	XX
Subfamily Murinae				
Species				
Tokudaia osimensis	XO	?	XO	XO

Nomenclature and taxonomic position according to Corbet and Hill (1991).
X̊ is an X and Ÿ is a Y chromosome with a mutation that affects sex determination.
XŸ? means that the existence of these females is not proven.
? means unknown.

which means that ovaries may develop in the presence of Sry. Further, in two species of the genus *Ellobius* the normal, fertile males seem to lack the Sry gene, which means that the Sry gene is not necessary for testis development.

Sex chromosomes

The species with bizarre mechanisms for sex determination are listed in Table 19.1 and, below, I will discuss briefly the various genera and species.

Akodon

This genus of South American field-mice comprises 40–50 species and their chromosomes have been studied mainly by Osvaldo Reig, Nestor Bianchi and their coworkers in Argentina.

Some species have puzzling sex-chromosomes with a variation in the size and morphology of the X. One variety of X is indistinguishable from the Y chromosome and the question was: is it a Y or a deleted X? Recently, Bianchi and colleagues have shown that it is a Y, because the Y-specific genes Zfy and Sry are present in these females of *Akodon azarae*. Thus, in this species, various female types exist: XX with two normal X chromosomes, XXp− with one normal and one deleted X, and XY. Another two *Akodon* species, *A. mollis* and *A. varius*, also have XY females.

The reason for sex reversal in *Akodon* is unknown, but it must be because some gene(s) in the Y chromosome does not work properly and prevents the Sry gene from starting testis formation. These females may be designated XY* in contrast to the X*Y females of the lemmings (see below): an asterisk (*) indicates a mutation which affects sex determination and causes sex reversal. XY* females in *Akodon* are always the offspring of XY* females, and the sex ratio in their litters is one male to two females (i.e. 1:2), due to the death of Y*Y zygotes.

Dicrostonyx

The genus *Dicrostonyx* has a circumpolar distribution in the tundra zones of Eurasia and North America. The systematics of collared lemmings is complex and controversial. Gordon Corbet and John Hill list nine species tentatively, one of which is present in mainland Siberia. Different forms of this species, *Dicrostonyx torquatus*, have been studied thoroughly by Emily Gileva and coworkers from a cytogenetic and taxonomic point of view. The genus *Dicrostonyx* offers excellent opportunities to study the role of chromosomal changes in speciation, because chromosomal polymorphism is common both within and between populations. The genetic diversity is much greater than is indicated by morphological characteristics.

The chromosomal mechanism for sex determination is similar to that of the wood lemming (*Myopus*, see below): three sex-chromosome types of female and one type of male occur, and one of the female types is indistinguishable from that of the male.

However, the chromosome situation in collared lemmings is more complex compared to the wood lemming. For example, the sex-chromosome constitution of the subspecies, *Dicrostonyx t. torquatus*, may be explained by a series of translocations between the sex chromosomes and autosomes, starting with a 'primitive' form living further to the east in Siberia.

There is one important difference between *Dicrostonyx* and *Myopus*. Both X*- and Y- carrying eggs are formed in X*Y females of *Dicrostonyx*, because they have not evolved the double non-disjunction mechanism preceding meiosis in X*Y females as in *Myopus* (see below). Because of the early death of YY zygotes, *Dicrostonyx* X*Y females have a 1:2 sex-ratio among their offspring. Theoretically, the litter size of X*Y females should be reduced by one-quarter compared with XX and X*X females, but, due to a higher ovulation rate in X*Y females, their reproductive output is not significantly lower than that of females with two X-chromosomes.

Myopus

The wood lemming *Myopus schisticolor* (Figure 19.1) is the species that is studied the most thoroughly of those listed in Table 19.1. It is the only species in its genus and has a wide distribution from Norway in the west to Kamchatka in the east. In contrast to its more famous relative, the Norway lemming, it prefers mossy forests and not the tundra and alpine birch region.

It was a great surprise when we discovered in 1976 that some female

Figure 19.1. A pair of wood lemmings, *Myopus schisticolor*. (Photograph by Karl Fredga and Mattias Klum.)

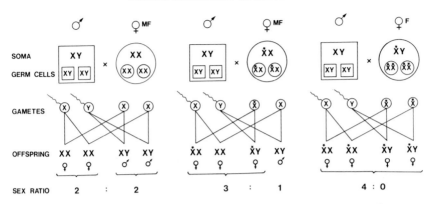

Figure 19.2. Theoretical breeding scheme of the wood lemming. Matings between males (XY) and females of different sex-chromosome constitutions will give rise to different sex-ratios in the offspring. Due to double non-disjunction in X*Y females only one type of egg, X*, is produced. Thus, only daughters are born to these females. X* is an X chromosome with a mutation which affects sex determination. MF, females producing both males and females; F, females producing females only. (From Fredga *et al.* (1977).)

wood lemmings had an apparently normal male (XY) karyotype. The presence of a Y chromosome had always been associated with male phenotypes. Our interest in the wood lemmings arose from two facts described by the Finnish biologists Olavi Kalela and Tarvo Oksala in 1966: (i) the sex ratio was skewed (1:3) in favour of females; and (ii) some females produced daughters only. The basic findings about the wood lemming may be summarized as follows (see Figure 19.2):

A mutation in an X chromosome affects sex determination. This X chromosome is designated X*

All males are normal XY

There are three types of females with respect to the sex chromosomes: XX, X*X and X*Y. There is no difference in external morphology between the three types of females

The two kinds of X chromosome can be distinguished under the microscope by the G-banding pattern of the short arm

There are two types of females with respect to their offspring: some give birth to sons and daughters (MF type), some give birth to daughters only (F type)

It is the X*Y females that are of F type. Owing to a mechanism of double non-disjunction in the fetal ovary, X*-carrying eggs only are formed

The phenomena listed above lead to the exceptional sex-ratio, with about three times as many females as males in both wild populations and in laboratory stocks

The special sex-chromosome constitution of the wood lemming provides an exceptional opportunity to investigate the interaction between genes involved in sex determination. In addition, individuals with numerical sex-chromosome aberrations are relatively common, usually owing to failure of the double non-disjunction mechanism in X*Y females. Among others, the following sex-chromosome combinations and phenotypes have been recorded: XXY, sterile male; X*XY, sterile male, fertile female, and true hermaphrodite; X*YY, fertile female; XO and X*O, fertile females. The various phenotypic expressions of the X*XY individuals may be explained by non-random inactivation of the X chromosomes. If X* is inactivated the fetus develops into a sterile male, if X is inactivated it develops into a fertile female, and if the activation pattern is different between the right and left side a true lateral hermaphrodite is formed.

XO is the most common type of numerical sex-chromosome aberrations in the wood lemming, producing fertile females with normal litter size. It is possible that they, like the XO mouse, have a reduced reproductive lifespan owing to premature exhaustion of the supply of oocytes, but in the wild it is unlikely that the wood lemming females would reach that age.

The rare X*YY females are of particular interest. They demonstrate that two Y-chromosomes cannot override the female-determining effect of X*. All fetuses that have an X* chromosome and no X develop into females, and the X*YY females are fully fertile.

Theoretically, the three types of normal female are expected to occur in the same proportions, but this is not the case. We have found that, both in natural populations and in laboratory colonies, the frequency of X*Y females is about 45% instead of 33%. This may, at least to some extent, be explained by differences in maturation rate, onset of reproduction and pregnancy rate of the different female types. Extensive field studies have demonstrated that in all these attributes the X*Y is superior to X*X, which is superior to XX. The litter size is age dependent but is the same for all three sex-chromosome types.

Why do X*Y females become sexually mature earlier than the others? Birth weight and ovary development may be important factors. We compared the birth weight of X*Y and X*X daughters in litters of X*Y females. In 8 out of 10 litters the X*Y young were heavier on average, and in 7 of these litters *all* the X*Y were heavier than their X*X litter mates. In addition, there are clearcut differences in the development of ovaries and oocytes as between three-day-old X*Y and X*X young. Ovarian volume, both somatic and germinal, was nearly twice as large in X*X compared to X*Y, and the number of oocytes was twice as many in the X*X. However, mean oocyte volume was higher in the X*Y, and these females had a markedly higher proportion of growing oocytes in relation to the total number than is seen in X*X females, 6.5% compared with 3%.

Microtus

Two species of voles belong to the category of 'bizarre' rodents.

Microtus cabrerae is endemic to the Iberian peninsula and it is possible that XY females exist in this species. Variation in length of the sex chromosomes, caused by the variable size of heterochromatic segments, is common and three types of X and five types of Y were described by Burgos and colleagues in 1988. The few females claimed to be XY had two different and uncommon types of Y. It is possible that a mutation in the Y chromosome causes sex reversal and that the presumed XY females are XY* as in *Akodon*, but further studies are needed.

Microtus oregoni is a small vole living along the North American west coast. It has interested cytogeneticists since the 1950s, because of its low chromosome number and unique sex-chromosome constitution. Both sexes of this species are gonosomic mosaics having different chromosome numbers in germ and somatic cells, as Susumu Ohno and his colleagues showed in the 1960s. Female somatic cells have 17 chromosomes and one X-chromosome only. When primordial germ-cells differentiate into oogonia, selective non-disjunction takes place. Germ cells with no X chromosomes (OO) die off and only XX oogonia survive. Thus, one type of egg only, with one X-chromosome, is produced.

Male somatic cells have 18 chromosomes and XY sex-chromosomes. When primordial germ cells differentiate into spermatogonia, once again non-disjunction takes place. XXY and OY germ cells are produced, but of these only OYs differentiate into definite spermatogonia. The reason for this is unknown, but, as a result, two types of sperm are produced, one with a Y chromosome and the other with no sex-chromosome at all. The former is the male-determining gamete, the latter is the female. The XY zygote develops into a male, the XO into a female lacking a paternal X-chromosome. It is unknown how this complex mechanism for sex determination has developed.

Ellobius

The taxonomy of the genus *Ellobius* is complex, and species nomenclature has changed from time to time. At present, five species are recognized by Russian scientists: *Ellobius fuscocapillus, E. lutescens, E. tancrei, E. talpinus* and *E. alaicus*. These subterranean mole voles have a wide range of more or less allopatric distribution in the mountains and steppes of Central Asia. Figure 19.3 shows the external appearance of *E. lutescens* (a) and *E. tancrei* (b).

The range of chromosome numbers is large in the genus *Ellobius*, from 17 to 54, and within one of the species, *E. tancrei*, there is a 'Robertsonian fan' with variation from 32 to 54 (diploid). But it is their sex chromosomes which are of greatest interest, because different mechanisms for sex determination have evolved. The chromosome situation is summarized in Table 19.2 which is based mainly on investigations by Nikolai Vorontsov, Elena Lyapunova, Karl Fredga and coworkers.

E. fuscocapillus has normal sex chromosomes: females are XX and males are XY. The X is of 'original' size ($\approx 5\%$ of the female haploid set) and the Y is considerably smaller.

E. lutescens has one X-chromosome only and identical karyotypes in both sexes (XO/XO). Among the 17 chromosomes the X is the smallest but of 'original' size. No difference between the X of males and that of females has been detected in spite of many different chromosome-banding techniques being used.

E. tancrei has two X-chromosomes in both males and females and identical karyotypes (XX/XX). The X is of 'original' size. Most likely *E. talpinus* and *E. alaicus* also have XX males and females.

For *Ellobius lutescens*, at least five different possibilities have been put forward to explain the sex-determining mechanism and the origin of the unpaired chromosome. The most likely is that the testis-determining factor has been translocated from an original Y-chromosome to the male X-chromosome, and that the translocated part is too small to be detected under the microscope. The condition in *E. tancrei* could also be explained in a similar way. Thus, the sex chromosomes in females and males of *E. lutescens* might be designated XO/X^{y}O, and in *E. tancrei* XX/X^{y}X.

Since Sry is regarded as being the gene responsible for mammalian sex determination, we intended to test the hypotheses outlined above by *in situ* hybridization with Sry probes. But first we had to check the presence of Sry in males, and its absence in females, by Southern blots (see Chapter 18 by Jennifer Graves). Then it was shown by Walther Vogel

Figure 19.3. Mole voles. (a) *Ellobius lutescens*; (b) *Ellobius tancrei*. (Photograph by Karl Fredga.)

and coworkers that *E. lutescens* males (and females) lack the Sry gene, and also the Zfy gene.. So we decided to investigate males and females of *E. tancrei*, and Vogel and colleagues have now shown that these also lack the Sry and Zfy genes. Thus, these two *Ellobius* species seem to determine sex without Sry. So far, we have had no opportunity to test *E. fuscocapillus* males, which is critical because of their normal XY sex-chromosome constitution.

It is still possible that *Ellobius* species have a very specific type of Sry,

Table 19.2 *Sex chromosomes and chromosome numbers in mole voles, genus* Ellobius

	Sex chromosomes		
Species	Female	Male	2*n*
E. fuscocapillus	XX	XY	36
E. lutescens	XO	XO	17
E. tancrei	XX	XX	32–54
E. talpinus	XX	XX	54
E. alaicus	XX	XX	52

Based on Vorontsov and Lyapunova (1989) and Fredga and Lyapunova (1991).

but probes from man, mouse, rabbit and marsupial all gave negative results in *E. lutescens* and *E. tancrei*. The probes identified male-specific bands clearly in Southern blots from a variety of mammals, including rodents like *Myopus schisticolor* and *Microtus arvalis* which belong to the same subfamily as *Ellobius*. So, sex determination in *Ellobius* is still a mystery.

Another remarkable fact is that germ cells with more than one X-chromosome are able to mature in the testicular environment of *E. tancrei*. As a rule, male mammals with more than one X-chromosome are sterile. The most well-known example is XXY Klinefelter's syndrome in humans; it is also known to occur in 12 other species. They all have a male phenotype but are infertile due to testicular hypoplasia.

Tokudaia

Tokudaia osimensis, the Amami spinous country rat, has been studied by Japanese researchers. The species is represented by various forms in the islands between Japan and Taiwan. In the subspecies *T.o. osimensis* from Amami and Tokuno, both males and females have an odd number of chromosomes and the sex-chromosome constitution XO. At metaphase I of male meiosis, the X chromosome appears as a univalent, just as in *Ellobius lutescens*. The mechanism for sex determination seems to be similar to that of *E. lutescens*. Another subspecies, *Tokudaia osimensis muenninki*, from Okinawa island, has an ordinary XX/XY sex-chromosome mechanism.

Table 19.3 *Presence* (+) *or absence* (−) *of some Y-linked genes expected to be involved in testis determination (Sry) or testis differentiation (Zfy)*

	Akodon			Myopus				Ellobius			
								lutescens		tancrei	
	F	M	M	F	F		M	F	M	F	M
Gene	XX	XY*	XY	XX	X*X	X*Y	XY	XO	XO	XX	XX
Zfx	+	+	+	+	+	−	+	+	+	+	+
Zfx*				−	+	+	−				
Zfy	−	++	++	−	−	++	++	−	−	−	−
Sry	−	++	++	−	−	+	+	−	−	−	−

+ + Indicates that many copies of the genes are present.
F, female; M, male.
X* and Y* (see Table 19.1).
Zfx* (see text).

Sry and Zfy genes

It is of great interest to know whether the Y-linked genes Sry and Zfy are present in XY females or not. There is strong evidence that Sry is indeed identical with the testis-determining factor, and, thus, is crucial for male development (see Chapter 18 by Jennifer Graves). Zfx and Zfy are present on the sex chromosomes of all eutherian mammals, and may have a role in spermatogenesis. Table 19.3 is a summary of our present-day knowledge concerning whether or not these 'bizarre' rodents have Sry, Zfx and Zfy. Some species have not been tested yet. In the wood lemming *Myopus* there are two allelic Zfx genes on the X and X* chromosomes which are designated Zfx and Zfx*, respectively. As a rule, eutherian mammals have only one copy of the Sry gene on their Y chromosomes, but some rodents are exceptions. At least 15 copies of Zfy sequences are distributed along the short arm of the wood lemming's Y chromosome, and about 10 copies are present in *Akodon azarae*. The fact that Zfy genes were amplified in both these species that have XY females led us to consider whether this had anything to do with the sex reversal. Bianchi and colleagues have now shown that this is not the case. Some other *Akodon* species without sex reversal also have about 10 copies of Zfy. Two other phylogenetically related species, *Oxymycterus rufus* and *Bolomys obscurus*, have about 24 copies and possess ordinary sex-chromosomes without sex reversal. Several of the *Akodon* species showing

Zfy amplification also have Sry amplification. Therefore, it is suggested that these two genes may map to the same region of the *Akodon* Y-chromosome and are coamplified. In humans, these two genes, ZFY and SRY, are located close to each other on the short arm of the Y chromosome (Yp11.3).

The following facts are clear from Table 19.3:

(1) Female sex can develop in spite of Sry being present. Examples are the XY females of *Akodon* and *Myopus*
(2) Male sex can develop in spite of Sry being absent. Examples are the XO and XX males of *Ellobius lutescens* and *E. tancrei*

The question is how is sex determined in these species? The *Akodon* case indicates that another Y-linked gene (or genes) is necessary for the Sry gene to work properly. The *Myopus* case shows that not only Sry but also a gene (or genes) on the X chromosome is necessary for testis formation. The *Ellobius* cases indicate that a hitherto unknown gene, or a strange variety of Sry, is responsible for male determination in these species.

Further reading

Bianchi, N. O., de la Chapelle, A., Vidal-Rioja, L. and Merani, S. (1989) The sex-determining zinc finger sequences in XY females of *Akodon azarae* (Rodentia, Cricetidae). *Cytogenetics and Cell Genetics*, **52**, 162–66.

Bianchi, N. O., Bianchi, M. S., Pamilo, P., Vidal-Rioja, L. and de la Chapelle, A. (1992) Evolution of zinc finger-Y and zinc finger-X genes in Oryzomyne–Akodontine rodents (Cricetidae). *Journal of Molecular Evolution*, **34**, 54–61.

Bianchi, N. O., Bianchi, M. S., Tolvanen, R. and de la Chapelle, A. (1992) 'The sex determining region Y gene (*Sry*) in *Akodon* (Cricetidae) species with XY females.' Abstracts, p. 55. 11th International Chromosome Conference, Edinburgh 1992.

Bondrup-Nielsen, S., Ims, R. A., Fredriksson, R. and Fredga, K. (1993) Demography of the wood lemming (*Myopus schisticolor*). In *The Biology of Lemmings*, pp. 493–507. Eds. N. C. Stenseth and R. A. Ims. Academic Press, UK.

Borisov, Y. M., Lyapunova, E. A. and Vorontsov, N. N. (1991) Evolution of the karyotype in the genus *Ellobius* (Microtinae, Rodentia). *Sovjet Genetics*, **27**, 375–83.

Burgos, M., Jiménez, R. and Diaz de la Guardia, R. (1988) XY females in *Microtus cabrerae* (Rodentia, Microtidae): a case of possibly Y-linked sex reversal. *Cytogenetics and Cell Genetics*, **49**, 275–7.

Corbet, G. B. and Hill, J. E. (1991) *A World List of Mammalian Species*. Third ed. Natural History Museum Publications, London and Oxford University Press, Oxford.

Fredga, K. (1983) Aberrant sex chromosome mechanisms in mammals. Evolutionary aspects. *Differentiation*, **23** (Suppl), 523–30.

Fredga, K. (1988) Aberrant chromosomal sex-determining mechanisms in mammals, with special reference to species with XY females. *Philosophical Transactions of the Royal Society of London, series B*, **322**, 83–95.

Fredga, K. and Lyapunova, E. A. (1991) Fertile males with two X chromosomes in *Ellobius tancrei* (Rodentia, Mammalia). *Hereditas*, **115**, 86–7.

Fredga, K., Fredriksson, R., Bondrup-Nielsen, S. and Ims, R. A. (1993) Sex ratio, chromosomes and isozymes in natural populations of the wood lemming, (*Myopus schisticolor*). In *The Biology of Lemmings*, 465–91. Eds. N. C. Stenseth and R. A. Ims. Academic Press, UK.

Fredga, K., Gropp, A., Winking, H. and Frank, F. (1976) Fertile XX- and XY-females in the wood lemming *Myopus schisticolor*. *Nature*, **261**, 225–7.

Fredga, K., Gropp, A., Winking, H. and Frank, F. (1977) A hypothesis explaining the exceptional sex ratio in the wood lemming (*Myopus schisticolor*). *Hereditas*, **85**, 101–4.

Gileva, E. A., Benenson, I. E., Konopisteva, L. A., Puchkov, V. F. and Makarantes, I. A. (1982) XO females in varying lemming *Dicrostonyx torquatus*: reproductive performance and its evolutionary significance. *Evolution*, **36**, 601–9.

Just, W., Rau, W. and Vogel, W. (1992) 'Sex determination in *Ellobius lutescens* without Sry?' Abstracts, p. 57. 11th International Chromosome Conference, Edinburgh 1992.

Just, W., Rau, W., Fredga, K. and Vogel, W. (1993) 'Fertile males of *E. lutescens* and *E. tancrei* without Sry and Zfy.' In Abstracts of Seventeenth International Congress of Genetics, p. 234. Birmingham, UK, 15–21 August 1993.

Kalela, O. and Oksala, T. (1966) Sex ratio in the wood lemming, *Myopus schisticolor* (Lilljeb.), in nature and in captivity. *Annales Universitatis Turkuensis, Series AII*, **37**, 1–24.

Lau, Y.-F. C., Yang-Feng, T. L., Elder, B., Chan, K., Fredga, K. and Wiberg, U. H. (1992) Unusual distribution of Zfy and Zfx sequences on the sex chromosomes of the wood lemming, a species exhibiting XY sex reversal. *Cytogenetics and Cell Genetics*, **60**, 48–54.

Ohno, S., Jainchill, J. and Stenius, C. (1963) The creeping vole (*Microtus oregoni*) as a gonosomic mosaic. I. The OY/XY constitution of the male. *Cytogenetics*, **2**, 232–9.

Ohno, S., Stenius, C. and Christian, L. (1966) The XO as the normal female of the creeping vole (*Microtus oregoni*). In *Chromosomes Today* 1, pp. 182–7. Eds. C. D. Darlington and K. R. Lewis. Oliver & Boyd, Edinburgh and London.

Page, D. C., Mosher, R. Simpson, E. M., Fisher, E. M. C., Mardon, G., Pollack, J., McGillivray, B., de la Chapelle, A. and Brown, L. (1987) The sex-determining region of the human Y chromosome encodes a finger protein. *Cell*, **51**, 1091–104.

Vorontsov, N. N. and Lyapunova, E. A. (1989) Two ways of speciation. In *Evolutionary Biology of Transient Unstable Populations*. Ed. A. Fontdevila. Springer-Verlag, Berlin Heidelberg.

20

Environmental control of gonadal differentiation

CLAUDE PIEAU, MARC GIRONDOT,
GISÈLE DESVAGES, MIREILLE DORIZZI,
NOËLLE RICHARD-MERCIER and
PATRICK ZABORSKI

In most species of invertebrates and vertebrates, sex is determined by sex chromosomes as early as the formation of the zygote resulting from the union of gametes. 'Genotypic sex-determination' (GSD) obeys two main systems, XX/XY (male heterogamety) and ZZ/ZW (female heterogamety), but can also be polyfactorial due to the presence of many genes with minor effects in the zygote.

In some species, sex is not fixed in the zygote, but sexual differentiation can be influenced by the local environment during embryonic or larval development ('environmental sex-determination' (ESD)). Thus, in the marine worm, *Bonellia viridis*, nearly all developing larvae become female if they are raised in isolation, whereas they become male if they are raised in association with adult females. The sex of larvae of mermithid nematodes parasitizing insects (such as *Romanomermis culicivorax*) is influenced by the nourishment extracted from their hosts: they become female if resources are abundant, or male if they are reduced. Sex determination appears to be sensitive to photoperiod in the amphipod, *Gammarus duebeni*. Temperature-sensitive sex mutations in the free-living nematode *Caenorhabditis elegans* and the fruit fly *Drosophila melanogaster* could be considered to be particular cases of ESD.

In vertebrates, several environmental factors have been scrutinized for their potential influence on the sexual differentiation of gonads. These factors include: the pH of water (in some cichlid and poeciliid fish); the ionic ratio K^+/Ca^{++} of water (in the amphibian *Discoglossus pictus*); and temperature (in the three classes of ectothermic vertebrates). Only temperature has been shown irrefutably to influence gonadal differentiation. In the following pages, we describe the effects of temperature on sex determination and the possible mechanism of its action in vertebrates. Then, we compare data in species with

433

'temperature-dependent sex-determination' (TSD) to those in species with GSD.

Effects of temperature on the sexual differentiation of gonads in vertebrates

In mammals and birds, sex determination is genotypic according to the XX/XY and ZZ/ZW systems respectively. Both systems of GSD are found in fish, amphibians and reptiles. Moreover, in many species, sex determination is influenced by temperature (TSD, Table 20.1)

Fish

The first observation of the influence of temperature on gonadal differentiation was made in an hermaphroditic species, *Rivulus marmoratus*, which has gonads containing both male and female gametic tissues, the ovotestes. When embryos were maintained at cold temperatures (20 °C or below), the ovarian part of the ovotestes degenerated in many individuals, causing them to become phenotypic males.

Sexual differentiation appears also to be sensitive to temperature in gonochoristic species. The Atlantic silverside *Menidia menidia* is an interesting example. This species inhabits a large part of the coast of eastern North America. In the laboratory, a regime of cold fluctuating temperatures of 11 °C to 19 °C produces more phenotypic females than a regime of warmer temperatures of 17 °C to 25 °C. However, the progeny of different females differ in their responses to temperature, showing that temperature interacts with genetic factors. Moreover, it has been shown that temperature sensitivity varies greatly among populations at different latitudes. Some populations display TSD clearly, whereas others display GSD, providing evidence of adaptative variation in environmental and genotypic sex-determination in this fish.

Amphibians

Several older works, beginning in 1914, described a masculinizing effect of heat treatment during larval development for some anuran species and an urodele species, all known to have GSD. There was no known marker for identification of the sexual genotype and the progeny of individuals were not studied. So, the complete and functional sex-reversal of gonadal phenotype was not demonstrated.

More recently, the effects of temperature on sexual differentiation have

Table 20.1 *Different systems of sex determination in vertebrates*

Classes; orders; suborders	GSD		TSD
	XX/XY	ZZ/ZW	
Mammals	+		
Birds		+	
Reptiles			
Crocodilians			+
Turtles	+	+	+
Squamates			
Lizards	+	+	+
Snakes		+	
Amphibians			
Anurans	+	+	+
Urodeles	+	+	+
Fish			
Teleosts	+	+	+

GSD, genotypic sex-determination; TSD, temperature-dependent sex-determination; XX/XY, male heterogamety; ZZ/ZW, female heterogamety.

been studied in two close species of salamanders, *Pleurodeles waltl* and *P. poireti*. Both display a ZZ/ZW mechanism of GSD. At ambient laboratory temperature (20 ± 2 deg. C), the sex ratio is one male to one female. When larvae of *P. waltl* are reared at 30 °C, the sex ratio is biased clearly towards males (70%). At 32 °C, 100% of individuals become phenotypic males. In *P. poireti*, rearing the larvae at 30 °C yields a bias towards females (66%). Therefore, high temperatures (30–32 °C) produce opposite effects in these species: ZW genotypic females of *P. waltl* become phenotypic males, whereas ZZ genotypic males of *P. poireti* become phenotypic females. Sex reversal of these individuals has been demonstrated irrefutably through different studies: genetic (analysis of offspring); cytogenetic (characterization of differential transcription loops during meiosis on the W and Z chromosomes); enzymatic (analysis of the electrophoretic pattern of peptidase-1, a dimeric sex-linked enzyme); and immunological (analysis of the expression of the serologically defined H-Y antigen).

The thermosensitive period for gonadal differentiation has been delimited in *P. waltl* by shifting the larvae at different stages, from room temperature to high temperatures (male-producing) and returning them to room temperature after different times. This period is long, approximately two months duration. As seen above, a large increase in temperature (more than 10 deg. C above the ambient temperature) is required to reverse the sexual phenotype of all individuals. Therefore, it can be expected that ESD either does not occur or is a rare occurrence in nature.

Reptiles

In many species of reptiles, the incubation of eggs at different temperatures always results in approximately 50% phenotypic males and 50% phenotypic females (Figure 20.1A). Sex determination is genotypic for snakes and for some species of turtles and lizards. In other species, including crocodilians, turtles and lizards, gonadal differentiation has been shown to be influenced by temperature. Generally, both sexes and sometimes intersexes differentiate only within a narrow range of temperature, the 'transitional range of temperature' (TRT). Within this range, the temperature which, theoretically, gives 50% phenotypic males and 50% phenotypic females has been called the 'critical', the 'threshold' or the 'pivotal' temperature. This pivotal temperature varies according to species but, even in a given species, it may be slightly different from one clutch to another. Above and below the TRT, 100% phenotypic males or 100% phenotypic females are obtained. In some species, temperatures below TRT are masculinizing whereas temperatures above TRT are feminizing (pattern 1, observed in many turtles, Figure 20.1B). In other species, the opposite occurs (pattern 2, observed in crocodilians and lizards, Figure 20.1C). In a third pattern, high and low temperatures are feminizing and intermediate temperatures are masculinizing: there are two TRTs and therefore two pivotal temperatures (pattern 3, observed in some species of turtles, crocodilians and lizards, Figure 20.1D).

The thermosensitive period for gonadal differentiation has been established in some species, using shifts of temperature from male-producing temperatures to female-producing temperatures and *vice versa*. This period corresponds to the first steps of sexual differentiation of the gonads (see below), generally in the 'middle' third of embryonic development. For turtles, it lasts approximately two weeks.

The production of phenotypic males, phenotypic females and intersexes (with ovotestes as in the European pond turtle, *Emys orbicularis*, see below) as well as the different responses between clutches around the pivotal temperature, clearly show genetic differences between individuals.

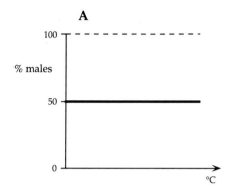

A

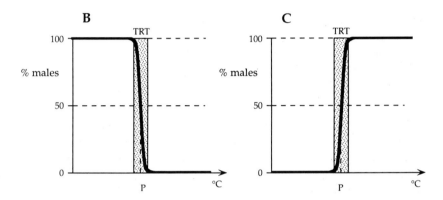

B **C**

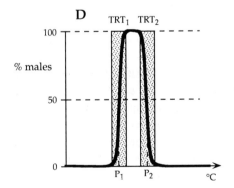

D

Figure 20.1. Sexual differentiation as a function of the incubation temperature of eggs in reptiles: (A) no effect of temperature; (B) low temperatures are masculinizing, high temperatures are feminizing; (C) low temperatures are feminizing, high temperatures are masculinizing; (D) low and high temperatures are feminizing, intermediate temperatures are masculinizing. (TRT, TRT_1, TRT_2, transitional ranges of temperature; P, P_1, P_2, pivotal temperatures.)

Whether these differences correspond to classical mechanisms of GSD is debatable. Heteromorphic sex-chromosomes are lacking in species with TSD, except perhaps one (*Staurotypus salvinii*). Analysis of the progeny is difficult, since, in many species, sexual maturity is reached several years after hatching. Polymorphic sex-linked enzymes are not known. However, sex-specific DNA fragments appear to exist in males of the sea turtles *Chelonia mydas* and *Lepidochelys kempi*. Likewise, the expression of the serologically defined H-Y antigen (SD-H-Y), a sex-specific molecule which characterizes heterogametic sex in many species, agrees in *C. mydas* with an XX/XY system, and with a ZZ/ZW system in *E. orbicularis*. A good concordance between gonadal sexual phenotype and expression of SD-H-Y has been found at the pivotal temperature in *E. orbicularis*, indicating that GSD might be operating at this temperature.

Mechanism for action of temperature on the sexual differentiation of gonads

In elucidating the mechanism for action of temperature on the sexual differentiation of gonads, the main problem is to find the molecular target on which temperature exerts its effects. Since, in many species, phenotypic males, phenotypic females and intersexes are produced within a very narrow range of temperatures, it could be expected that masculinizing or feminizing substances with a threshold effect are produced in the gonads and that such production is temperature-dependent. Individual genetic differences would account for slight differences in the temperature at which this threshold is reached. Steroid hormones are good candidates for such substances. Indeed, many previous studies in other vertebrates have shown that steroids are synthesized early in the gonads and that partial or even complete sex-reversal can be obtained by treatments with exogenous androgens or oestrogens. To illustrate how temperature affects steroid production, and hence sex, we describe our work on steroid synthesis in the embryonic gonads and the effects of steroids on gonadal sexual differentiation in reptiles exhibiting TSD. First, we summarize the main features of gonadal development in reptiles, taking the European pond turtle (*E. orbicularis*) as an example.

Development of gonads in embryos of E. orbicularis

In this species the pivotal temperature is 28.5 °C, and incubation of eggs below 28 °C yields 100% phenotypic males, whereas 100% phenotypic females are obtained above 29.5 °C. Figure 20.2 is a schematic drawing of the development of gonads corresponding to the age and the developmental stages of embryos, and Figure 20.3 is an illustration of the structure of gonads at some of these stages. Gonads begin to form between stages 12 and 14. On each side of the intestinal mesentery, the coelomic epithelium covering the medio-ventral surface of the mesonephros thickens and is colonized by the migrating germ cells. These gonad anlagen form the so-called 'genital ridges', and the surface epithelium becomes the so-called 'germinal epithelium' (Figure 20.3A). This epithelium proliferates and gives thin cords of epithelial cells extending into the underlying mesenchyme. Between stages 14 and 16, the gonad anlagen consist of two distinct parts: the 'germinal epithelium' at the surface and the inner part (often called 'medulla') which contains epithelial cords, the so-called 'sex cords' (Figure 20.3B). As this structure is observed at both male- and female-producing temperatures, the gonads are usually considered to be undifferentiated at this time. However, sex cords are slightly thinner at feminizing temperatures than at masculinizing ones. The thermosensitive period (stages 16 to 22, Figure 20.2) corresponds to the first steps of sexual differentiation of the gonads. In differentiating testes, the germinal epithelium flattens, and germ cells leave this epithelium and migrate in the sex cords which enlarge and become testicular cords or tubes (Figure 20.3C). In differentiating ovaries, the sex cords become thin, whereas the germinal epithelium thickens due to multiplication *in situ* of germ cells. A cortex is thus formed (Figure 20.3D). At stage 22, some germ cells enter into meiosis.

During the thermosensitive period, gonadal differentiation can be sex reversed by temperature shifts from a masculinizing to a feminizing temperature or *vice versa*. After this period, gonadal differentiation is no longer reversible (Figure 20.2): in testes, testicular cords (future seminiferous tubules) become more and more individualized (Figure 20.3E); in ovaries, the cortex continues to thicken and at the end of the embryonic life (stages 25 and 26) primary follicles containing growing oocytes are formed in its internal part (Figure 20.3F).

In some individuals, around the pivotal temperature, gonads differentiate as ovotestes with testicular cords in the medulla and an ovarian-like cortex at the surface. After hatching, the cortex regresses at least partially.

Steroid synthesis in embryonic gonads

The enzymes involved in steroidogenesis appear to be present in the gonads at the early stages of their development and biochemical differences

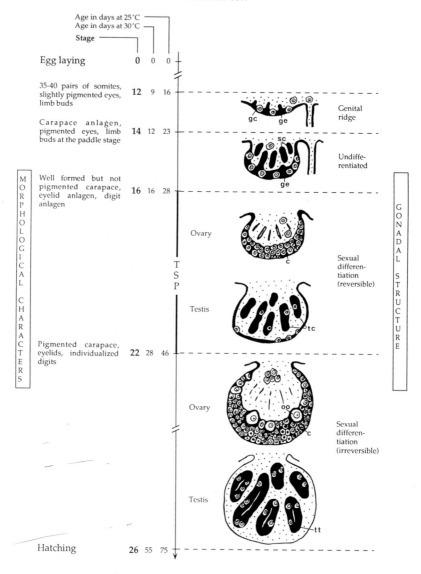

Figure 20.2. Development of gonads as a function of age and developmental stage in *Emys orbicularis* embryos. (TSP, thermosensitive period; c, cortex; gc, germ cells; ge, germinal epithelium; oo, oocyte; sc, sex cord; tc, testicular cord; tt, testicular tubule.)

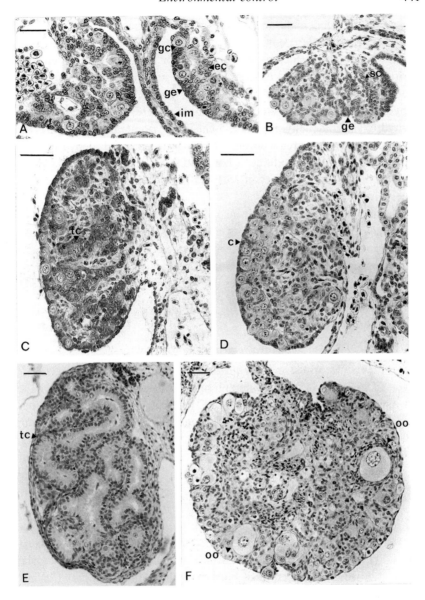

Figure 20.3. Transverse sections through gonads of *Emys orbicularis* embryos incubated at 25 °C, 28.5 °C or 30 °C: (A) genital ridges, stage 14 at 25 °C; (B) undifferentiated gonad, stage 15 at 30 °C; (C) differentiating testis, stage 20 at 28.5 °C; (D) differentiating ovary, stage 20 at 28.5 °C; (E) testis, stage 25⁺ at 25 °C; (F) ovary, stage 25⁺ at 30 °C. (Bar represents 50 μm. c, cortex; ec, epithelial cells; gc, germ cells; ge, germinal epithelium; im, intestinal mesentery; oo, oocyte; sc, sex cord; tc, testicular cord. For a description of the developmental stages, see Figure 20.2.)

are found between male and female gonads at these stages. Thus, the gonads of *E. orbicularis* embryos incubated at 25 °C (male-producing) or at 30 °C (female-producing), either during or after the thermosensitive period for gonadal sexual differentiation, are able to metabolize different radiolabelled steroids used as substrates (pregnenolone, progesterone, dehydroepiandrosterone, androstenedione). Moreover, the activity of 3β-hydroxysteroid dehydrogenase-5-ene-4-ene isomerase (3β-HSD), the enzyme converting pregnenolone to progesterone and dehydroepiandrosterone to androstenedione, is higher in testes at 25 °C than in ovaries at 30 °C. This higher activity of 3β-HSD in testes has also been found in other turtle species.

In embryonic gonads of *E. orbicularis*, steroidogenesis appears to be complete, up to the production of androgens (testosterone, 5α-dihydrotestosterone) and oestrogens (oestrone and oestradiol-17β). The amounts of oestrogens detected by radioimmunoassays are very low. However, they are higher in differentiating ovaries at 30 °C than in differentiating testes at 25 °C during the early stages of the thermosensitive period.

The involvement of oestrogens

Experiments performed in *E. orbicularis* have shown that oestrogens are involved in gonadal sexual differentiation. If oestrone or oestradiol-17β are injected into eggs incubated at 25 °C (masculinizing temperature) just before or at the beginning of the thermosensitive period, the gonads differentiate into ovaries instead of testes. The development of testicular cords is inhibited, whereas that of an ovarian cortex is stimulated. Conversely, epithelial cords or tubes, similar to potential seminiferous cords, differentiate in the medulla of the gonads if tamoxifen (an antioestrogen binding to oestrogen receptors) is injected, at the same stages, into eggs incubated at 30 °C (feminizing temperature). Thus, tamoxifen prevents the inhibitory action of oestrogens on testicular cord development. This interpretation is confirmed by the presence of testicular cords following the injection of both tamoxifen and oestradiol-17β into eggs incubated at 25 °C. Tamoxifen is an antagonist of oestrogens for maintenance of testicular cords. However, it induces formation of an ovarian cortex at 25 °C. Such an agonistic action of tamoxifen has been reported *in vivo* and *in vitro* in mammalian tissues and cells.

The feminization of gonads by exogenous oestrogens at a maleproducing temperature has been obtained in several other species of turtles, the alligator *Alligator mississippiensis* and the lizard *Eublepharis macularius.* In all cases, the sensitive period for hormonal effects on gonadal sexual differentiation coincides with the thermosensitive period. Oestrogen treatments after this period do not inhibit testicular cords, but can stimulate vestiges of the germinal epithelium at the surface of the gonads. At these places, germ cells divide and enter into meiosis, indicating that the germinal epithelium has its own oestrogen receptors.

Treatment with tamoxifen failed to masculinize the gonads of embryos incubated at a female-producing temperature in another species of turtle (*Trachemys scripta*) and in *A. mississippiensis.* This could be due either to a strong agonistic action of tamoxifen in these species or to a different conformation of the product used or its metabolized derivatives.

A key role for aromatase

Cytochrome P-450 aromatase (P-450$_{AROM}$) is the enzyme responsible for the aromatization of androstenedione into oestrone and of testosterone into oestradiol-17β. A fine correlation between changes in gonadal aromatase activity and changes in gonadal structure has been found in *E. orbicularis* embryos. Aromatase activity is much higher in ovaries than in testes. When a cortex is present at the surface of the testes, the thicker the cortex the higher the aromatase activity. At the beginning of the thermosensitive period, aromatase activity is low, but it is somewhat higher for turtle embryos incubated at 30 °C compared with those at 25 °C. Afterwards, at 25 °C it remains very low up to hatching, whereas at 30 °C it increases considerably (compare A and B, Figure 20.4). If eggs are incubated at 25 °C up to stage 18 and then exposed to 35 °C (highly feminizing) for different times, the response for gonadal aromatase activity is exponential and is parallel to that observed in embryos incubated at 30 °C (Figure 20.4C). If eggs are shifted from 25 °C to 35 °C at stage 23, aromatase activity does not increase (Figure 20.4D). Altogether, results show that the sensitive period for the effects of temperature on gonadal aromatase activity corresponds to the thermosensitive period for gonadal sexual differentiation and that temperature does not act directly on aromatase activity, but on the regulation of P-450$_{AROM}$ synthesis.

In other systems (human adipose cells, human and rat ovarian granulosa cells), changes in aromatase activity are correlated with changes in the levels of mRNA encoding P-450$_{AROM}$. In *E. orbicularis*, temperature-induced changes in the synthesis of P-450$_{AROM}$ could also result from changes in the levels of mRNA encoding the enzyme. According to this view, the expression of the P-450$_{AROM}$ gene would be temperature-dependent.

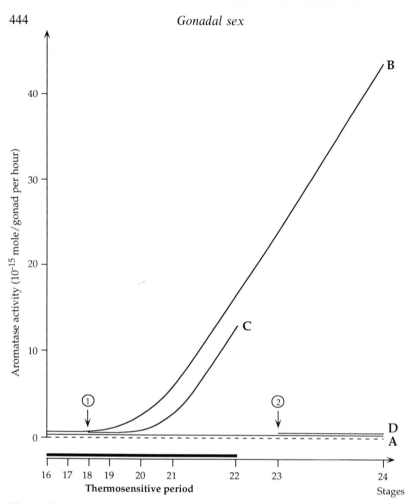

Figure 20.4. Aromatase activity in developing gonads of *Emys orbicularis* embryos obtained from eggs incubated at different temperatures: (A) eggs incubated at 25 °C; (B) eggs incubated at 30 °C; (C) eggs first incubated at 25 °C and then shifted to 35 °C at stage 18 (arrow 1); (D) eggs first incubated at 25 °C and then shifted to 35 °C at stage 23 (arrow 2). (For a description of the developmental stages, see Figure 20.2.)

Whether temperature acts on the expression of the P-450$_{\text{AROM}}$ gene itself or that of another upstream gene remains to be determined.

Comparisons with other vertebrates

Partial modification and even reversal of the sexual phenotype of the gonads have been obtained under the effects of sex steroids in all classes of vertebrates except eutherian mammals. In birds, the importance of oestrogens and the P-450$_{AROM}$ enzyme in gonadal sexual differentiation has been demonstrated clearly. First, exogenous oestrogens induce sex-reversal in genotypic males (ZZ). Second, tamoxifen causes total masculinization of the right gonad and partial masculinization of the left gonad in genotypic females (ZW). Third, aromatase inhibitors induce testicular development, up to the production of spermatozoa, in genotypic females (ZW).

Up to now, attempts to reverse gonadal phenotype by administration of steroids *in vivo* have failed in eutherian mammals. It is possible that a particular system of steroid metabolism normally protects the embryo against maternal steroids and does not allow exogenous steroids to reach its gonads. Recent data dealing with the effects of the anti-Müllerian hormone (AMH) on aromatase activity and gonadal differentiation *in vitro* are of particular interest. AMH inhibits aromatase activity in differentiated fetal ovaries, and it induces structures that are like testicular cords in undifferentiated gonads of rat genotypic females. This masculinization could result from the decrease in the level of gonadal endogenous oestrogens.

All these results suggest strongly that the regulation of expression of the P-450$_{AROM}$ gene plays a key role in sexual differentiation of gonads in all vertebrates (Figure 20.5).

Data for adult mammals show that in adipose cells and ovarian granulosa cells, the regulation of expression of the P-450$_{AROM}$ gene is multifactorial. Some factors such as cyclic AMP and gonadotrophins activate its transcription, other factors such as prolactin and epidermal growth factor repress it. For testicular differentiation in mammals, the testis-determining factor (SRY or homologs) could repress transcription of the P-450$_{AROM}$ gene itself, or could activate transcription of the AMH gene or that of another gene, the product of which represses the P-450$_{AROM}$ gene (cascade reaction). In birds, ovarian differentiation could be governed by a gene carried by the W chromosome. This gene would encode a factor activating the transcription of the P-450$_{AROM}$ gene, directly or indirectly *via* cyclic AMP. Naturally, this factor would be synthesized before repressing factors, such as AMH or the product of the homolog of the SRY gene. As shown above, in species with TSD, a thermosensitive factor could regulate the transcription of the P-450$_{AROM}$ gene or that of another gene intervening upstream in a cascade system.

It is well known that steroid hormones activate the transcription of some genes, whereas they repress the transcription of others. Structural

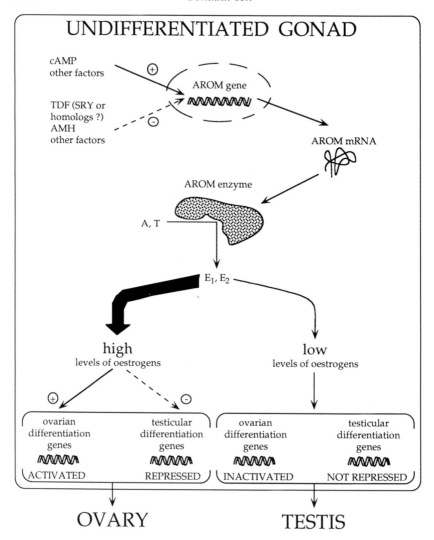

Figure 20.5. Postulated involvement of cytochrome P-450 aromatase (AROM) in the sexual differentiation of gonads in vertebrates. (+, Activation of transcription; −, repression of transcription; A, androstenedione; E_1, oestrone; E_2, oestradiol-17β; T, testosterone; AMH, anti-Müllerian hormone; cAMP, adenosine 3′:5′-cyclic monophosphate; SRY, sex region of the Y chromosome; TDF, testis-determining factor.)

genes for both ovarian and testicular differentiation are present in all individuals. When the level of oestrogens is high, ovarian differentiation genes are activated, whereas testicular differentiation genes are repressed. When the level of oestrogens remains low, testicular differentiation genes are not repressed whereas ovarian differentiation genes are in-activated (Figure 20.5).

It is clear that no frontier exists between GSD and ESD. In all cases, sexual differentiation of gonads results from cascade reactions. The same molecules, in particular steroids, appear to be implicated. However the mechanism of regulation of the expression of some major genes, such as that encoding P-450$_{AROM}$ will vary somewhat in the different systems.

Acknowledgements

We thank Mrs L. Guillon for her help with the manuscript.

Further reading

Reviews

Bogart, M. H. (1987) Sex determination: a hypothesis based on steroid ratios. *Journal of Theoretical Biology*, **128**, 349–57.

Bull, J. J. (1980) Sex determination in reptiles. *Quarterly Review of Biology*, **55**, 3–21.

Bull, J. J. (1983) *Evolution of Sex Determining Mechanisms*. The Benjamin/Cummings Publishing Company, Inc., Menlo Park, Ca.

Dournon, C., Houillon, C. and Pieau, C. (1990) Temperature sex-reversal in amphibians and reptiles. *International Journal of Developmental Biology*, **34**, 81–92.

Janzen, F. J. and Paukstis, G. L. (1991) Environmental sex determination in reptiles: ecology, evolution, and experimental design. *Quarterly Review of Biology*, **66**, 149–79.

Korpelainen, H. (1990) Sex ratios and the conditions required for environmental sex determination in animals. *Biological Reviews*, **65**, 147–84.

Raynaud, A. and Pieau, C. (1985) Embryonic development of the genital system. In *Biology of the Reptilia*, pp. 149–300, Vol. 15, Development B. Eds. C. Gans and F. Billett. John Wiley and Sons, New York.

Original articles

Bull, J. J. and Vogt, R. C. (1979) Temperature-dependent sex determination in turtles. *Science*, **206**, 1186–8.

Bull, J. J., Gutzke, W. H. N. and Crews, D. (1988) Sex reversal by estradiol in three reptilian orders. *General and Comparative Endocrinology*, **70**, 425–8.

Charnier, M. (1966) Action de la température sur la sex-ratio chez l'embryon d'*Agama agama* (*Agamidae*, Lacertilien). *Comptes Rendus de la Société de Biologie, Paris*, **160**, 620–2.

Conover, D. O. and Heins, S. W. (1987) Adaptative variation in environmental and genetic sex determination in a fish. *Nature*, **326**, 496–8.

Conover, D. O. and Kynard, B. E. (1981) Environmental sex determination: interaction of temperature and genotype in a fish. *Science*, **213**, 577–9.

Desvages, G. and Pieau, C. (1991) Steroid metabolism in gonads of turtle embryos as a function of the incubation temperature of eggs. *Journal of Steroid Biochemistry and Molecular Biology*, **39**, 203–13.

Desvages, G. and Pieau, C. (1992) Aromatase activity in gonads of turtle embryos as a function of the incubation temperature of eggs. *Journal of Steroid Biochemistry and Molecular Biology*, **41**, 851–3.

Desvages, G. and Pieau, C. (1992) Time required for temperature-induced changes in gonadal aromatase activity and related gonadal structure in turtle embryos. *Differentiation*, **52**, 13–18.

Dorizzi, M., Mignot, M.-T., Guichard, A., Desvages, G. and Pieau, C. (1991) Involvement of oestrogens in sexual differentiation of gonads as a function of temperature in turtles. *Differentiation*, **47**, 9–17.

Ferguson, M. W. J. and Joanen, T. (1982) Temperature of egg incubation determines sex in *Alligator mississippiensis*. *Nature*, **296**, 850–3.

Gutzke, W. H. N. and Chymiy, D. B. (1988) Sensitive periods during embryogeny for hormonally induced sex determination in turtles. *General and Comparative Endocrinology*, **71**, 265–7.

Pieau, C. (1971) Sur la proportion sexuelle chez les embryons de deux Chéloniens (*Testudo graeca* L. et *Emys orbicularis* L.) issus d'oeufs incubés artificiellement. *Comptes Rendus de l'Académie des Sciences, Paris*, **272(D)**, 3071–4.

Pieau, C. (1972) Effets de la température sur le développement des glandes génitales chez les embryons de deux Chéloniens, *Emys orbicularis* L. et *Testudo graeca* L. *Comptes Rendus de l'Académie des Sciences, Paris*, **274(D)**, 719–22.

Pieau, C. (1974) Différenciation du sexe en fonction de la température chez les embryons d'*Emys orbicularis* L. (Chélonien); effets des hormones sexuelles. *Annales d'Embryologie et de Morphogenèse*, **7**, 365–94.

Wibbels, T. and Crews, D. (1992) Specificity of steroid hormone-induced sex determination in a turtle. *Journal of Endocrinology*, **133**, 121–9.

Yntema, C. L. (1976) Effects of incubation temperatures on sexual differentiation in the turtle, *Chelydra serpentina*. *Journal of Morphology*, **150**, 453–62.

Zaborski, P., Dorizzi, M. and Pieau, C. (1982) H-Y antigen expression in temperature sex-reversed turtles (*Emys orbicularis*). *Differentiation*, **22**, 73–8.

Zaborski, P., Dorizzi, M. and Pieau, C. (1988) Temperature-dependent gonadal differentiation in the turtle *Emys orbicularis*: concordance between sexual phenotype and serological H-Y antigen expression at threshold temperature. *Differentiation*, **38**, 17–20.

Afterwords

The editors felt it was important to end this volume with a statement about human sex differences. Since we could not reach agreement, we offer the reader two different perspectives on the views of Aristotle, as quoted in the Foreword.

21

A man's a man for a' that

R. V. SHORT

The preceding chapters illustrate both how much, and yet how little, we have learned about the differences between the sexes since Aristotle's penetrating observations of 2300 years ago, cited in the Foreword, or Charles Darwin's ideas on sexual selection, first published just over 100 years ago in *The Descent of Man, and Selection in Relation to Sex*. Darwin was in no doubt about male superiority:

Difference in the mental powers of the two sexes. With respect to differences of this nature between man and woman, it is probable that sexual selection has played a highly important part. I am aware that some writers doubt whether there is any such inherent difference; but this is at least probable from the analogy of the lower animals which present other secondary sexual characters.

Woman, owing to her maternal instincts, displays those qualities towards her infants in an eminent degree; therefore it is likely that she would often extend them towards her fellow-creatures. Man is the rival of other men; he delights in competition, and this leads to ambition which passes too easily into selfishness. These latter qualities seem to be his natural and unfortunate birthright. It is generally admitted that with woman the powers of intuition, of rapid perception, and perhaps of imitation, are more strongly marked than in man.

The chief distinction in the intellectual powers of the two sexes is shown by man's attaining to a higher eminence, in whatever he takes up, than can woman – whether requiring deep thought, reason, or imagination, or merely the use of senses and hands. If two lists were made of the most eminent men and women in poetry, painting, sculpture, music (inclusive both of composition and performance), history, science, and philosophy, with half-a-dozen names under each subject, the two lists would not bear comparison.

Scientists today are remarkably reluctant to do research on, or even speculate about, the nature and magnitude of any innate differences in behaviour or intellectual ability that might distinguish men from women, for fear of being labelled sexist. Presumably, feminists would be incensed

451

by the dogmatic assertions of Aristotle and Darwin about human sex differences. But perhaps Aristotle and Darwin were right. After all, nobody questions the significant differences in average weight, height and strength between men and women, testimony no doubt to our polygynous past. It would be surprising if these physiological differences were not accompanied by some behavioural ones: man, the hunter, would have needed very different skills and abilities compared to woman, the gatherer and childbearer.

The problem arises when society makes value judgements about our sex differences, and rewards male aggression whilst penalizing female maternal instincts, for example. As has been pointed out by the World Health Organization, women constitute one-third of the world's labour force, are responsible for two-thirds of the hours worked, receive only 10% of the world's income, and own less than 1% of the world's property. Women are also disadvantaged in most competitive sports; indeed it would be interesting to devise a series of Olympic contests in which women could outperform men. We must not confuse the highly desirable goal of equality of *rights* for men and women with the mistaken notion that the sexes are equal in *ability*. The early philosophers regarded woman as the tally-half of man; they could see that each sex had attributes that the other lacked, and the advantage of marriage was that the sum was greater than its parts. By dividing responsibilities between the sexes, we might increase their mutual interdependence and hence strengthen the pair bond. But which sex differences are truly innate, and which are acquired culturally? And do our genes necessarily determine our destiny?

One of the most compelling pieces of evidence to suggest that there are indeed innate, genetically determined gender differences in humans comes from the studies of the Spiros in the 1970s. They analysed the behaviour of boys and girls reared in the gender-neutral cultural environment of an Israeli kibbutz. Boys and girls were raised in a communal nursery from the first week of life, and the two sexes continued to live together throughout infancy, childhood and adolescence, apart from brief daily visits with their parents. All facilities were communal, and any sex-role identification in dress or behaviour was discouraged.

In spite of this, sex differences in behaviour soon began to appear. Boys played more strenuously than girls, and liked to pretend they were driving machines, whereas girls preferred more artistic games, and liked to play with dolls and prams. As they grew up, the boys became more egotistical, and the girls became more socially sensitive. With the advent of puberty came a sense of shame and embarrassment, with the girls refusing to take

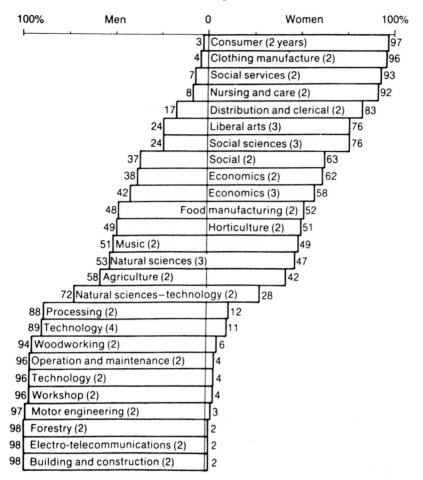

Figure 21.1. Sex distribution of applicants for courses in Swedish secondary schools, Autumn 1980. (From Wistrand B (1981).)

showers with the boys. The girls started to keep to themselves, and their living area was tidier than that of the boys. The boys became more aggressive, and tended to dominate classroom discussions. On reaching adulthood, the women sought contact with children and adopted feminine styles of clothing, whilst the men began to take on leadership roles and to dominate communal meetings. Although Melford Spiro had set out to demonstrate that culture determines human nature, he was reluctantly forced to conclude that it was human nature that determines culture.

It is instructive to look at a country like Sweden, where the principle of total equality between the sexes has been fully sanctioned since the late

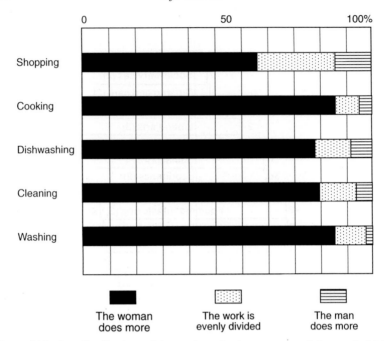

Figure 21.2. Sex distribution of domestic tasks between married or cohabiting heterosexual couples, 1975. (From Wistrand B (1981).)

1960s. Nevertheless, this has done almost nothing to alter either the marked sexual skewing of career choices by school children (Figure 21.1), or the lifestyles of married couples (Figure 21.2). Is this just cultural inertia, or does it reflect an innate sex difference in behaviour? The nature-nurture conundrum is almost insoluble as far as the control of human behaviour is concerned and the truth must lie somewhere between the two extremes.

As far as our sexual behaviour is concerned, we cannot help being impressed by the fact that some genetically male but phenotypically female children with 5α-reductase deficiency who have been reared as girls nevertheless spontaneously adopt a male role as their androgen levels rise at puberty. We have the harrowing personal diary of Herculine Barbin, born near La Rochelle, France, in 1838 and raised as a girl in a convent. She became increasingly hirsute at puberty, and touchingly and tenderly describes how she fell deeply in love with one of the other girls in the convent, and started to sleep with her. The sudden onset of acute genital pains forced her reluctantly to seek medical help. The doctor, discovering a greatly enlarged clitoris, reported the case to the

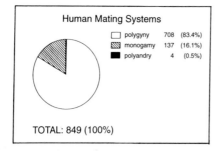

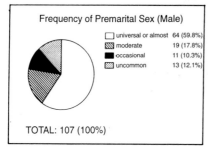

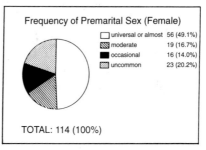

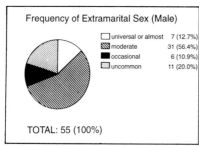

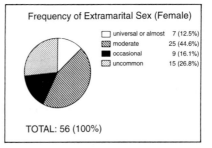

Figure 21.3. Human mating systems, and frequency of male and female premarital and extramarital sex in traditional cultures around the world prior to Western influences. (From Smith, R. L. (1984) Human sperm competition. In *Sperm Competition and the Evolution of Animal Mating Systems*, pp. 601–59. Ed. R. L. Smith. Academic Press, London.)

Monseigneur, who banished her from the convent, separating her from her lover. She was reassigned as a male at the age of 22 years, and as Hercule Barbin was sent to Paris to work on the railway. Hercule committed suicide by asphyxiation in a charcoal stove eight years later. The autopsy findings leave us in no doubt that this, too, was a case of 5α-reductase deficiency.

So our hormones can overrule cultural imprinting, and can overturn our assigned gender at birth. But we must not forget that some 5α-reductase-deficient individuals continue to live as females for the whole of their lives, so even our hormones do not always determine our destiny.

Regardless of what science has to say about human sex differences, we can be sure that most human societies will continue to practise some degree of assortative mating, in which women usually marry men who are taller and older than themselves. This sexual selection will ensure that existing physical dimorphisms are perpetuated in the future. And notwithstanding all the Judaeo-Christian cultural imprinting with the concept of no premarital sex and lifelong monogamy after marriage, in most cultures many men will have sex with more than one woman during their lifetimes, and a smaller proportion of women will have sex with more than one man (Figure 21.3). If proof of this were needed, we have only to look at the widespread incidence of sexually transmitted diseases, and the rapid global spread of AIDS; if humans were monogamous, sexually transmitted diseases would disappear in a generation. So there is a measure of truth to Mrs Amos Pinchot's dreamtime revelation:

> *Hoggamus, higgamus*
> *Men are polygamous*
> *Higgamus, hoggamus*
> *Women monogamous.*

Further reading

Harcourt, A. H., Harvey, P. H., Larson, S. G. and Short, R. V. (1981) Testis weight, body weight and breeding systems in primates. *Nature*, **293**, 138–9.
McDougall, R. (1980) *Herculine Barbin. Being the recently dicovered memoirs of a Nineteenth-Century French Hermaphrodite.* Parthenon, New York.
Wistrand, B. (1981) *Swedish Women on the Move.* Swedish Institute, Stockholm.

22

A few chromosomal differences among friends

EVAN BALABAN

I have always approached discussions of human sex differences with an element of reserve. This reticence does not stem from any fear of being labelled sexist, but rather from my inability to distinguish primary from secondary causes when relating biological to societal sex differences. It is amazing how little European male discourse on the subject of sex differences has changed since the time of Aristotle, who was a great believer in biology as a primary cause. Citing Darwin and more recent work by the anthropologist Melford Spiro, my colleague Roger Short has argued why this point of view may be right: a modern Aristotelian view of sex differences. An alternative view is equally deserving of consideration.

In the modern Aristotelian view, the seeds of our present condition were sown in prehistory. Early human populations were subject to strong evolutionary forces, and this evolutionary past has left its mark on our behavioural present via 'instincts': maternal, territorial, aggressive and sexual. Yet it is important to distinguish between any particular way in which we reached our present condition (history) and the inevitability of our present condition, which Aristotle and Darwin failed to do.

Most evolutionary thinking about social or mating behaviour recognizes explicitly the roles of the environment and of chance in modulating whether populations of organisms are (for instance) monogamous or polygamous. Such models emphasize the inherently plastic nature of sex roles: biology equals flexibility. Arguments which use present-day sex roles to infer past selection pressures (such as prehistoric societies where the men hunted and the women gathered food and cared for children) and then use the inferred past to justify the inevitability of the present, limit our vision. By interpreting biology as fate, we lose valuable insights into an interesting and important

457

part of our collective history: why things did not indeed turn out some other way.

Biology does not mean inevitability, and if we do not yet know the extent to which our measurement of sex differences is a function of (i) the different social environments inhabited by men and women, (ii) cultural bias, and (iii) the actual biological differences between men and women, then we may want to be more wary of the idea that many current societal sex differences are fundamental and transcendent. If studies of human behaviour tell us anything, it is that our notions of maleness and femaleness do not appear to correspond to simple features like the roundness or smoothness of Mendel's peas.

Modern Aristotelians frequently cite data like those gathered in Sweden and the Israeli kibbutzes to prove the folly of trying social engineering on human nature. One problem with these studies has been their willingness to substitute belief for analysis. They accept the ideology of equality espoused by such societies as fact, rather than subjecting the social behaviour of individuals to close scrutiny. Does the failure of Utopian social experiments tell us more about the inevitability of human nature, or about the inertia of cultural systems and the difficulty of isolating social movements from the cultural contexts which precede and surround them? As Melford Spiro points out in a disclaimer to the conclusions in his 1979 book *Gender and Culture: Kibbutz Women Revisited*:

Any attempt to assess the possible determinants of the counterrevolutionary changes that have occurred in the kibbutz movement in such institutions as marriage, the family, and sex-role differentiation is beset with formidable difficulties. The problem is too complex, the data too limited, and our methods of investigation were too primitive to permit an unequivocal interpretation.

The Aristotelian view of inevitable differences in the abilities of individuals has always posed a problem for democratic societies. How do we deal with the fact that all individuals are not created equal? We all agree in theory that equality in rights should not be confused with equality in attributes. But, in practice, equal rights extract an economic price, and there are inevitable conflicts between 'rights' and 'abilities'. Many of our legal systems try to resolve such issues by seeking natural precedents; the view that there is something inevitable about sex differences can thus take on the force of law. It is, therefore, all the more essential that we attain a clearer understanding of the extent to which our measurement of sex differences is contaminated by our preconceptions.

Biologists agree that genes can influence all aspects of human sex differences, yet we disagree on the strength of the causal links between particular sex differences and particular sets of genes, and the extent to which certain sex differences will always be a part of our biological heritage. Inevitability allows one falsely to believe that it is possible to see into the future. But if you think that the future depends on a complicated interaction between genes, environments and chance, then all bets are off. Scenarios of sexual selection producing a continuing soap opera of eternally philandering males and their perpetually faithful yet disillusioned spouses may represent our current social psyche more than some hard kernel of biological reality. When the mechanisms for determining sex differences are complex, biology always has a rich repertoire of possibilities.

What if most women really do like rearing children more than men? Or what if just the opposite is true but men in many societies are trapped by their macho ethos? If something happens to be true in a given place at a given time, the argument that it is true because it is innate or natural or the product of an evolutionary adaptation does not necessarily help us to make social decisions. Biology does not change the value of social facts. Biologists may have talent as creative forces which generate intellectual grist for society's fears and fancies about sex differences, but our science is a poor source of wisdom for making ethical decisions.

Biology driven by social considerations can be just as nefarious as social considerations driven by biology. Recent public opinion surveys in North America indicate that people are more tolerant of homosexuality if being gay is some inevitable consequence of the biological cards one is dealt. At the same time a spate of new studies on presumed biological correlates of sexual orientation have appeared, including papers claiming a genetic influence on homosexuality (inferred either from the correlation of sexual orientation among twins, siblings and adoptees, or from the correlation of DNA markers in homosexual siblings), and studies claiming structural differences in the brains of homosexual and heterosexual men. Many biologists, myself included, are openly critical of the technical details employed in these studies and in the propriety of their conclusions. Whether the claims of such studies turn out to be true or false, their intrusion into the social arena sets a dangerous precedent. Human rights should not rest on the minutiae of how complex human traits like sexual orientation unfold. Whatever mixture of chance, environments and genes makes individuals different, society still has the same choices. It can either respect individual differences or try to eradicate them.

The study of sex differences has an import which transcends its ability to inform us about our past or our future. Sex-difference research has fostered human freedom in an unprecedented way via the development of a technology capable of regulating reproduction. It is the hope of all of us that this continues to be the case, and it is our collective responsibility to ensure that such technology is not abused.

Whatever you believe about the causation of human sex differences, one fact is indisputable. There are very few societies around the world where females run the government, make the law and run the churches. If we had the voice of a female Aristotle speaking to us from a similarly privileged position, would its characterization of sex differences have looked any different? Would western societies have developed along different lines? I look forward to a day when we will know whether the opinions of Aristotle and Darwin remain contemporary because of historical circumstance or because of a few chromosomal differences among friends.

Further reading

Bleier, R. (1988) Sex differences research: science or belief? In *Feminist Approaches to Science*, pp. 147–64. Ed. R. Bleier. Pergamon Press, New York.

Byne, W. and Parsons, B. (1993) Human sexual orientation: the biologic theories reappraised. *Archives of General Psychiatry*, **50**, 228–39.

Fausto-Stirling, A. (1992) *Myths of Gender; Biological Theories about Women and Men* (2nd edition). Basic Books/Harper Collins, New York.

Lewontin, R. (1993) *Biology as Ideology*. Harper Perennial, New York.

Index

Page numbers in *italics* refer to figures and tables.